PLASMA ELECTRONICS

Second Edition

PLASMA ELECTRONICS

Applications in Microelectronic Device Fabrication

Second Edition

Toshiaki Makabe

Zoran Lj. Petrovic

CRC Press
Taylor & Francis Group
Boca Raton London New York

CRC Press is an imprint of the
Taylor & Francis Group, an **informa** business

CRC Press
Taylor & Francis Group
6000 Broken Sound Parkway NW, Suite 300
Boca Raton, FL 33487-2742

First issued in paperback 2016

Version Date: 20140710

ISBN 13: 978-1-138-03415-0 (pbk)
ISBN 13: 978-1-4822-2205-0 (hbk)

Visit the Taylor & Francis Web site at
http://www.taylorandfrancis.com

and the CRC Press Web site at
http://www.crcpress.com

Contents

Preface to the First Edition

Over the past three decades low-temperature plasma applications have been extended from primarily lighting to the fabrication of microelectronic devices and new materials, far exceeding our expectations. Radio frequency plasmas ranging from 10^5 Hz to 10^9 Hz, in particular, are now used to process metallic, semiconductor, and dielectric materials for the fabrication of ultra large-scale integrated (ULSI) circuits and to deposit various kinds of functional thin films and to modify the surface properties. Without plasma-produced ions and dissociated neutral radicals for etching and deposition on wafers, microelectronics manufacturing for ULSI circuits would simply be unfeasible.

The advent of ULSI fabrication has greatly changed how the field of plasma science is approached and understood. Low-temperature non-equilibrium plasmas are sustained mainly by electron impact ionization of a feed gas driven by an external radio frequency power source. These low-temperature plasmas acquire characteristics intrinsic to the feed gas molecules as determined by their unique electron-collision cross-section sets. This uniqueness means that plasmas must be understood using quantum, atomic, and molecular physics. The disparate time and spatial scales involved in low-temperature plasma processing (submeter to nanometer and seconds to nanoseconds) makes plasma processing an inherently stiff problem. The characteristics of low-temperature plasmas contrast markedly with highly ionized equilibrium plasmas that are ensembles of charged particles whose behavior can be understood through their long-range Coulomb interactions and collective effects and characterized by plasma frequency, Debye length, and electron temperature. The fundamental collision and reaction processes occurring both in gas phase and on surface in low-temperature plasmas are the bases for understanding their behavior and exploiting them for practical applications.

In the emerging nanotechnology era, device design, reliability, and the design of integrated plasma processes for their fabrication are tightly coupled. Being able to predict feature profile evolution under the influence of spatiotemporally varying plasmas is indispensable for the progress of topdown nanotechnology. Prediction of plasma damage and its mitigation are also crucial. Prediction & mitigation are 2 process and it will be performed through a series of vertically integrated numerical modeling and simulations ranging from the reactor scale to those cognizant of device elements subtending the plasma.

Motivated by the important role of plasmas in technology and the need for simulations to understand the associated complex processes, we emphasize in this book academic fusion among atomic and molecular physics, surface physics, the Boltzmann transport theory, electromagnetic theory, and computational science as plasma electronics. We do this to describe and predict the space and time characteristics of low-temperature plasmas and associated processing intrinsic to specific feed gases. An underlying theme throughout this work is computer-aided plasma analysis and synthesis, with emphasis on computational algorithms and techniques.

This book is based on a series of lectures presented at Keio University as part of its graduate program. The university's interest in the subject matter and feedback were essential parts of developing this text. We believe that the book is well suited as an instrument for self-instruction through its topical exercises and problems arranged in each chapter.

It is a pleasure to acknowledge our debt to David Graves and Robert Robson, who have helped in a variety of ways during the long period of our research life composing plasma electronics. Finally, we are indebted to T. Yagisawa for figure preparation and his attention to detail.

Toshiaki Makabe and Zoran Lj. Petrovic
Keio University

Preface to the Second Edition

Low temperature plasma has become a dry processing tool predictive of processes based on the plasma's fundamental collision physics and surface chemistry, and more sophisticated plasma processes have been propagated for manufacturing ultra-large-scale integrated (ULSI) circuits and combined devices of electronics and photonics in this decade. During the same period, new applications of the plasma process have resulted in micro/nano fabrications for devices associated with microanalysis, micro-reactors, and micro-electromechanical systems (MEMS), based on the high potential for efficiency and resource saving in the era of information communication technology and environmental society. This second edition reflects the fast progress in the quantitative understanding of the non-equilibrium, low-temperature plasma and the surface processing and the progress in the predictive modeling of the plasma and the process. Low-temperature plasma is otherwise known as a collisional plasma, in contrast with a collisionless plasma. As expected from the short-ranged collisional interaction, low-temperature plasma produces a large amount of long-lived excited species and neutral radicals that express a proper active function through the quantum characteristics peculiar to each of the feed gas molecules, in addition to the classical spatiotemporal plasma structure and function between electrons and positive (and negative) ions.

The low-temperature plasma process and the numerical predictive methods are compactly implemented in the form of 21 new figures, 18 new tables, 9 new problems, and 3 exercises in the second edition. We have added a chapter about the development of atmospheric-pressure plasma, in particular, microcell plasma, with a discussion of its practical application to improve surface efficiency, including the quantitative discussion of heat transfer and heating of the media and the reactor. Sections about the phase transition between the capacitive and inductive modes in an ICP, and about MOS-transistor and MEMS fabrications have been newly included. Using original works performed at Keio University, I made the revisions to the text. It is a pleasure to note my appreciation for the collaboration of Takashi Yagisawa and my graduate students, and to express gratitude to my long-time collaborator and coauthor, Zoran Petrovic.

Tokyo, in the Autumn 2013

Toshiaki Makabe
Keio University

About the Authors

Toshiaki Makabe received his BSc, MSc, and Ph.D. degrees in electrical engineering all from Keio University. He became a Professor of Electronics and Electrical Engineering in the Faculty of Science and Technology at Keio University in 1991. He also served as a guest professor at POSTECH, Ruhr University Bochum, and Xi'an Jiaotong University. He was Dean of the Faculty of Science and Technology and Chair of the Graduate School from 2007 to 2009. Since 2009, he is the Vice-President of Keio University in charge of research. He has published more than 170 papers in peer-reviewed international journals, and has given invited talks at more than 80 international conferences in the field of non-equilibrium, low-temperature plasmas and related basic transport theory and surface processes. He is on the editorial board of Plasma Sources Science and Technology, and many times he has been a guest editor of the special issue about the low-temperature plasma and the surface process of the Japanese Journal of Applied Physics, Australian Journal of Physics, Journal of Vacuum Science and Technology A, IEEE Transactions on Plasma Science, and Applied Surface Science, etc. He received the awards; "Fluid Science Prize" in 2003 from Institute of Fluid Science, Tohoku University, "Plasma Electronics Prize" in 2004 from the Japan Society of Applied Physics, "Plasma Prize" in 2006 from the American Vacuum Society, etc. He is an associate member of the Science Council of Japan, and a foreign member of the Serbian Academy of Sciences and Arts. He is a fellow of the Institute of Physics, the American Vacuum Society, the Japan Society of Applied Physics, and the Japan Federation of Engineering Societies.

Zoran Lj. Petrovic obtained his Master's degree in the Department of Applied Physics, Faculty of Electrical Engineering in the University of Belgrade, and earned his Ph.D from Australian National University. He is the Head of the Department of Experimental Physics in the Institute of Physics, University of Belgrade. He has taught postgraduate courses in microelectronics, plasma kinetics and diagnostics and was a visiting professor in Keio University (Yokohama, Japan). He has received the Nikola Tesla award for technological achievement and the Marko Jaric Award for Great Achievement in Physics. He is a full member of the Academy of Engineering Sciences of Serbia and Serbian Academy of Sciences and Arts where he chairs the department of engineering science. Zoran Petrovic is a fellow of American Physical Society, vice president of the National Scientific Council of Serbia and president of the Association of Scientific Institutes of Serbia. He is a member of editorial boards of Plasma Sources Science and Technology and Europena Physical Journal D. He has authored or co-authored over 220 papers in leading international scientific journals, and has given more than 90 invited talks at professional conferences. His research interests include atomic and molecular collisions in ionized gases, transport phenomena in ionized gases, gas breakdown, RF and DC plasmas for plasma processing, plasma medicine, positron collisions and traps and basic properties of gas discharges.

List of Figures

List of Tables

CHAPTER 1

Introduction

1.1 PLASMA AND ITS CLASSIFICATION

The plasma state is defined as the fourth state of matter as distinct from the solid, liquid, and gas phases. It consists of free positive and negative charges with electrical quasi-neutrality in addition to feed gas components. The word "plasma" was first introduced by Langmuir in 1928 and derives from the Greek $\pi\lambda\alpha\sigma\mu\alpha$ meaning "something formed" [1]. The strength of the plasma state is described in terms of the ionization degree $H_{DI} = n_p/N$, where $n_e\,(= n_p)$ and N is the number density of the charged particles and neutral molecules in a plasma. Strongly ionized plasma is realized in $H_{DI} \geq 10^{-3}$. In particular, electron transport in a strongly ionized plasma is subject to the long-range Coulomb interaction with surrounding electrons and positive ions, and the smooth trajectory makes it impossible to identify changes in momentum by collision during flight (Figure 1.1a). Therefore, strongly ionized plasma is termed collisionless plasma (Table 1.1). On the other hand, the characteristics of a weakly ionized plasma are properly represented by the short-range interaction between the electrons and the neutral molecules in the feed gas under external electric or magnetic fields. In such a collision-dominated plasma, the electron trajectory is characterized by both flight in the field and collision with surrounding molecules (Figure 1.1b). In comparison, a neutral molecule has a straight trajectory under a short-range collision in gases (Figure 1.1c).

In the gas phase, a plasma state can be produced electrically, thermally, or optically through the ionization of neutral molecules in the feed gas. In particular, nonequilibrium plasma or low-temperature plasma, which is produced by the collisional ionization of free electrons under an external electrical power, has a property that the electron energy is much higher than that of the neutral gas. We herein distinguish low-temperature plasma in the generation mechanism from thermal plasma produced by the thermal ionization of neutral molecules.

Low-temperature plasmas from micrometer to meter size are artificially maintained even in an electrodeless reactor as well as between metallic or

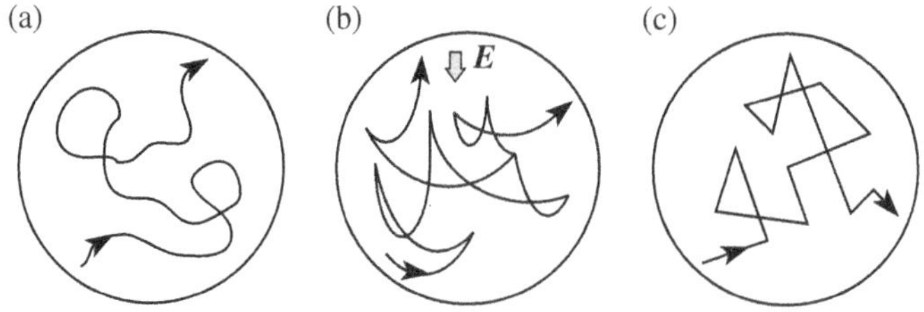

FIGURE 1.1 Electron trajectory in three different media. In strongly ionized plasma (a), in weakly ionized plasma (b), and in neutral gas (c).

TABLE 1.1 System Comparison between a Low Temperature Collisional Plasma and Collisionless Plasma

	Low-Temperature Collisional Plasma	**Collisionless Plasma**
Classification	Weakly ionized plasma	Strongly ionized plasma
Degree of ionization	Low($<10^{-3}$) under quasi-neutralty	High($\geq 10^{-3}$) under quasi-neutrality
Elemental process	Two-body collision b/w e + Molecule (short-range interaction)	Many-body collision b/w charged-particles (long-range Coulomb interaction)
Collision probability	Set of collision cross sections based on molecular quantum structure	Coulomb logarithm based on plasma density and the temperature
Temperature	$T_g \ll (T_e$ and $T_p)$	$T_g \sim (T_e$ and $T_p)$
Function of	Reduced electric field $\boldsymbol{E}/N(\boldsymbol{r}, t)$	Space potential $V_s(\boldsymbol{r}, t)$
System equations		
Governing eq.	Maxwell equation	Maxwell equation
Kinetic eq.	Boltzmann equation	Collisionless Boltzmann equation

dielectric electrodes by using electrical power sources ranging from direct current to radiofrequency levels (tens of kHz to several hundred MHz and GHz).

1.2 APPLICATION OF LOW-TEMPERATURE PLASMA

Low-temperature plasma technology differs from that of collisionless plasma in that a low-temperature plasma is produced and maintained in a collision-dominated region and exhibits proper characteristics and functions intrinsic to the quantum state of the feed gas molecules. This is one of the primary

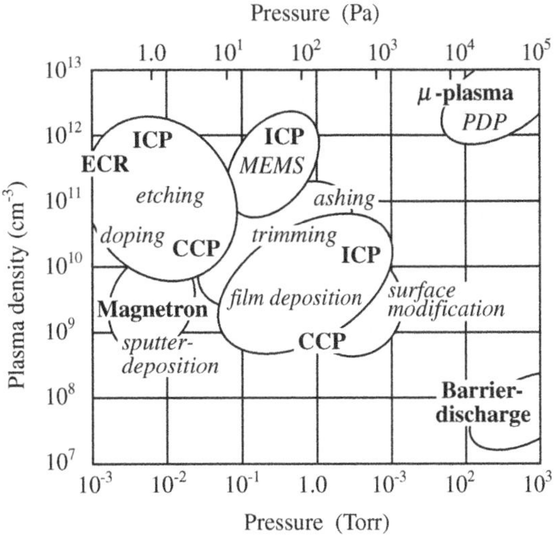

FIGURE 1.2 Low-temperature plasma and material processes.

advantages of low-temperature plasmas for material processing and device fabrication requiring a variety of surface processes and very different reactions among materials adjacent to each other. The technology assisted by low-temperature plasma is generally referred to as plasma processing and is classified into plasma-enhanced chemical vapor deposition, plasma etching, sputtering, ashing, surface modification, and so on, on a scale of size ranging from nanometers to meters (Figure 1.2). Typical geometrical arrangements of an ultra-large-scale integrated circuit demonstrated partly in Figure 1.3 are manufactured through more than 100 plasma processes.

The principal plasma processes for material fabrication are etching and deposition, which are highly competitive. In an era of nanotechnology, plasma technology has developed into a combination of top-down and bottom-up processes. The bottom-up (chemical) approach is effectively achieved under plasma-enhanced conditions.

1.3 ACADEMIC FUSION

In the past two decades, great progress has been made in the application of low-temperature plasma [2]. Most currently used models of low-temperature radiofrequency plasma were proposed between the mid-1980s and the early 1990s [3]. Further achievements in plasma technology at the design level will lead to greater miniaturization of ultra-large-scale integrated (ULSI) circuits in microelectronics and nanoelectronics and to the further functionalization of new materials. These developments will be synchronized with the progress of computers utilizing high-speed and high-performance ULSI chips. We stand

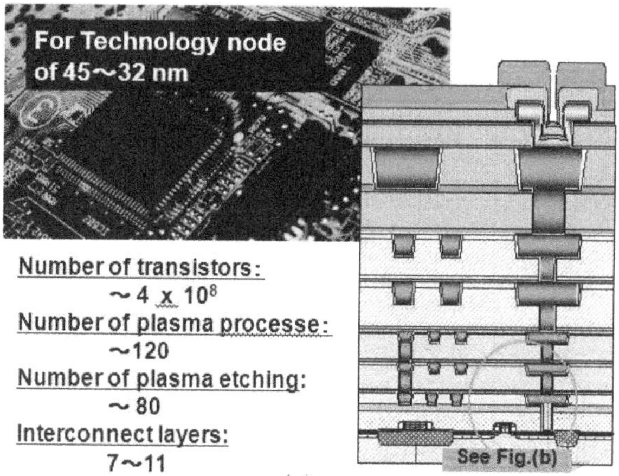

(a)

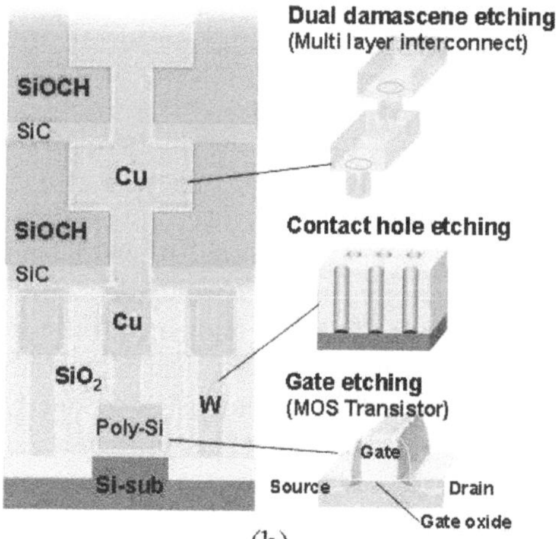

(b)

FIGURE 1.3 Schematic diagram of ULSI (a) and the structure to be etched by a reactive plasma (b). *From Makabe, T, and Tatsumi, T. Plasma Sources Science and Technology, "Workshop on atomic and molecular collision data for plasma modelling: database needs for semiconductor plasma processing", Volume 20, Number 2, 2011. With permission.*

at the advent of the design era of sophisticated plasmas based on atomic and molecular physics both in the gas phase and in the surface phase.

The low-temperature plasma and related technologies will join the exciting and practical fields of academic fusion reconstructed by computational science

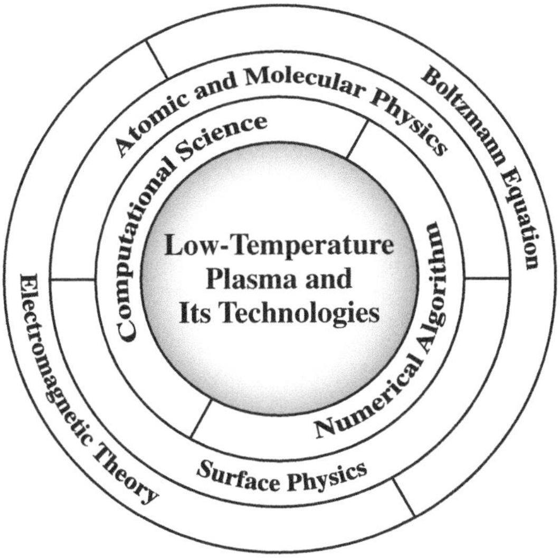

FIGURE 1.4 Academic fusion for low-temperature plasmas.

with quantum atomic and molecular physics, surface physics and chemistry, Boltzmann equation of particles, and Maxwell's electromagnetic theory as is shown in Figure 1.4.

To address the design issue, it is essential to prepare a two- or three-dimensional image of the plasma structure in a reactor and feature profile on a substrate surface by experimental observations or computational modeling.

References

[1] Goldston R.J. and Rutherford P.H. 1995. *Introduction to Plasma Physics.* Bristol: IOP.

[2] American Institute of Physics. 2003. 50 years of science, technology, and the AVS (1953–2003), *J. Vac. Sci. Technol. A (Special Issue)* 21:5.

[3] Makabe, T., Ed. 2002. *Advances in Low Temperature RF Plasmas, Basis for Process Design.* Amsterdam: Elsevier.

CHAPTER 2

Phenomenological Description of the Charged Particle Transport

In this chapter we attempt to lay out the foundation of the phenomenology of charged particle transport in gases. We are primarily concerned with electrons, although ions are also covered briefly. We leave the more detailed description of transport theory for one of the later chapters; here we describe aspects of transport that may be understood from a phenomenological consideration of the charged particles either in the real or in velocity space. We derive the corresponding velocity distributions and transport (swarm) properties and some of their basic relationships. We also derive general equilibrium distributions of charged particles in real and velocity space in thermal equilibrium.

2.1 TRANSPORT IN REAL (CONFIGURATION) SPACE

Typically, without the external electric field, a charged particle (we discuss mainly the case of electrons here) swarm will acquire a Gaussian spatial profile in density. This profile is the result of the particle's random thermal motion and collisions with gas molecules, provided that initially all particles start from the same position. Long-range Coulomb interactions between charged particles have no influence on the development of the swarm in a weakly ionized gas. As a result, electrons will have a net movement in the negative direction of the field superimposed on their random motion by collisions in all directions under an external field. When a field is added, the Gaussian spatial

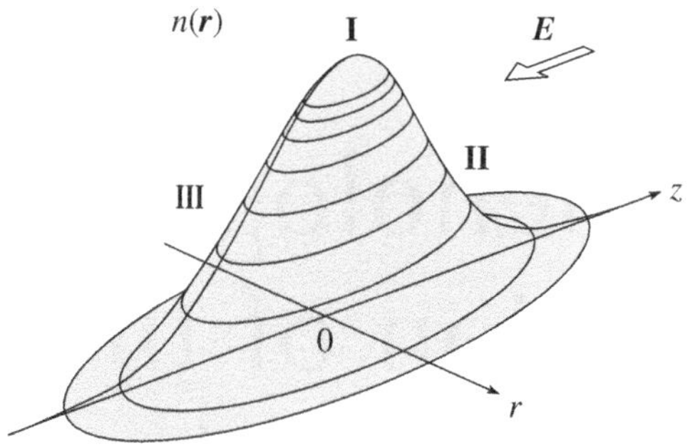

FIGURE 2.1 Spatial profile of electron density for a swarm released from a single point in space and allowed to develop under the influence of electric field and collisions with background molecules.

profile will be slightly shifted and even skewed in the direction opposite to that of the field.

A swarm of electrons and its development are considered in a direct current (DC) electric field $\boldsymbol{E}(=-E\mathbf{k})$, where $\boldsymbol{k}$ is a unit vector along the z-axis. For the purpose of this chapter, we assume that the transport properties are uniformly distributed throughout the electron swarm, and we mainly follow the semiquantitative analysis of Parker and Lowke [1].

2.1.1 Momentum Balance of Electrons

The effect of field will also result in the net flux of electrons ($\boldsymbol{\Gamma}$) in the direction opposite to the field. In other words, a net effective velocity ($\boldsymbol{v}_0$) may be used to describe the average motion of a swarm of electrons with number density distribution $n(\boldsymbol{r})$. If a swarm introduced into an electrical field acquires a uniform distribution, it will attain an average, *drift* velocity ($\boldsymbol{v}_d$). When we release a swarm of particles from a single point as described above, then there will be a spatial density distribution of charged particles $n(\boldsymbol{r})$, as shown in Figure 2.1. Thus the density gradient will result in flux due to diffusion ($-\boldsymbol{D}\partial n/\partial \boldsymbol{r}$) in the direction opposite to that of the gradient. The total flux of particles in one-dimensional space along a DC electric field $\boldsymbol{E} = -E\boldsymbol{k}$ is then given by

$$\boldsymbol{\Gamma} = n\boldsymbol{v}_0 = n\boldsymbol{v}_d - D\frac{\partial n}{\partial z}\boldsymbol{k}, \tag{2.1}$$

where D is the diffusion coefficient.

The momentum balance of electrons will be applied to derive the transport parameters. Electrons gain velocity and consequently average momentum only along the axis of the electrical field. Then, the total change of the momentum of the swarm per unit time is equal to the impulse $e\boldsymbol{E}$. The momentum change is caused by the collision with background gas, mainly the elastic collision with the rate of R_m in a low-energy electron. The momentum balance equation is therefore written as

$$\frac{d}{dt}(mnv_z\boldsymbol{k}) = e(-\boldsymbol{E})n - mn\left(\boldsymbol{v}_d - \frac{D}{n}\frac{\partial n}{\partial z}\boldsymbol{k}\right)R_m, \tag{2.2}$$

where m and $e(>0)$ are the mass and charge of the electron, respectively. In the stationary state, $dv_z/dt = 0$, the total differential on the left-hand side of Equation 2.2 is separated into two parts, one of which is zero. Applying transformation $v_z = \partial z/\partial t$, we separate the Equation 2.2 is into two with respect to the electron number density $n(\boldsymbol{r})$ and its spatial derivative

$$\begin{aligned} 0 &= (-e\boldsymbol{E} - m\boldsymbol{v}_d R_m)\, n, \\ 0 &= \left(mv_z^2 - mDR_m\right)\frac{\partial n}{\partial z}. \end{aligned} \tag{2.3}$$

It is justified (as shown later) to assume that the z component of the velocity is not substantially perturbed by the electric field-induced drift, and therefore $\frac{1}{3}v^2 = v_x^2 = v_y^2 \sim v_z^2$. In that case we may simplify the second part of Equation 2.3 to give the approximate value of the diffusion coefficient as

$$\boldsymbol{v}_d = -\frac{e\boldsymbol{E}}{mR_m}. \tag{2.4}$$

$$D = \frac{v^2}{3R_m}. \tag{2.5}$$

Problem 2.1.1
Derive Equations 2.4 and 2.5.

2.1.2 Energy Balance of Electrons

The former derivation of the properties of the electron swarm relied only on momentum balance as given by Equation 2.2. Now, we consider the energy balance starting from the concept of the mean energy $\langle\varepsilon_m\rangle$. The change of the total energy of the system of electrons with $n(\boldsymbol{r})$ is given by

$$\frac{d}{dt}(n\langle\varepsilon_m\rangle) = e(-\boldsymbol{E})n \cdot \left(\boldsymbol{v}_d - \frac{D}{n}\frac{\partial n}{\partial z}\boldsymbol{k}\right) - n\frac{2m}{M}\langle\varepsilon_m\rangle R_m, \tag{2.6}$$

where M is the mass of the gas molecules. The first term on the right-hand side represents the energy gain by the movement along the field axis in unit time, and the second term represents the collisional energy losses, given by the sum of the elastic loss, $2m\langle\varepsilon_m\rangle/M$. When we substitute Equations 2.4 and 2.5 in the energy balance of Equation 2.6, we obtain for the stationary state

$$\frac{e^2E^2}{mR_m} - \frac{2eE\langle\varepsilon_m\rangle}{3mR_m}\frac{1}{n}\frac{\partial n}{\partial z} - \frac{2m}{M}\langle\varepsilon_m\rangle R_m = 0. \tag{2.7}$$

We consider solving this equation under certain conditions, particularly in regard to the spatial profile of the electron number density $n(z)$.

1. $\partial n/\partial z = 0$: This corresponds to the point **I** in Figure 2.1, that is, to the maximum of the density distribution. From Equation 2.7 we then obtain
$$\langle \varepsilon_{m0} \rangle = \langle \varepsilon_m \rangle|_{\frac{\partial n}{\partial z}=0} = \frac{M}{2m^2}\left(\frac{eE}{R_m}\right)^2. \tag{2.8}$$
2. $\partial n/\partial z \neq 0$: This corresponds to all points except for **I** in Figure 2.1. For the region with a negative density gradient (region **II** in Figure 2.1) the mean energy is higher than that at the peak (point **I**):
$$\langle \varepsilon_m \rangle = \langle \varepsilon_{m0} \rangle + \Delta\varepsilon.$$

So far we have assumed that the collisional rate was independent of the electron energy. However, this is normally not the case. Thus we use a Taylor expansion around the mean energy at the peak **I** for the collision rate:

$$R_m(\langle \varepsilon_m \rangle) = R_{m0} + \frac{\partial R_m}{\partial \varepsilon}|_0 \Delta\varepsilon,$$

which in combination with Equations 2.7 gives

$$\langle \varepsilon_m \rangle = \langle \varepsilon_{m0} \rangle \left[1 - \frac{2\langle \varepsilon_{m0} \rangle}{3eE\left(1 + 2\frac{\partial R_m}{\partial \varepsilon}|_0 \frac{\langle \varepsilon_{m0} \rangle}{R_{m0}}\right)} \frac{1}{n}\frac{\partial n}{\partial z}\right]. \tag{2.9}$$

Exercise 2.1.1
Discuss the value of the mean electron energy in two cases, (a) $\partial n/\partial z < 0$ and (b) $\partial n/\partial z > 0$, as compared with the mean energy at the peak of the swarm $\langle \varepsilon_{m0} \rangle$. For case (a), the gradient of the electron number density shown in Figure 2.1 is negative as in region **II**, and therefore $\langle \varepsilon_m \rangle > \langle \varepsilon_{m0} \rangle$. However, for case (b), the gradient of the electron number density is positive (see region **III** in Figure 2.1) and therefore $\langle \varepsilon_m \rangle < \langle \varepsilon_{m0} \rangle$. Thus, unless the collision rate R_m changes very rapidly, we may expect the mean energy of electrons to increase toward the front of the swarm where flux due to diffusion adds to the drift-induced component. The spatial profile of energy may have consequences on the transport coefficients, provided that some energy-dependent processes also exist, which is generally the case.

Problem 2.1.2
Derive expression 2.9 for the mean electron energy $\langle \varepsilon_m \rangle$.

In the presence of density gradients in the swarm, random motion under collisions with gas molecules will result in diffusion. So far we have considered diffusion only in one dimension along the direction of the electric field. When, however, we consider three-dimensional space defined by the Cartesian coordinates $\boldsymbol{r}(x, y, z)$ and the corresponding unit vectors $(\boldsymbol{i}, \boldsymbol{j}, \boldsymbol{k})$, while keeping the same direction of the electric field $\boldsymbol{E} = -E\boldsymbol{k}$, the electron number density develops as $n(x, y, z)$. The conditions of the zero gradient in one direction now become lines along the surface of the density profile. The total flux of electrons in Equation 2.1 are revised for the three-dimensional case as

$$\boldsymbol{\Gamma} = nv_d\boldsymbol{k} - D_0\left(\frac{\partial n}{\partial x}\boldsymbol{i} + \frac{\partial n}{\partial y}\boldsymbol{j}\right) - D_0\left(1 - \frac{\xi}{1+2\xi}\right)\frac{\partial n}{\partial z}\boldsymbol{k}, \tag{2.10}$$

where

$$\xi = \frac{\langle\varepsilon_{m0}\rangle}{R_{m0}}\frac{\partial R_m}{\partial\varepsilon}\bigg|_0. \tag{2.11}$$

As a result of the influence of the electric field, the diffusion along the z-axis (direction of the field) has special properties and is in general different from the diffusion in perpendicular directions. The diffusion of electrons (and charged particles in general) becomes anisotropic with two components of the diffusion tensor $\mathbf{D}$: D_L (longitudinal diffusion coefficient) and D_T (transverse diffusion coefficient). These two coefficients may be given as

$$D_L = D_0\left(1 - \frac{\xi}{1+2\xi}\right), \tag{2.12}$$

and

$$D_T = D_0 = \frac{v^2}{3R_m}, \tag{2.13}$$

which are valid only for approximations that were involved in the derivation of the formulae. In the case where the collision rate R_m is independent of the mean energy, the diffusion will be isotropic. The same is true for the zero electric field.

In this section we have seen that the transport of electrons (the same is true for ions) may be described with the aid of transport coefficients. We have given approximate relations for the drift velocity and components of the diffusion tensor. In general these properties will depend on the mean electron energy. However, it is not practical to use mean energy for tabulating the data for transport coefficients. It is better to use an external parameter associated with the value of the external electric field $\boldsymbol{E}$. The reduced electric field is defined as the ratio between the magnitude of the electric field $\boldsymbol{E}$ and the neutral gas number density N, that is, E/N. It is shown that the reduced field is proportional to the energy gained by electrons between two collisions. E/N is expressed in units known as Townsend (in honor of the founder of gaseous electronics), which is defined as 1 Td $= 10^{-21}\,\mathrm{Vm}^2 = 10^{-17}\,\mathrm{Vcm}^2$ (see Figure 2.2).

Problem 2.1.3

By substituting Equation 2.9 into Equation 2.6, derive the longitudinal D_L (Equation 2.12) and transverse D_T (Equation 2.13) coefficients. Assuming analytic dependence of the collision rate R_m on energy, find the ratio D_L/D_T

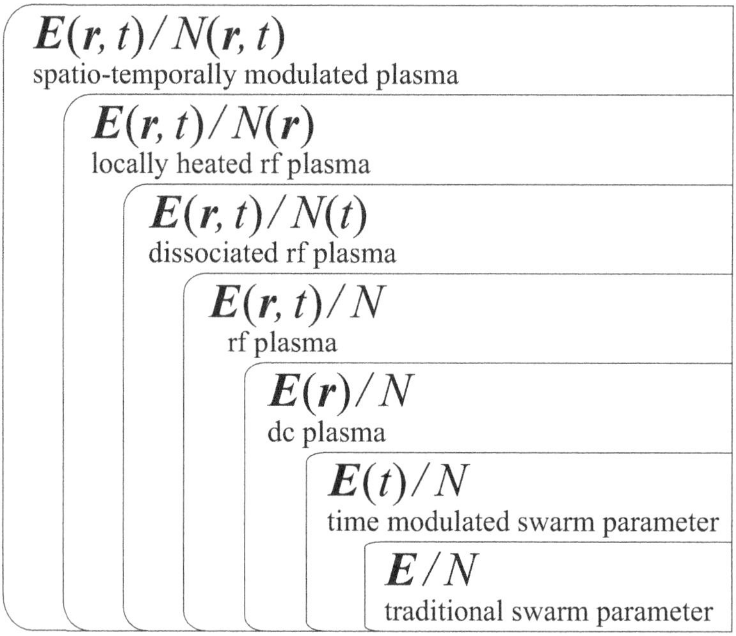

FIGURE 2.2 Hierarchical structure of the reduced electric field, E/N.

in analytic form. Discuss the conditions when the ratio is equal to, less than, or greater than 1. (A hint: choosing $R_m \sim (\langle\varepsilon_m\rangle/\langle\varepsilon_{m0}\rangle)^{(\ell+1)/2}$ *is particularly convenient.)*

Problem 2.1.4

Pressure p *may be used to represent the gas number density* N*; find the relationship between* E/N *and* E/p *given that*

$$E/N[\mathrm{Vcm}^2] = 1.036 \times 10^{-19} T_g[\mathrm{K}] E/p[\mathrm{Vcm}^{-1}\mathrm{Torr}^{-1}],$$

where T_g *is gas temperature. Here, at* $T_g = 273$ *K,* E/P *of 1* Vcm^{-1}*Torr is equal to 2.823 Td (*$= 10^{-17}\,\mathrm{Vcm}^2$*). Prove that the mean energy gain between two collisions is proportional to* E/N*.*

2.2 TRANSPORT IN VELOCITY SPACE

An ensemble of electrons may be studied in velocity space. Whereas in the real space the spatial density distribution $n(\boldsymbol{r})$ was used to describe the swarm, in velocity space we may consider the probability distribution of velocities. The velocity distribution $g(\boldsymbol{v})$ is defined so that the quantity $g(\boldsymbol{v})d\boldsymbol{v}$ is the probability of finding particles with velocity $\boldsymbol{v}$ within the small element $d\boldsymbol{v}$. Due to the symmetry with respect to the v_z-axis, it is advantageous to use a polar coordinate system (v, θ, φ). Thus $d\boldsymbol{v} = v^2 \sin\theta \; d\theta \; d\varphi$. The magnitude of the velocity $v = |\boldsymbol{v}|$ is associated with the energy of electrons $\varepsilon = mv^2/2$, and in the case of thermal equilibrium the distribution is given by the so-called Maxwellian distribution:

$$g_M(v) = \left(\frac{m}{2\pi kT}\right)^{3/2} \exp\!\left(-\frac{mv^2}{2kT}\right). \tag{2.14}$$

At the same time, we may define the velocity distribution function $g(\boldsymbol{v}, \boldsymbol{r}, t)$, which describes swarm development with respect to both velocity and real space and with respect to time.

2.2.1 Electron Velocity Distribution and Swarm Parameters

Thermal equilibrium, however, does not occur for low-density swarms of charged particles in an electric field. Normally, an electric field affects electrons, which gain energy but are not able to dissipate it to give translational motion to gas molecules or ions, because elastic collisions are not particularly effective in energy transfer to the target due to the very different masses; that is, $m/M \ll 1$. Thus electrons in swarms and in low-density collisional plasmas are normally not in thermal equilibrium. Two distinct conditions may occur for the electron swarm in an electric field. In the case that the properties of the swarm are uniform in the real space, we will have a quasi-equilibrium. Then, the energy gained by the field is dissipated in collisions, both elastic and inelastic. As a result, properties of electrons given by electron energy distribution do not change within the swarm. In such a case (i.e., under hydrodynamic conditions), it will be possible to separate spatial and velocity distributions. However, close to the initial stage or boundary wall or soon after the release the properties of the electrons may not be uniform in space or time or both. These nonequilibrium conditions are not expressed in terms of thermodynamic equilibrium; however, in relation to the balance between energy gain and loss, the swarm will develop until the balance is achieved or the swarm is lost. Such cases are described as a nonlocal (in space) or not relaxed electron swarm transport.

In this section we focus on relaxed spatially uniform swarms that are in the hydrodynamic regime. Thus the velocity distribution provides a complete description of the swarm. Figure 2.3 shows a velocity distribution in velocity space. Here, we take advantage of the azimuthal symmetry of the swarm in velocity space and present the distribution as a function of the axial (axis of electric field) velocity v_z and radial (perpendicular) velocity v_r. The electric field produces asymmetry of the distribution with a peak shifted in one direction. The small difference between the components in the direction and opposite to the direction of the field produces a net drift of particles. At the same time the velocities directed in opposite directions occur at all points throughout the swarm profile. In the bulk of the spatial swarm profile, loss of particles moving outside a small section is compensated by gain from the nearby

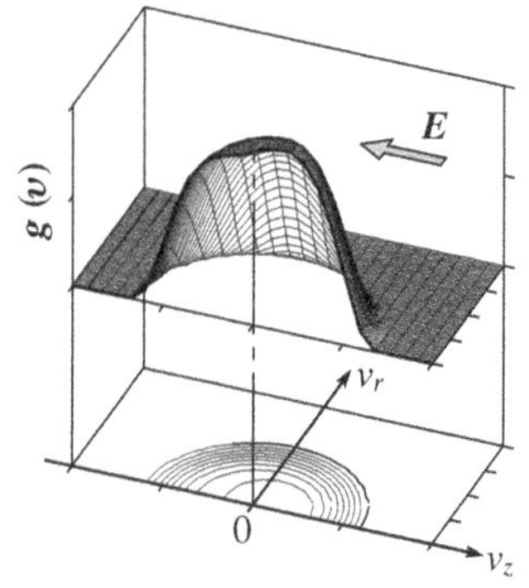

FIGURE 2.3 Typical electron velocity distribution $g(\boldsymbol{v})$ as a function of axial v_z and radial v_r velocities under a DC electric field.

sections that have a similar density; this, however, is not the case at the edges. The velocity distribution thus contains information about both drift and diffusion.

Problem 2.2.1

The drift velocity and mean energy of electrons in argon under the uniform field, $E = 100$ Vcm^{-1}, and at pressure of 1 Torr (at temperature of 300 K) are $v_d = 2.3 \times 10^7$ cms^{-1} and 8 eV, respectively. Show, then, that the mean random velocity v_r and drift velocity v_d satisfy the relation $(v_d/v_r) \ll 1$.

In Figure 2.4 we show the effect of the electric field on the velocity distribution $g(\boldsymbol{v})$ at each velocity $\boldsymbol{v}$ with and without the field. When the electric field $\boldsymbol{E}$ is turned on, a small component of velocity $\Delta\boldsymbol{V}$ is added to all electrons in the direction opposite to the field. In the simplified model the circular contour representing random (thermal) isotropic motion of electrons is slightly shifted as if its center were moved to the position $\mathbf{O}'$ from $\mathbf{O}$. The resulting velocity is $\boldsymbol{v}'$, which is given by: $\boldsymbol{v}' - \boldsymbol{v} \cong \Delta\boldsymbol{V}\cos\theta$ (where θ is the angle between $\boldsymbol{v}'$ and $-\boldsymbol{E}$). We can now estimate the effect of the electric field on the electron velocity distribution as

$$g(\boldsymbol{v}') = g(\boldsymbol{v} + \Delta\boldsymbol{V}\cos\theta) \cong g_0(v) + \Delta\boldsymbol{V}\cos\theta\frac{dg_0(v)}{dv}, \tag{2.15}$$

where we applied the condition $|\,\boldsymbol{v}\,| \gg |\,\Delta\boldsymbol{V} cos\,\theta\,|$. The distribution $g(\boldsymbol{v}')$ is composed of an isotropic part in all directions and having, in general, a non-Maxwellian term and a small anisotropic term. Both terms are influenced by the electric field. This expansion is analogous to a two-term expansion in spherical harmonics of the velocity distribution, where the second term is the anisotropic term (see Chapter 5).

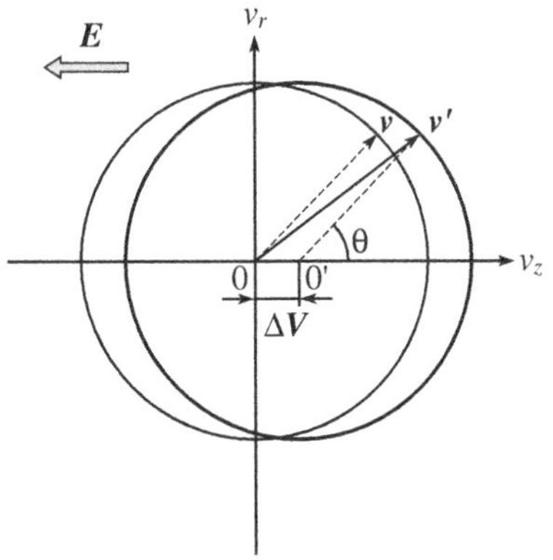

FIGURE 2.4 Contour plot of electron velocity distribution $g(\boldsymbol{v})$ as a function of axial v_z and radial v_r velocities. When the electric field is turned on, a small component of velocity $\Delta \boldsymbol{V}$ is added to all electrons in the direction opposite to the field $\boldsymbol{E}$.

Now we consider the momentum conservation of electrons with velocity distribution $g(\boldsymbol{v})$ in gases under an electric field. First, let us define $Q_m(v)$ as the momentum transfer cross section of the electron by elastic scattering with the background gas molecule with number density N. Then, for one group of electrons with velocity $\boldsymbol{v}$, the collision frequency of the momentum exchange is $NQ_m(v)v$, and the total momentum dissipated in collisions in the direction opposite to the field $\boldsymbol{E}$ is $mv\cos\theta NQ_m(v)v$. The total momentum exchange, therefore, is obtained by averaging over all velocities by using the distribution, and in the steady state we have

$$\frac{\partial(nm < \boldsymbol{v} >)}{\partial t} = e(-\boldsymbol{E})n - \int nm\boldsymbol{v}NQ_m(v)vg(\boldsymbol{v})d\boldsymbol{v} = 0. \tag{2.16}$$

Here, $g(\boldsymbol{v})$ is normalized as $\int g(\boldsymbol{v})d\boldsymbol{v} = 1$, having $d\boldsymbol{v} = 2\pi \sin\theta d\theta v^2 dv$. When we substitute Equation 2.15 into Equation 2.16, we obtain the following for the second term in Equation 2.16:

$$\begin{aligned}&\int nmv^2NQ_m(v)\left[g_0(v) + \Delta \boldsymbol{V}\cos\theta\frac{dg_0(v)}{dv}\right]\cos\theta 2\pi\sin\theta d\theta v^2 dv \\ &= \frac{4\pi}{3}\int nmv^4NQ_m(v)\Delta \boldsymbol{V}\frac{dg_0(v)}{dv}dv.\end{aligned}$$

After we apply partial integration in order to replace the derivative of the velocity distribution and take into account the symmetry of the distribution

$g_0(v)$, we obtain

$$-\frac{4\pi}{3}\int m\frac{d}{dv}\left[v^4 NQ_m(v)\Delta\boldsymbol{V}\right]g_0(v)dv = e\boldsymbol{E}. \tag{2.17}$$

This is an integral equation that is satisfied for an arbitrary field $\boldsymbol{E}$ and cross sections $Q_m(v)$ only if $\Delta\boldsymbol{V} = -\left[e\boldsymbol{E}/mNQ_m(v)v\right]$. In this case we write the expansion for $g(\boldsymbol{v})$ as

$$g(\boldsymbol{v}) = g_0(v) - \frac{eE\cos\theta}{mNQ_m(v)v}\frac{dg_0(v)}{dv}. \tag{2.18}$$

The second term in Equation 2.18 shows the effect of an external field on the velocity distribution. The cosine term is the lowest-order development of the actual distribution in the velocity space in addition to the symmetric $g_0(v)$ term. Usually the cross section for the momentum transfer of elastic scattering is much greater than the sum of cross sections for other inelastic collisions. So far we have taken into account only elastic collisions. The energy balance of electrons will consist of energy loss in such collisions, as we mentioned before, $2m\langle\varepsilon\rangle/M$. Therefore, the total energy loss in elastic collisions in unit time is expressed as

$$\frac{2m}{M}\frac{1}{2}mv^2 NQ_m(v)vg(\boldsymbol{v})d\boldsymbol{v}.$$

The energy gain, on the other hand, is determined by the motion of electrons along the electric field axis:

$$eEv\cos\theta g(\boldsymbol{v})d\boldsymbol{v}.$$

If we make the balance and apply the expansion 2.18 we obtain

$$\begin{aligned}&\int\frac{m}{M}mv^2 NQ_m(v)v\left[g_0(v) - \frac{eE\cos\theta}{mNQ_m(v)v}\frac{dg_0(v)}{dv}\right]d\boldsymbol{v}\\ &= \int eEv\cos\theta\left[g_0(v) - \frac{eE\cos\theta}{mNQ_m(v)v}\frac{dg_0(v)}{dv}\right]d\boldsymbol{v}.\end{aligned} \tag{2.19}$$

After integration over the polar angles θ and φ and by equating the terms inside the integrals, we obtain the following first-order differential equation:

$$4\pi\frac{m}{M}mv^2 NQ_m(v)vg_0(v) = -\frac{4\pi}{3}eEv\frac{eE}{mNQ_m(v)v}\frac{dg_0(v)}{dv}.$$

The solution will be obtained by direct integration without making any assumptions on the velocity dependence of the cross section $Q_m(v)$:

$$g_0(v) = A\exp\left[-\frac{3m}{M}\int_0^v\left(\frac{NQ_m(v)v}{eE/m}\right)^2 vdv\right], \tag{2.20}$$

where A should be determined from the normalization $\int g_0(v)d\boldsymbol{v} = 1$. Now we perform the conversion to electron energy distribution $f(\varepsilon)$:

$$g_0(v)4\pi v^2 dv = f(\varepsilon)d\varepsilon; \quad \frac{1}{2}mv^2 = \varepsilon. \tag{2.21}$$

The Maxwellian distribution in Equation 2.14 in thermal equilibrium is rewritten as a function of energy ϵ:

$$f_M(\varepsilon) = A\sqrt{\varepsilon}\exp\left(-\frac{3\varepsilon}{2\langle\varepsilon\rangle}\right), \tag{2.22}$$

and the mean energy is given by $\langle\varepsilon\rangle = 3kT_e/2$. In a more general case when there is an electric field but for a constant cross section Q_m, we obtain from Equation 2.20 the so-called Druyvestyn distribution:

$$f_D(\varepsilon) = A\sqrt{\varepsilon}\exp\left[-0.548\left(\frac{\varepsilon}{\langle\varepsilon\rangle}\right)^2\right]. \tag{2.23}$$

Exercise 2.2.1
Plot the distributions $f(\epsilon)$, $f_M(\epsilon)$, and $f_D(\epsilon)$ for electrons in Ar at 100 Td. Here the mean energy $< \epsilon >$ is 6.47 eV.
Figure 2.5 shows the Maxwellian, Druyvestyn, and nonequilibrium (f_0) energy distributions for electrons at the same mean energy. Note the differences among the three distributions.

Now we proceed to determine the transport coefficients based on the energy or velocity distributions in velocity space. The drift velocity is defined in this chapter as the average velocity of a swarm; that is,

$$\boldsymbol{v}_d = \int_v \boldsymbol{v}g(\boldsymbol{v})d\boldsymbol{v}\ . \tag{2.24}$$

Combination of expression 2.18 and Equation 2.24 gives

$$\begin{aligned} \boldsymbol{v}_d &= \int_{v,\theta} v\cos\theta\left[g_0(v) - \frac{e\boldsymbol{E}\cos\theta}{mNQ_m(v)v}\frac{dg_0(v)}{dv}\right]2\pi v^2\sin\theta d\theta dv \\ &= -\frac{4\pi}{3}\frac{e\boldsymbol{E}}{m}\int\frac{v^2}{NQ_m(v)}\frac{dg_0(v)}{dv}dv. \end{aligned} \tag{2.25}$$

When we make the transition to the energy distribution in Equation 2.21 we obtain

$$\boldsymbol{v}_d = -\frac{1}{3}\sqrt{\frac{2}{m}}e\boldsymbol{E}\int\frac{\varepsilon}{NQ_m(\varepsilon)}\frac{d}{d\varepsilon}\left(\frac{f(\varepsilon)}{\sqrt{\varepsilon}}\right)d\varepsilon, \tag{2.26}$$

where the value of $(2/m)^{1/2}eE$ of an electron with 1 eV is 5.93×10^7 cm^{-1}. The transverse diffusion coefficient D_T may be derived on the basis of the

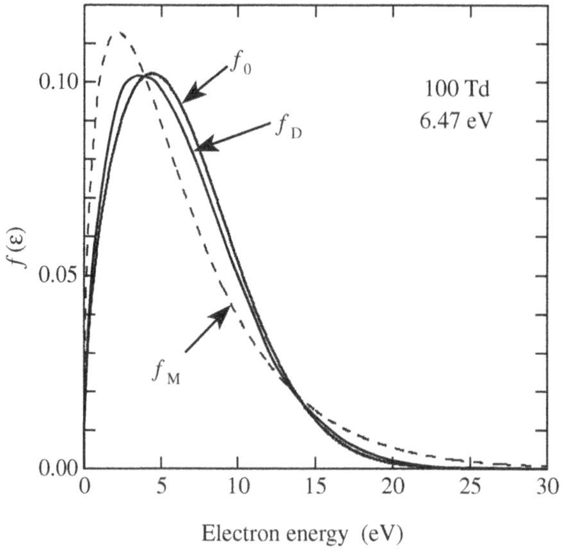

FIGURE 2.5 Comparison among $f(\epsilon)$, $f_M(\epsilon)$, and $f_D(\epsilon)$ with the same mean energy.

standard formula $\lambda\langle v\rangle/3$ without field (i.e., for neutral molecules), where λ is the mean free path between two collisions given by $\langle (NQ_m)^{-1}\rangle$. Then, by using $g(\boldsymbol{v})$, D_T is expressed in an integral form:

$$D_T = \frac{1}{3}\int \frac{v}{NQ_m(v)} g(\boldsymbol{v})d\boldsymbol{v} = \frac{4\pi}{3}\int \frac{v^3}{NQ_m(v)} g_0(v)dv. \tag{2.27}$$

The expression by the energy distribution is

$$D_T = \frac{1}{3}\sqrt{\frac{2}{m}}\int \frac{\sqrt{\varepsilon}}{NQ_m(\varepsilon)} f(\varepsilon)d\varepsilon. \tag{2.28}$$

In addition to elastic collisions, there are a number of inelastic processes that may occur. The most significant is the ionization whereby electron collision with a molecule of the background gas produces the pair of a new electron and positive ion. The inelastic processes have a threshold energy as the molecular quantum structure and they are usually characterized by the rate (number of events per second per electron). The ionization rate is calculated from the ionization cross section $Q_i(\epsilon)$ as

$$\begin{aligned} R_i &= \int_v NQ_i(v)vg(\boldsymbol{v})d\boldsymbol{v} \\ &= \int_{v,\theta} NQ_i(v)v\left[g_0(v) - \frac{e\boldsymbol{E}\cos\theta}{mNQ_m(v)v}\frac{dg_0(v)}{dv}\right]2\pi v^2 \sin\theta d\theta dv \\ &= \int_v NQ_i v(v)g_0(v)4\pi v^2 dv \\ &= \sqrt{\frac{2}{m}}\int_\varepsilon NQ_i(\varepsilon)\sqrt{\varepsilon}f(\varepsilon)d\varepsilon. \end{aligned} \tag{2.29}$$

Another traditional way to describe ionization events is to define the number of ionization events that one electron makes while crossing a unit distance α (also known as Townsend's ionization coefficient), which can be easily defined as

$$\alpha = \frac{R_i}{v_d}. \tag{2.30}$$

Normalization of the energy distribution $f(\varepsilon)$ used in all calculations of transport coefficients is given by $\int f(\varepsilon)d\varepsilon = 1$ (see Equation 2.21). The process that leads to production of a negative ion and loss of the original electron is the electron attachment described by the cross section $Q_a(\epsilon)$. These two processes, ionization and electron attachment, change the number of electrons and thus are labeled as nonconservative. All these coefficients or rates are directly used in deriving electron transport (fluid) equations.

There are also a number of inelastic but conservative (number-conserving) processes (Q_k) that are associated with a wide range of possible energy losses (thresholds). Typically these energy losses ε_k are quite large, and these processes are much more efficient in controlling the mean energy of electrons than elastic processes, which have an energy loss of approximately $2m\langle\varepsilon\rangle/M$. On the other hand, the momentum loss (exchange) is most influenced by the elastic scattering, whose cross section is usually much greater than the sum of the inelastic processes, $Q_m(\varepsilon) \gg Q_i(\varepsilon) + Q_a(\varepsilon) + \Sigma Q_k(\varepsilon)$. Therefore, we measure the contribution of the elastic scattering to the electron transport in terms of the momentum transfer. Rates and spatial excitation coefficients are defined in the same way as for ionization. However, these coefficients do not directly enter the transport equations, although they may be used in some corrections and energy balance equations.

Thus momentum transfer is dominated by elastic collisions, whereas energy balance is primarily controlled by inelastic processes (with the exception of rare gases and some metallic vapors that have no inelastic processes at low energies). However, it is not practical to use mean energy as the swarm parameter because it cannot be measured directly, so we traditionally define a property of electron swarms known as characteristic energy $\varepsilon_k = eD_T/\mu$, where $\mu = v_d/E$ is the mobility. Characteristic energy has the dimension of energy and is usually within 30% of the mean energy (although sometimes the differences may be up to a factor of two or more). This quantity is usually expressed in units eV or, even more often, by the analogous quantity D/μ, which is expressed in volts. The definition of characteristic energy is given by combining Equations 2.26 and 2.28:

$$\varepsilon_k = e\frac{D_T}{\mu} = e\frac{D_T}{v_d}E = \frac{-eE\frac{1}{3}\sqrt{\frac{2}{m}}\int \frac{\sqrt{\varepsilon}}{NQ_m(\varepsilon)}f(\varepsilon)d\varepsilon}{\frac{1}{3}\sqrt{\frac{2}{m}}eE\int \frac{\varepsilon}{NQ_m(\varepsilon)}\frac{d}{d\varepsilon}\left(\frac{f(\varepsilon)}{\sqrt{\varepsilon}}\right)d\varepsilon},$$

which for the thermal Maxwellian distribution in Equation 2.22, in particular, becomes

$$\varepsilon_k = \frac{\int \frac{\varepsilon}{NQ_m(\varepsilon)}\exp\left(-\frac{3}{2}\frac{\varepsilon}{\langle\varepsilon\rangle}\right)d\varepsilon}{\frac{3}{2\langle\varepsilon\rangle}\int \frac{\varepsilon}{NQ_m(\varepsilon)}\exp\left(-\frac{3}{2}\frac{\varepsilon}{\langle\varepsilon\rangle}\right)d\varepsilon} = \frac{2}{3}\langle\varepsilon\rangle \ \ [\text{eV}]. \tag{2.31}$$

In the case of the thermal equilibrium where $\langle\varepsilon\rangle = 3kT/2$, we obtain the following relation:

$$\frac{D_T}{\mu} = \frac{kT}{e}. \tag{2.32}$$

This is also known as Einstein's relation, because Einstein derived an identical equation for the Brownian motion of particles. Moreover, Nernst derived the equation for the electrolytic transport of ions, and Townsend derived it for the electrons in gas discharges even before Einstein, so this relation is often referred to as the Nernst–Townsend relation. This relation is valid only for thermal equilibrium, although it is often used, especially in plasma modeling, for nonequilibrium conditions. In that case, however, it is possible to use corrected forms that are more appropriate.

Problem 2.2.2

Discuss the momentum transfer loss of inelastic collision, and compare the influence of the collisional effect between the energy loss and the momentum loss.

2.2.2 Ion Velocity Distribution and Mean Energy

Ions, in principle, follow the same laws that determine the electron transport in gases under the influence of electric or magnetic fields. The basic differences are that the mass of ions M_p is of the same order of magnitude as that of gas molecules M, and that the charge q may not be equal to e and is either positive or negative (here we use the subscript p to denote positive ions, but the conclusions are generally valid for the transport of negative ions as well, only the collisional processes are of a different nature). The energy exchange in elastic collisions is generally significant in an ion swarm.

Let us consider an ion swarm with uniform density in real space in a steady state in gases under an external field $\boldsymbol{E} = E\boldsymbol{k}$. At low kinetic energy of ions, the collision between the ion and neutral molecule is subject to an induced-dipole interaction, that is, constant mean free time τ, which is given by

$$\tau = \frac{1}{1.105}\left(\frac{\epsilon_0}{\pi}\right)^{1/2}\frac{1}{qN}\left(\frac{M_pM}{M_p+M}\right)^{1/2}\alpha^{-1/2} ,$$

where α is the polarizability of the ion in the dipole interaction with the neutral molecule [2]. The governing equation of the velocity distribution $G(\boldsymbol{v})$ of ions is then given in the form of the relaxation equation using the BGK approximation [3, 4],

$$\frac{dG(\boldsymbol{v})}{dt} = \frac{qE}{M_p}\frac{dG(\boldsymbol{v})}{dv_z} = -\frac{G(\boldsymbol{v})-G^0(\boldsymbol{v})}{\tau} . \tag{2.33}$$

Here $G^0(\boldsymbol{v})$ is the distribution just after a collision and is approximated by a two-temperature displaced Maxwellian velocity distribution,

$$G^0(\boldsymbol{v}) = \frac{\beta_\perp}{\pi}\left(\frac{\beta_\parallel}{\pi}\right)^{1/2}\exp\left[-\beta_\perp v_r^2 - \beta_\parallel(v_z - < v_z > + qE\tau/M_p)^2\right] , \tag{2.34}$$

where $\beta_\perp = 1/(2 < v_x^2 >)$, and $\beta_\parallel = 1/2[< v_z^2 > - < v_z >^2 - (qE\tau/M_p)^2]$. Equation 2.33 is then solved as

$$\begin{aligned} G(\boldsymbol{v}) &= \frac{\beta_\perp}{2\pi(qE\tau/M_p)}\exp\left(-\beta_\perp v_r^2 - \frac{v_z - < v_z > + qE\tau/M_p}{qE\tau/M_p} + \frac{1}{4\beta_\parallel(qE\tau/M_p)^2}\right) \\ &\times \operatorname{erfc}\left[2\beta_\parallel^{1/2}\frac{qE\tau}{M_p}\left(\frac{1}{4\beta_\parallel(qE\tau/M_p)^2} - \frac{v_z - <v_z> + qE\tau/M_p}{2(qE\tau/M_p)}\right)\right] , \end{aligned} \tag{2.35}$$

where erfc is a complementary error function.

The momentum transfer of an ion in the elastic collision is $M(\boldsymbol{v}_p - \boldsymbol{V})/(M_p+M)$ where $\boldsymbol{V}$ is the velocity of the gas molecule. When we introduce the relative velocity $\boldsymbol{v}_\gamma = \boldsymbol{v}_p - \boldsymbol{V}$, we can write the momentum balance as

$$\int M_p\frac{M}{M_p+M}v_{\gamma z}\boldsymbol{k}(1-\cos\omega)N\sigma_m(v_\gamma,\omega)v_\gamma G(\boldsymbol{v}_\gamma)d\boldsymbol{v}_\gamma = q\boldsymbol{E} , \tag{2.36}$$

where σ_m is the differential cross section, ω is the angle of scattering in the center of mass system, and $G(\boldsymbol{v}_\gamma)$ is the relative velocity distribution of ions.

As the collision between the ion and the molecule at low energy is subject to the induced-dipole interaction, $\int(1-\cos\omega)N\sigma_m(v_\gamma,\omega)v_\gamma d\Omega$ is a constant (R_m), and Equation 2.36 reduces to

$$\frac{M_pM}{M_p+M}R_m\int v_{\gamma z}\boldsymbol{k}G(\boldsymbol{v}_\gamma)d\boldsymbol{v}_\gamma = q\boldsymbol{E} .$$

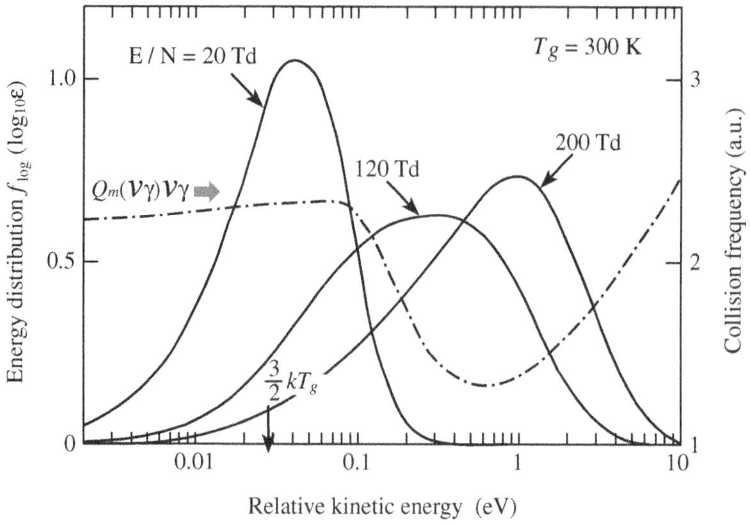

FIGURE 2.6 Energy distribution at three different E/N and elastic momentum transfer cross section of O^- in O_2.

The integral represents the mean velocity, that is, drift velocity $\boldsymbol{v}_{dp}$ in velocity space, which is then equal to

$$\boldsymbol{v}_{dp} = \frac{M_p + M}{M_p M} \frac{q\boldsymbol{E}}{R_m}. \tag{2.37}$$

When an ion elastically affects a molecule and is isotropically scattered in the center-of-mass system, the averaged energy loss per collision is given by

$$\Delta\varepsilon_l = \frac{M_p M}{(M_p + M)^2} \left[M_p \langle v_p^2 \rangle - M \langle V^2 \rangle \right]. \tag{2.38}$$

Problem 2.2.3
Derive the relation 2.38.

In a steady state, the energy balance of the ion swarm in unit time is given by

$$q\boldsymbol{E} \cdot \boldsymbol{v}_{dp} - \Delta\varepsilon_l R_m = qEv_{dp} \left[1 - \frac{M_p M}{(M_p + M)^2} \left(M_p \langle v_p^2 \rangle - M \langle V^2 \rangle \right) \frac{M_p + M}{M_p M} \frac{1}{v_{dp}^2} \right] = 0, \tag{2.39}$$

which is satisfied when

$$(M_p + M) v_{dp}^2 + M \langle V^2 \rangle - M_p \langle v_p^2 \rangle = 0.$$

By using this relation we obtain the expression of the mean ion energy, which is known as Wannier's formula [2]:

$$\langle \varepsilon_p \rangle = \frac{1}{2} M_p \langle v_p^2 \rangle = \frac{1}{2}(M_p + M) v_{dp}^2 + \frac{1}{2} M \langle V^2 \rangle. \tag{2.40}$$

The physical meaning of Equation 2.40 is that the mean energy of ions is separated into three parts: one due to the effect of the electric field given by the drift velocity (first term), one due to the apparent kinetic energy of the encounter molecule (second term), and the thermal energy equal to $3kT_g/2$ (third term).

Problem 2.2.4

When an ion swarm has a shifted Maxwellian velocity distribution in gases with a Maxwell distribution of T_g, derive that the relative velocity distribution between the ion and the molecule $G(\boldsymbol{v}_r)$ also has the Maxwellian velocity distribution.

2.3 THERMAL EQUILIBRIUM AND ITS GOVERNING RELATIONS

In the case of thermal equilibrium, all processes are balanced by opposite processes, so it is possible to make some quite general distributions both in real and velocity space. In this section, we deal primarily with electrons inasmuch as extensions to other charged particles are trivial. One should bear in mind that in thermal equilibrium all particles should in principle have the same temperature and that the temperature remain constant throughout the whole region, which is in equilibrium. This rule of homogeneous temperature is often difficult to satisfy when it comes to plasmas. We therefore introduce the term "local thermodynamic equilibrium" (LTE), which means that the conditions for thermal equilibrium are satisfied at each point in the plasma but that the value of the effective temperature varies with position. In the case of LTE, the temperature is merely a parameter used to apply the laws of thermodynamic equilibrium to a plasma, but this approximation is nevertheless quite effective.

2.3.1 Boltzmann Distribution in Real Space

When there is an external electric field, the flux of charged particles consists of drift and diffusion components. Let us consider the z-axis along the field direction. In a weak field, the drift velocity $\boldsymbol{v}_d$ is proportional to $\boldsymbol{E}$ and is written as $\boldsymbol{v}_d = \mu \boldsymbol{E}$ using the mobility μ. Therefore, the flux in the z direction is expressed as

$$\begin{aligned} \boldsymbol{\Gamma} &= n\langle \boldsymbol{v} \rangle \\ &= n\boldsymbol{v}_d - D\nabla n = n\mu \boldsymbol{E} - D\frac{\partial n}{\partial z}\boldsymbol{k}. \end{aligned} \tag{2.41}$$

Given that, in thermal equilibrium under randomization, the mean velocity must be zero $\langle \boldsymbol{v} \rangle = 0$, Equation 2.41 leads to a differential equation that connects the spatial density distribution of charged particles $n(z)$ and the spatial potential $V(z)$:

$$\frac{\mu}{D}E = \frac{-\mu}{D}\frac{\partial V(z)}{\partial z} = \frac{1}{n}\frac{\partial n}{\partial z}. \tag{2.42}$$

This equation is solved by direct integration if we define the boundary condition such that for $z = 0$, the potential is $V(0) = 0$ and the density is $n(0) = n_0$. In this case the density distribution is

$$n(z) = n_0 \exp\left(-\frac{\mu}{D}V(z)\right). \tag{2.43}$$

We substitute the Einstein relation 2.32 into Equation 2.43 and obtain

$$n(z) = n_0 \exp\left(-\frac{eV(z)}{kT}\right), \tag{2.44}$$

which is known as Boltzmann's distribution law. This equation describes how the population of charged particles in thermal equilibrium with effective energy kT will be distributed under the space potential $V(z)$. Thus, even for a potential that corresponds to an energy much larger than the thermal energy, there will be a certain number of electrons, although this number is very small. The Boltzmann distribution is often applied to determine the population of excited molecules at different discrete energy levels, in which case it is actually applied to bound electrons and their population at discrete bound levels of atoms or molecules.

2.3.2 Maxwell Distribution in Velocity Space

The probability that a charged particle will be found in a small volume of velocity space defined by $(v_x, v_x + dv_x), (v_y, v_y + dv_y)$, and $(\mathrm{v_z}, \mathrm{v_z} + \mathrm{dv_z})$ is given by

$$g(v_x, v_y, v_z)dv_x dv_y dv_z\,, \tag{2.45}$$

where $g(v_x, v_y, v_z)$ is the velocity distribution. Under conditions of thermal equilibrium, the distribution $g(v_x, v_y, v_z)$ must satisfy the following conditions:

(a) Each of the components of the velocity distribution must be fully independent of each other, and

(b) The velocity distribution must have the same magnitude in all directions, and the mean values of velocities in all directions must be zero.

In Figure 2.7 we show a typical distribution function in two dimensions. When we apply conditions (a) and (b), we expand the distribution into three independent functions

$$\begin{aligned} g(v_x, v_y, v_z)dv_x dv_y dv_z &= G(v_x)dv_x G(v_y)dv_y G(v_z)dv_z \\ &= g(v^2)dv_x dv_y dv_z\,, \end{aligned} \tag{2.46}$$

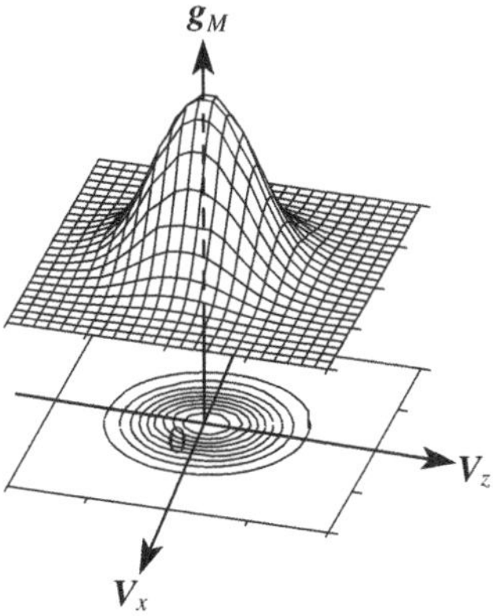

FIGURE 2.7 Equilibrium velocity distribution in v_x–v_y space.

where the magnitude of the velocity $\boldsymbol{v}$ is defined by $v^2 = v_x^2 + v_y^2 + v_z^2$, and the distribution in terms of v^2 is then defined as

$$g(v^2) = G(v_x)G(v_y)G(v_z). \tag{2.47}$$

We must choose the function g to satisfy the functional Equation 2.47.

Exercise 2.3.1
Show that the solution to Equation 2.47 is

$$g\left(v_x^2\right) = a^2 \exp\left(-\alpha v_x^2\right), \quad (\alpha > 0). \tag{2.48}$$

Let us observe a special case of one-dimensional motion where $v_y = v_z = 0$. In this case, Equation 2.48 is written as

$$\begin{aligned} g\left(v_x^2\right) &= G(v_x)G(0)G(0) \\ &= a^2 G(v_x), \end{aligned}$$

where $a = G(0)$ and, by the same argument, $g(v_y^2) = a^2 G(v_y)$ and $g(v_z^2) = a^2 G(v_z)$. Therefore, Equation 2.47 is rewritten as

$$g(v^2) = g\left(v_x^2 + v_y^2 + v_z^2\right) = \frac{1}{a^6} g\left(v_x^2\right) g\left(v_y^2\right) g\left(v_z^2\right). \tag{2.49}$$

Note that the functional Equation 2.49 satisfies Equation 2.48 as the solution of each component.

Therefore, we obtain

$$g(v^2) = g\left(v_x^2 + v_y^2 + v_z^2\right) = A \exp\left(-\alpha\left(v_x^2 + v_y^2 + v_z^2\right)\right), \tag{2.50}$$

where the value of the constants A and α must be determined. When we normalize the velocity distribution $g(v_x, v_y, v_z)$ to 1,

$$\begin{aligned} 1 &= \int g(v_x^2+v_y^2+v_z^2)dv_x dv_y dv_z \\ &= A\int_{-\infty}^{\infty} \exp(-\alpha v_x^2)dv_x \int_{-\infty}^{\infty} \exp(-\alpha v_y^2)dv_y \int_{-\infty}^{\infty} \exp(-\alpha v_z^2)dv_z \\ &= A\left(\sqrt{\frac{\pi}{\alpha}}\right)^3, \end{aligned}$$

and therefore

$$A = \left(\sqrt{\frac{\alpha}{\pi}}\right)^{3/2}.$$

Now we apply the distribution $g(v_x, v_y, v_z)$ to determine the mean kinetic energy of charged particles with mass m at temperature T according to the conditions of thermodynamic equilibrium

$$\begin{aligned} \frac{3}{2}kT &= \frac{1}{2}m\langle v^2\rangle = \left(\frac{\alpha}{\pi}\right)^{3/2}\frac{m}{2}\int (v_x^2+v_y^2+v_z^2)e^{-\alpha(v_x^2+v_y^2+v_z^2)}dv_x dv_y dv_z \\ &= \left(\frac{\alpha}{\pi}\right)^{3/2}\frac{3m}{2}\frac{1}{2\alpha}\left(\frac{\pi}{\alpha}\right)^{3/2}, \end{aligned}$$

which leads directly to

$$\alpha = \frac{m}{2kT} \quad \text{and} \quad A = \left(\frac{m}{2\pi kT}\right)^{3/2}. \tag{2.51}$$

We use the constants given by Equation 2.51 to obtain the final form of the Maxwellian distribution from Equation 2.50,

$$g_M = g(v_x, v_y, v_z) = \left(\frac{m}{2\pi kT}\right)^{3/2} \exp\left(-\frac{m(v_x^2+v_y^2+v_z^2)}{2kT}\right). \tag{2.52}$$

In the previous derivation of Maxwell's velocity distribution, we used the following integrals: $\int_0^\infty e^{-\alpha x^2}dx = 1/2\sqrt{\pi/\alpha}$, and $\int_0^\infty x^2 e^{-\alpha x^2}dx = 1/2\alpha\sqrt{\pi/\alpha}$. In place of Maxwell's velocity distribution, the Maxwellian speed distribution $F(v)$ is defined as

$$\begin{aligned} F(v)dv &= \int_\theta \int_\varphi g_M(v^2)v^2 sin\,\theta d\theta d\varphi dv \\ &= 4\pi v^2 g_M(v^2)dv\,. \end{aligned} \tag{2.53}$$

The following averaged velocities are defined for the Maxwellian speed distribution: (i) most probable speed $\tilde{v}$, (ii) root mean square velocity $\sqrt{\langle v^2\rangle}$, and (iii) mean speed $\langle v\rangle$. These are obtained as follows.

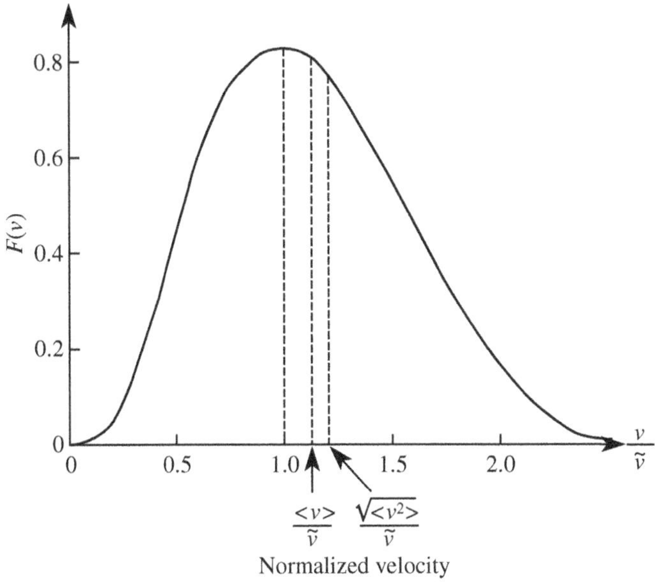

FIGURE 2.8 Normalized Maxwell speed distribution and the relationships among these averaged velocities.

1. The most probable speed $\tilde{v}$ is obtained from the condition $dF/dv = 0$, and its value is

$$\tilde{v} = \left(\frac{2kT}{m}\right)^{1/2}. \tag{2.54}$$

2. The root mean square velocity $(\langle v^2 \rangle)^{1/2}$ is the square root of the mean value of v^2:

$$\sqrt{\langle v^2 \rangle} = \sqrt{\int v^2 F(v) dv} = \sqrt{\frac{3kT}{m}}. \tag{2.55}$$

3. The mean speed is obtained as

$$\langle v \rangle = \int v F(v) dv = \sqrt{\frac{8kT}{\pi m}}. \tag{2.56}$$

We can easily compare the magnitudes of these three mean values as

$$\tilde{v} < \langle v \rangle < \sqrt{\langle v^2 \rangle}. \tag{2.57}$$

Figure 2.8 shows the plot of the normalized Maxwell speed distribution $F(v)$ and gives the values of the three possible choices for averaging the velocity. Different physical quantities may be averaged with different powers of speed and will be associated with different effective velocities. It is clear that the mean speed may be used to establish the momentum transfer, that the mean square speed is relevant for consideration of the mean energy, and that the most probable speed is the peak of the probability distribution. Although one may directly produce averages of physical phenomena, associating them with mean values of speed provides a better physical insight into the phenomenon under consideration.

References

[1] Parker, J.H., Jr. and Lowke, J.J. 1969. *Phys. Rev.* 181:290–301.

[2] Wannier, G.H. 1953. *Bell Syst. Tech. J.* 32:170.

[3] Bhatnagar, P.L., Gross, E.P., and Krook, M. 1954. *Phys. Rev.* 94:511.

[4] Whealton, J.H. and Woo, S.B. 1971. *Phys. Rev. A* 6:2319.

CHAPTER 3

Macroscopic Plasma Characteristics

3.1 INTRODUCTION

Plasma is a compound phase with electric quasi-neutrality, generally consisting of electrons, positive ions, and neutral molecules. In particular, strongly ionized plasma consists of electrons and positive ions. The phase is called collisionless plasma, as the electron (ion) has few short-range binary collisions with neutral molecules, and the plasma system is subject to long-range Coulomb interactions. When the short-range two-body collision is major, the system is collisional, that is, collision dominated. Collisionless plasma shows unique characteristics. We briefly describe the characteristics [1–3].

3.2 QUASI-NEUTRALITY

The presence of electrons and ions with density n_e and n_p will produce an electric field subject to Poisson's equation,

$$div\boldsymbol{E} = e\frac{n_p - n_e}{\epsilon_0}, \tag{3.1}$$

where $e(>0)$ is the elemental charge and ϵ_0 is the permittivity (dielectric constant) of vacuum. A plasma where the density of positive ions is exactly equal to that of electrons has no electric field. Naturally, based on the great difference in mass between the electron and positive ion, a quasi-neutral state with $n_p \sim n_e$ is maintained in plasmas on a macroscopic scale. This causes a finite positive plasma potential with respect to the wall earthed to the ground.

Notice that a low-temperature plasma externally excited by direct current (DC) or rf power source has a plasma production region and holds a structure

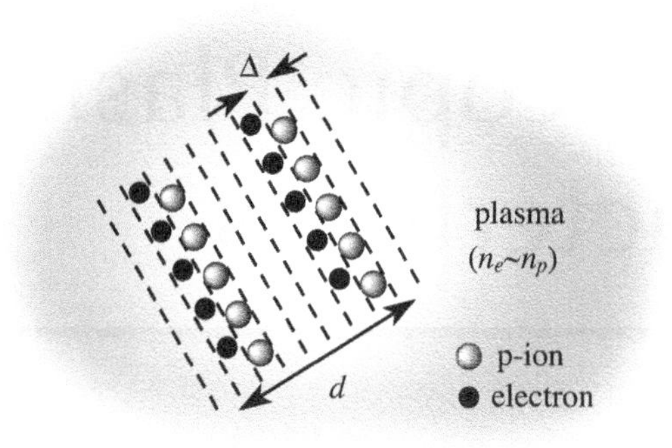

FIGURE 3.1 Plasma fluctuation and the charge separation in a collisionless plasma with macroscopic quasi-neutrality.

bifurcating into a positive ion sheath with a high field to produce an electron–ion pair and a bulk plasma with quasi-neutrality.

3.3 CHARGE-SEPARATION IN PLASMAS

3.3.1 Spatial Scale of Charge-Separation

In a bulk plasma, quasi-neutrality will be macroscopically kept in space and time. The quasi-neutral condition, however, will be locally disturbed by the intrinsic random motion of the charged particles in a plasma. We first assume the presence of a localized space–charge layer as shown in Figure 3.1 as a result of the random fluctuation of charged particles in a bulk plasma with macroscopic quasi-neutrality. In one-dimensional space, the local field and the potential difference caused by the two layers is given by Poisson's Equation 3.1 as

$$E = \frac{en\Delta}{\varepsilon_0} \quad \text{and} \quad V = \frac{en\Delta d}{\varepsilon_0},$$

where $n(\sim n_e \sim n_p)$, Δ is the thickness of each of the layers, and d is the distance between layers.

In a collisionless plasma, when the potential energy of the charged particle in the layers is less than the random thermal energy, kT_e or kT_p,

$$e\frac{en\Delta d}{\varepsilon_0} < kT_e, \tag{3.2}$$

the local disturbance of the charge-separation will be kept. Considering the scale of $\Delta \sim d$,

$$d < \left\{ \frac{\varepsilon_0 k T_e}{n e^2} \right\}^{1/2} \equiv \lambda_D. \tag{3.3}$$

Here, λ_D defined in Equation 3.3 gives a microscopic maximum spatial scale for the charge-separation and is known as the electron Debye (shielding) length or Debye radius.

Exercise 3.3.1

Calculate the Debye length when plasma density and electron temperature in a collisionless plasma are $10^{16}\,\mathrm{m}^{-3}$ and 3 eV, respectively.

$$\lambda_D = \left\{ \frac{\varepsilon_0 k T_e}{n e^2} \right\}^{1/2} \sim 7.43 \times 10^3 \times \left(\frac{kT[\mathrm{eV}]}{n[\mathrm{m}^{-3}]} \right)^{1/2} [\mathrm{m}]. \tag{3.4}$$

That is,

$$\lambda_D = 7.43 \times 10^3 \times \left(\frac{3}{10^{16}} \right)^{1/2} = 1.29 \times 10^{-4}\ [\mathrm{m}].$$

3.3.2 Time Scale for Charge-Separation

As there exists a huge difference in mass between the electron and the positive ion, the electron layer in Figure 3.1 will go toward the positive ion layer under the Coulomb force. In a collisionless plasma, the local drift motion of the electron layer passes through the massive positive layer by inertia, and the pair of charged layers is oppositely reformed. The electron layer will continue to vibrate around the massive positive ions in the absence of two-body collision between the neutral molecules. The temporal motion is described by the harmonic oscillation,

$$m \frac{d^2}{dt^2} \Delta(t) = -\frac{e^2 n}{\varepsilon_0} \Delta(t). \tag{3.5}$$

The oscillation is known as the electron plasma oscillation or Langmuir oscillation. The frequency of the electron plasma oscillation, electron plasma frequency, is numerically given as

$$\omega_e = \left(\frac{e^2 n}{m \varepsilon_0} \right)^{1/2} \sim 56.4 \left(n[\mathrm{m}^{-3}] \right)^{1/2}\ [\mathrm{s}^{-1}], \tag{3.6}$$

The time scale of the charge-separation in a collisionless plasma τ_{es}, given by ω_e^{-1}, has the relation with λ_D,

$$\frac{\lambda_D}{\tau_{es}} = \left(\frac{kT_e}{m_e} \right)^{1/2} \approx V_e, \tag{3.7}$$

where V_e is the electron thermal speed.

Problem 3.3.1

Discuss that the plasma frequency ω_p of the ion with charge Ze is defined in the same way as the electron,

$$\omega_p = \left(\frac{Z^2 e^2 n_p}{M_p \varepsilon_0}\right)^{1/2} = Z\left(\frac{m}{M_p}\right)^{1/2} \omega_e. \tag{3.8}$$

Also discuss that practical plasma oscillation is given by $(\omega_e^2 + \omega_p^2)^{1/2} = \omega_e(1 + z^2 m_e/M_p)^{1/2}$.

Problem 3.3.2

Estimate the values of ω_e and ω_p in collisionless Ar plasma with density 10^{16} m^{-3}.

Problem 3.3.3

A plasma is externally irradiated by an electromagnetic wave with frequency of ω. Derive the condition that the external disturbance does not immerse deeply in the plasma.

3.4 PLASMA SHIELDING

3.4.1 Debye Shielding

When a material is inserted into a plasma, the surface of the material is immediately charged regardless of metal or insulator, and it is electrically shielded from the surrounding plasma (see Figure 3.2). We consider the spatial scale of the shielding caused by a spherical metal inserted into a plasma with zero space potential. The potential distribution close to the metal is obtained by Poisson's equation as a function of radial distance r,

$$\nabla^2 V(r) = \frac{1}{r^2}\frac{d}{dr}\left(r^2 \frac{dV(r)}{dr}\right) = -\frac{e}{\varepsilon_0}(n_p - n_e). \tag{3.9}$$

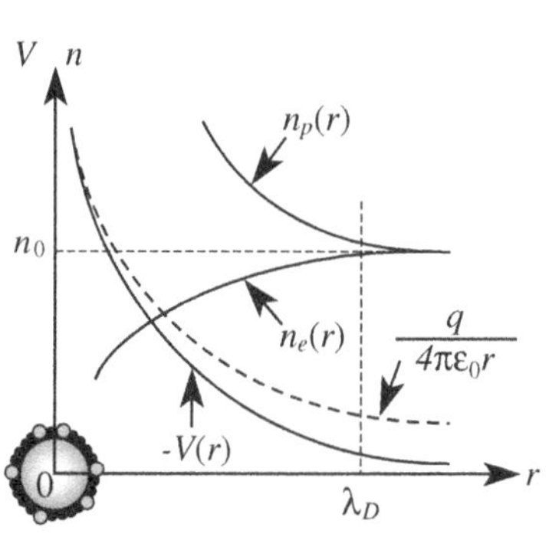

FIGURE 3.2 Plasma shielding of a charged sphere.

There are a number of electrons and ions in the shielded region in front of the sphere. The number density of electrons and ions, $n_e(\boldsymbol{r})$ and $n_p(\boldsymbol{r})$, follows the Boltzmann distribution in space, discussed in Section 2.3,

$$n_p = n_0 \exp\left(-\frac{eV(r)}{kT_p}\right), \tag{3.10}$$

$$n_e = n_0 \exp\left(\frac{eV(r)}{kT_e}\right), \tag{3.11}$$

where n_0 is the quasi-neutral plasma density far from the influence of the impurity metal. At a long distance from the metallic surface satisfying $eV(r) \ll kT_e, kT_p$, the exponential terms in Equations 3.10 and 3.11 will be expanded by the Taylor series, and we obtain

$$\begin{aligned}\frac{1}{r^2}\frac{d}{dr}\left(r^2\frac{dV(r)}{dr}\right) &= -\frac{en_0}{\varepsilon_0}\left\{\exp\left(-\frac{eV(r)}{kT_p}\right) - \exp\left(\frac{eV(r)}{kT_e}\right)\right\} \\ &\approx \frac{e^2 n_0}{\varepsilon_0}\frac{T_p + T_e}{kT_pT_e}V(r).\end{aligned} \tag{3.12}$$

The general converging solution of Equation 3.12 is

$$V(r) = \frac{A}{r}\exp\left(-\frac{r}{\lambda_D}\right),$$

where A (> 0) is constant and λ_D is

$$\lambda_D^2 = \frac{\varepsilon_0 kT_pT_e}{e^2 n_0 (T_p + T_e)}. \tag{3.13}$$

The potential close to the metal with a surface charge q_0 is expressed by the Coulomb potential, $q_0/4\pi\epsilon_0 r$. Then, the final form is

$$V(r) = \frac{q_0}{4\pi\varepsilon_0 r}\exp\left(-\frac{r}{\lambda_D}\right). \tag{3.14}$$

Note that $V(r)$ is negative due to $q_0(< 0)$ under $m_e \ll M_p$. Equation 3.14 shows that the scale length of the shielding is on the order of the Debye length λ_D.

3.4.2 Metal Probe in a Plasma

A small metallic probe with variable potential is inserted into a plasma in a steady state. A static shielding sheath is formed there without plasma production or loss. We consider the voltage-current characteristics $I(V)$ when the potential at the probe is externally varied with respect to the wall of the plasma. They are known as probe characteristics in a plasma. Figure 3.3 shows typical probe characteristics.

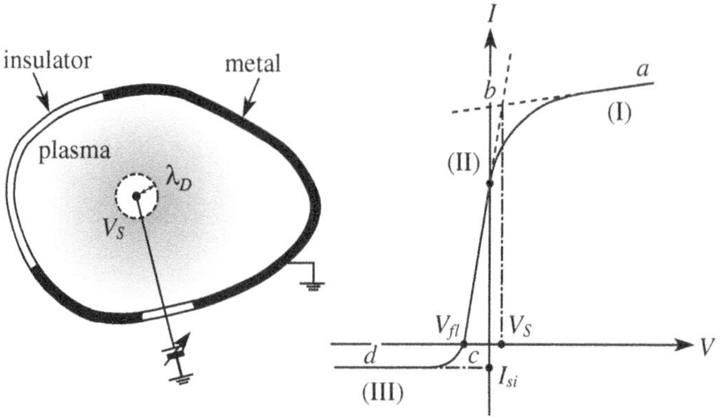

FIGURE 3.3 Ideal Langmuir probe characteristics in a plasma.

When the probe has a sufficiently large surface, a saturated electron current will be collected at $V > V_s$ where V_s is the space potential of the plasma. Practically due to the finite surface area, $I(V)$ characteristics at $V > V_s$ will increase gradually as shown in Figure 3.3. The probe potential at c where the current is zero is named the floating potential. The name comes from the zero net flux at the surface of an insulator inserted into a plasma.

The probe will be given an absorbing boundary that absorbs electrons and ions incident on the surface. At the region bc that satisfies $V < V_s$, the electron sheath will be formed in front of the probe $(V_s - V < 0)$. The electron incident right on the probe surface will be retarded by the field, and only the electron with kinetic energy $mv^2/2 > |e(V_s - V)|$ will be absorbed as the electron current. We consider that the electron has the Maxwellian velocity distribution with temperature T_e in the plasma and has no collision in the passive sheath in front of the probe. Then the electron probe current will be given by

$$\begin{aligned} I(V) &= en_eS\int_{v_x=-\infty}^{\infty}\int_{v_y=-\infty}^{\infty}\int_{v_z=\sqrt{\frac{2e|V_p|}{m}}}^{\infty} v_z\left(\frac{m}{2\pi kT_e}\right)^{3/2} \\ &\quad\times\exp\left\{-\frac{m(v_x^2+v_y^2+v_z^2)}{2kT_e}\right\}dv_x dv_y dv_z \\ &= en_eS\left(\frac{m}{2\pi kT_e}\right)^{3/2}\int_{-\infty}^{\infty}\exp\left(-\frac{mv_x^2}{2kT_e}\right)dv_x\int_{-\infty}^{\infty}\exp\left(-\frac{mv_y^2}{2kT_e}\right) \\ &\quad\times dv_y\int_{v_z}^{\infty}\exp\left(-\frac{mv_z^2}{2kT_e}\right)dv_z \\ &= en_eS\left(\frac{m}{2\pi kT_e}\right)^{3/2}\frac{2\sqrt{\pi}}{2\sqrt{\frac{m}{2kT_e}}}\frac{2\sqrt{\pi}}{2\sqrt{\frac{m}{2kT_e}}}\frac{1}{\frac{m}{kT_e}}\exp\left(-\frac{eV_p}{kT_e}\right), \end{aligned} \tag{3.15}$$

where S is the effective area of the probe. By considering the mean speed $\langle v_e \rangle = (8kT_e/\pi m)^{1/2}$ in Equation 3.15, the probe $I(V)$ characteristics are described by

$$I(V) = \frac{en_e \langle v_e \rangle}{4} S \exp\left(\frac{V - V_s}{kT_e}\right). \tag{3.16}$$

Equation 3.16 shows that the electron temperature T_e in the plasma is obtained by the gradient of $\ln I(V)$. At ab satisfying $V > V_s$ (electron saturated region), all electrons approaching the probe will be collected, the lower limit of the integral in Equation 3.15 will be zero, and

$$I = \frac{en_e \langle v_e \rangle}{4} S. \tag{3.17}$$

Equation 3.17 means that electron current consists of $n_e/2$ in one-dimensional position space and $\langle v_e \rangle/2$ in velocity space. In principle, by using the electron temperature T_e, the plasma density (electron density) will be given by the saturated curve, Equation 3.17, in the region of bc. The probe having one small metal is called the single probe or the Langmuir probe. Other types of probes are also used as a simple tool for plasma diagnostics.

Exercise 3.4.1

Discuss the pressure condition that the electron temperature T_e is estimated from the curve in region (II) in bc in Figure 3.3.

As a typical plasma we assume, $n_e = 10^{15}$ m^{-3}, $kT_e = 3.0$ eV. Then, the Debye length is

$$\lambda_e = 7.43 \times 10^3 \times \left(\frac{kT_e[\text{eV}]}{n_e[\text{m}^{-3}]}\right)^{1/2} = \frac{7.43 \times 10^3 \times \sqrt{3}}{\sqrt{10} \times 10^7} = 4.07 \times 10^{-4} \text{ [m]}.$$

The mean speed of electrons is

$$\langle v_e \rangle = (8kT_e/\pi m)^{1/2} = 6.71 \times 10^7 \times \sqrt{kT_e\,[\text{eV}]} \approx 1.16 \times 10^6 \text{ [ms}^{-1}\text{]}.$$

The collision rate R is roughly approximated at $10^7 p$[Pa s^{-1}], and the flight time of the electron in the static sheath in front of the probe is

$$\frac{\lambda_e}{\langle v_e \rangle} \approx \frac{4.07 \times 10^{-4}}{1.16 \times 10^6} \approx 3.51 \times 10^{-10} \ll \frac{1}{R} \left(\approx \frac{1}{10^7 p}\right).$$

Therefore, the collisionless condition is obtained as

$$p \ll \frac{1}{10^7} \times \frac{1}{3.51 \times 10^{-10}} \approx 2.85 \times 10^2 \text{ [Pa]}.$$

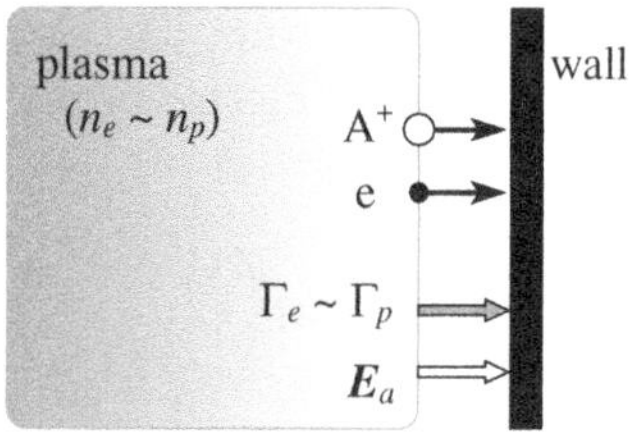

FIGURE 3.4 Fluxes of charged particles in front of a wall.

3.5 PARTICLE DIFFUSION

3.5.1 Ambipolar Diffusion

We consider a plasma without production in a vessel far from a plasma source. Electrons and ions in the plasma have a random motion with a kinetic energy. Even in the same kinetic energy (temperature) between electrons and ions, the electron flux incident on the wall as a result of the random motion will be much higher than the massive positive ion, and the relation $\Gamma_e \gg \Gamma_p$ is immediately attained (see Figure 3.4). Then, an electric field that accelerates positive ions and decelerates electrons toward the wall surface (i.e., ambipolar field) will be formed, and in a macroscopically steady state, the charged particles' flow to the wall with the relation

$$\Gamma_e \approx \Gamma_p \tag{3.18}$$

is realized.

We deal semiquantitatively with the ambipolar diffusion phenomena caused between the static plasma and the surrounding reactor wall. Each of the fluxes of electrons and ions with density, n_e and n_p, is given,

$$\boldsymbol{\Gamma}_e = n_e \langle \boldsymbol{v}_e \rangle = n_e \boldsymbol{v}_{de} - D_e \frac{dn_e}{dr}, \tag{3.19}$$

$$\boldsymbol{\Gamma}_p = n_p \langle \boldsymbol{v}_p \rangle = n_p \boldsymbol{v}_{dp} - D_p \frac{dn_p}{dr}, \tag{3.20}$$

where $\boldsymbol{v}_j$, $\boldsymbol{v}_{dj}$, and D_j are the mean velocity, drift velocity, and diffusion coefficient of electrons and positive ions, respectively. In a steady state as mentioned above, the electron and ion have the same velocity toward the wall, ambipolar diffusion velocity $\boldsymbol{v}_a$,

$$\boldsymbol{v}_a = -D_a \frac{1}{n} \frac{dn}{dr}, \tag{3.21}$$

where D_a is the ambipolar diffusion coefficient given by

$$D_a = \frac{\mu_p D_e + \mu_e D_p}{\mu_p + \mu_e}, \tag{3.22}$$

where μ_e and μ_p are the mobility of electrons and ions, respectively. In particular, at low reduced field the mobility may be constant. In thermal equilibrium of electrons and ions with Maxwellian velocity distribution, the ambipolar diffusion coefficient is expressed in terms of the electron and ion temperatures, T_e and T_p,

$$D_a = \frac{D_p\left(1 + \frac{T_e}{T_p}\right)}{\left(1 + \frac{\mu_p}{\mu_e}\right)}. \tag{3.23}$$

In the case of $T_e \gg T_p$,

$$D_a = D_p\frac{T_e}{T_p}. \tag{3.24}$$

In particular, in $T_e = T_p$,

$$D_a = 2D_p. \tag{3.25}$$

Problem 3.5.1
Derive the ambipolar diffusion coefficient, Equation 3.22, and show that the ambipolar diffusion field is written as

$$\boldsymbol{E}_a = \frac{D_p - D_e}{\mu_e + \mu_p}\frac{1}{n}\frac{dn}{dr}. \tag{3.26}$$

3.5.2 Spatial and Time Scale of Diffusion

We consider the time and spatial scale of particles in a collisionless or collisional plasma. In the case where the production and loss of the particle is negligible, the continuity equation of the particles is

$$\frac{\partial}{\partial t}n(\boldsymbol{r},t) = D\nabla^2 n(\boldsymbol{r},t). \tag{3.27}$$

The diffusion Equation 3.27 implies that the spatial distribution and time scale of the diffusion are determined by the geometry of the reactor. Here, we divide $n(\boldsymbol{r},t)$ into two independent functions $n(\boldsymbol{r})$ and $T(t)$. Then we have

$$\frac{1}{T(t)}\frac{dT(t)}{dt} = \frac{1}{n(\boldsymbol{r})}D\nabla^2 n(\boldsymbol{r}) \equiv -\frac{1}{\tau_D}. \tag{3.28}$$

Here, $\tau_D(>0)$ is the separation constant.

Problem 3.5.2
Discuss the physical meaning that the r.h.s. of Equation 3.28 has negative value.

Equation 3.28 is written as

$$T(t) = T_0\exp(-t/\tau_D), \tag{3.29}$$

and

$$\nabla^2 n(\boldsymbol{r}) = -\frac{1}{D\tau_D} n(\boldsymbol{r}) = -\frac{n(\boldsymbol{r})}{\Lambda^2}, \quad \text{where} \quad \Lambda^2 = D\tau_D. \tag{3.30}$$

Equation 3.30 shows that the spatial characteristics of particle diffusion Λ^2 are determined by the geometrical boundary condition. Λ is called the characteristic diffusion length and τ_D the diffusion decay time.

Problem 3.5.3

Calculate the characteristic diffusion length Λ in the case where the reactor geometry is (a) an infinite parallel plate (separation of d_0), (b) an infinite rectangle (two sides, d_0 and l_0), (c) a cylinder (radius r_0), and (d) a sphere (radius r_0), respectively.

$$\begin{aligned}\frac{1}{\Lambda^2} &= (\pi/d_0)^2; \text{ infinite plates,} \\ &= (\pi/d_0)^2 + (\pi/l_0)^2; \text{ infinite rectangle,} \\ &= (2.405/r_0)^2; \text{ cylindrical,} \\ &= (\pi/r_0)^2; \text{ spherical.}\end{aligned} \tag{3.31}$$

Problem 3.5.4

Neutral molecules diffuse in three-dimensional space without boundary. Calculate the number density of molecules $n(\boldsymbol{r}, t)$ in Equation 3.27. In particular, when the initial condition of the density is given by $n(\boldsymbol{r}, t = 0) = \delta(\boldsymbol{r})$, derive the density distribution

$$n(\boldsymbol{r}, t) = \left(\frac{1}{4\pi Dt}\right)^{3/2} \exp\left(-\frac{r^2}{4Dt}\right). \tag{3.32}$$

The above stochastic process is known as the Wiener process and the density distribution is the normal distribution $N(0, 2Dt)$ with mean value of 0 and variance of $2Dt$.

The characteristic diffusion length Λ_D is much influenced by the loss mechanism of the particle. For example, we consider the electron diffusion in electronegative gases with density N and with the electron attachment rate coefficient k_a. Then the electron continuity equation is

$$\frac{\partial}{\partial t} n(\boldsymbol{r}, t) = -k_a n(\boldsymbol{r}, t) N + D\nabla^2 n(\boldsymbol{r}, t). \tag{3.33}$$

Equation 3.33 is reduced to

$$\frac{1}{T(t)} \frac{dT(t)}{dt} = -k_a N + \frac{1}{n(\boldsymbol{r})} D\nabla^2 n(\boldsymbol{r}) \equiv -\frac{1}{\tau_D}. \tag{3.34}$$

As a result, the characteristic diffusion length changes from Equation (3.30) to

$$\Lambda^2 = \left(\frac{1}{D\tau_D} - \frac{k_a N}{D}\right)^{-1}. \tag{3.35}$$

Exercise 3.5.1
Molecules excited to an optically forbidden level are called metastables and have a long lifetime. The metastable will be de-excited without photo-emission when it reacts with the wall with absorbed molecules and exhausts its excess inner energy to the wall. Estimate the effective lifetime τ of metastables in a reactor of volume V_{ol} with reflection wall,

$$\tau = \left(\frac{2}{\pi}\right)^2 \frac{\Lambda^2}{D} + \frac{2V_{ol}(2-\beta)}{S_t v \beta}. \tag{3.36}$$

Here, the reflection coefficient γ_r at the wall is related to the surface sticking coefficient S_t by $\gamma_t = 1 - S_t$. v and β are, respectively, the thermal velocity and surface loss probability of the metastable.

3.6 BOHM SHEATH CRITERION

3.6.1 Bohm Velocity

A fluid description of a region between a plasma and material boundary (i.e., plasma sheath) will give a simple physical image from the ambipolar diffusion transport of the charged particle. As described in the previous section, the net current to the surface disappears as a result of the ambipolar diffusion in the steady state, and the surface is kept at the floating potential. The electron and ion dynamics are valid in the sheath under the following assumptions.

1. The frequency of the external power source is much higher than the collision rate $\omega \gg R$ (i.e., the collisionless sheath).

2. In the momentum continuity equation (see details in Chapter 5), the pressure gradient $dP(z)/dz = kTdn_e(z)/dz$ is dominant for electrons, whereas other terms except the pressure are predominant for ions.

3. The ion flux is constant, that is, $n_p(z)v_{dp}(z) = const$, without ionization and recombination in the collisionless sheath. Under these conditions, the momentum equations of electrons and ions in the steady state are, respectively,

$$m_e \frac{dP(z)}{dz} = -en_e(z)\frac{dV(z)}{dz}, \tag{3.37}$$

$$M_p n_p(z) v_{dp} \frac{dv_{dp}}{dz} = en_p(z)\frac{dV(z)}{dz}. \tag{3.38}$$

From the above ion flux continuity and Equation 3.38, we obtain

$$\begin{aligned} n_p(z) &= \frac{n_p(z_0)v_{dp}(z_0)}{v_{dp}(z)} \\ &= n_p(z_0)\frac{1}{\left\{1 - \frac{2e[V(z)-V(z_0)]}{M_p v_{dp}(z_0)^2}\right\}^{1/2}}. \end{aligned}$$

In a local thermal equilibrium (LTE) under Equation 3.37, the electron number density has the Boltzmann distribution (see Section 2.3). The region from a bulk plasma/sheath boundary at z_0 to the sheath terminal, consisting of electrons and ions, is expressed by Poisson's equation

$$\begin{aligned}\frac{dV(z)^2}{dz^2} &= -\frac{e}{\varepsilon_0}[n_p(z) - n_e(z)] \\ &= -\frac{en_e(z_0)}{\varepsilon}\left\{\left(1 - \frac{2e[V(z) - V(z_0)]}{M_p v_{dp}(z_0)^2}\right)^{-1/2} - \exp\left(-\frac{e[V(z_0) - V(z)]}{kT_e}\right)\right\}.\end{aligned}$$

In the sheath, positive ions are dominant; that is, $n_p(z) - n_e(z) > 0$. Accordingly,

$$\left(1 - \frac{2e[V(z) - V(z_0)]}{M_p v_{dp}(z_0)^2}\right)^{-1/2} - \exp\left(\frac{e[V(z) - V(z_0)]}{kT_e}\right) > 0. \tag{3.39}$$

In particular, at the position z close to the boundary between the bulk plasma and the sheath, $e[V(z) - V(z_0)] \ll kT_e$ will be satisfied. By using a Taylor expansion of the exp term in Equation 3.39,

$$\left(1 - \frac{2e[V(z) - V(z_0)]}{kT_e}\right) < \left(1 - \frac{2e[V(z) - V(z_0)]}{M_p v_{dp}(z_0)^2}\right).$$

Finally, we have the initial directional velocity of ions just entering the sheath from the bulk plasma (i.e., Bohm velocity):

$$v_{dp}(z_0) > \left(\frac{kT_e}{M_p}\right)^{1/2}. \tag{3.40}$$

3.6.2 Floating Potential

Here, we define the sheath edge at the position z_0' where v_{dp} is equal to $(kT_e/M_p)^{1/2}$.

$$V_{plasma} - V(z_0') = \frac{M_p v_{dp}(z_0')^2}{2e} = \frac{kT_e}{2e}. \tag{3.41}$$

Therefore, the electron density at z_0' is

$$n_e(z_0') = n_{e0}\exp\left(-\frac{e[V_{plasma} - V(z_0')]}{kT_e}\right) = 0.61n_{e0}. \tag{3.42}$$

When a small metal plate isolated electrically from the reactor is immersed in a bulk plasma with a plasma potential V_{plasma} and electron temperature T_e, the surface potential V_{fl}, called the floating potential, is given at the condition of zero net current $e\Gamma_e = e\Gamma_p$ as

$$V_{fl} = V_{plasma} - \frac{kT_e}{2e}\ln\left(\frac{M_p}{2.3m}\right). \tag{3.43}$$

Exercise 3.6.1

Derive the above expression.

Although $e\Gamma_e = e\Gamma_p$ is satisfied, positive ions have a strong directional flux, and electrons show an isotropic flux with random speed $< v_e >= (8kT_e/\pi m)^{1/2}$ (see Chapter 2). Therefore,

$$\frac{1}{4} n_{e0} \exp\left(-\frac{e[V_{plasma} - V_{fl}]}{kT_e}\right) < v_e > (= (n_p(\text{surf})v_{dp}(\text{surf})) = n_p(z_0^{'})\left(\frac{kT_e}{M_p}\right)^{1/2}.$$

By using Equation 3.42 under $n_e(z_0^{'}) \sim n_p(z_0^{'})$,

$$-\frac{e[V_{plasma} - V_{fl}]}{kT_e} = \ln\left(2.4\left(\frac{\pi m}{8M_p}\right)^{1/2}\right).$$

Finally we will obtain Equation 3.43.

Problem 3.6.1

A plasma is classified into a "collisional plasma" and "collisionless plasma" based on the density ratio of the ions and neutral molecule, i.e. degree of ionization (see page 263). The reduced field strength $E(\boldsymbol{r})/N$ is the dominant parameter for the collisional plasma, while the space potential $V(\boldsymbol{r})$ is of first importance to describe the collisionless plasma. Discuss the reason from the perspective of a general electron transport.

References

[1] Chen, F.F. 1984. *Introduction to Plasma Physics and Controlled Fusion*, Vol. 1. New York: Plenum.

[2] Golant, V.E., Zhilinsky, A.P., and Sakharov, I.E. 1980. *Fundamentals of Plasma Physics*. New York: John Wiley & Sons.

[3] Cherrington, B.E. 1979. *Gaseous Electronics and Gas Lasers*. Oxford: Pergamon.

CHAPTER 4

Elementary Processes in Gas Phase and on Surfaces

A large number of individual elementary processes are important in determining the kinetics of low-temperature plasma. The basic processes in the gas phase include collisions of electrons, ions, and fast neutrals with molecules of the feed gas and with surfaces. Collisions between charged particles, collisions between excited particles, and collisions of radicals with gas molecules are also important. In addition, one needs to consider ion–molecule reactions and different types of photon interactions with molecules, excited species, and surfaces. It is essential to understand the nature of elementary processes in order to investigate the collective phenomena and kinetics of plasmas as a whole. Electron collisions are particularly important, because they can produce new electrons and ions under most conditions and also produce the chemically active species required for numerous plasma applications.

Gas phase collisions are dominant in low-temperature plasmas. If we change the background feed gas in a low-temperature plasma reactor, the nature of the plasma will change considerably. In highly ionized plasmas, the nature of the gas is relatively unimportant (unless thresholds for ionization change dramatically), because they are dominated by charged particle collisions. In an earlier chapter we have seen that the motion of charged particles between two collisions is classical unless the gas density is very high. However, quantum effects do occur for a very brief period of time during the collision. Thus we must briefly review the quantum representation of particles and particle beams and quantum scattering theory [1.2]. Classical collision theory is also reviewed in this chapter, because it proves useful in some cases. We then identify the basic characteristics of different collision processes [3.4.5].

We analyze electron collisions over a very wide range of energies while briefly discussing ion and fast neutral scattering over ranges relevant for practical plasmas. Finally, we discuss collisions and reactions of neutrals and ions at energies close to room temperature, which is appropriate for low-temperature plasmas.

Numerous elementary processes take place at surfaces. In this chapter we consider surfaces as a source of charged particles both through collisions and arising from surface heating and surface reactions. A special group of surface processes associated with plasma etching are also described briefly here and in greater detail in Chapters 8 and 12.

4.1 PARTICLES AND WAVES

4.1.1 Particle Representation in Classical and Quantum Mechanics

An individual particle in classical physics is represented by its mass m, its energy ε, and its momentum $\boldsymbol{p}$. Assuming that the particle is moving along the z-axis, the components of the momentum are $\boldsymbol{p}(p_x = 0, p_y = 0,\ p_z = \sqrt{2m\varepsilon})$. In Figure 4.1 we show a classical representation of a moving particle. In quantum mechanics, the motion of a particle is described by a matter wave (de Broglie wave). Individual atomic and elementary particles are subject to quantum mechanical laws; therefore, their properties are subject to the Heisenberg uncertainty principle, which relates the uncertainty of the spatial coordinate $\Delta\boldsymbol{r}$ with the uncertainty of the momentum $\Delta\boldsymbol{p}$. The uncertainty relation can be written as

$$\Delta\boldsymbol{r} \cdot \Delta\boldsymbol{p} \geq \frac{h}{2\pi} (= \hbar), \tag{4.1}$$

where h is Planck's constant.

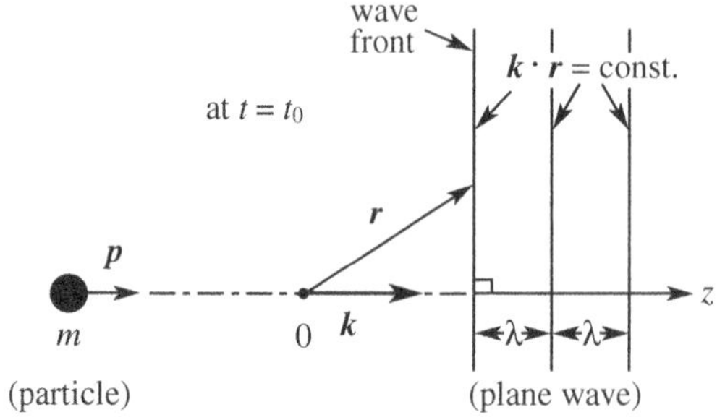

FIGURE 4.1 Classical and quantum (plane wave) description of a particle with the mass m and momentum $\mathbf{p}(0, 0, p_z)$.

In quantum mechanics, the energy and the momentum of the particle $(\varepsilon, \boldsymbol{p})$ are related to the angular frequency and the wave number of the plane wave $(\omega, \boldsymbol{k})$ through

$$\varepsilon = \hbar\omega \tag{4.2}$$

$$\boldsymbol{p} = \hbar\boldsymbol{k}. \tag{4.3}$$

A particle moving along the z-axis is represented as a plane wave

$$\Phi(\boldsymbol{r}, t) = A_0 \exp[j(\boldsymbol{k}\cdot\boldsymbol{r} - \omega t)], \tag{4.4}$$

as shown in Figure 4.1. Here, A_0 is the amplitude, and $\boldsymbol{k}\cdot\boldsymbol{r} = const$ expresses the wave front. This plane wave is fully defined by two pairs of quantities, $(\omega, \boldsymbol{k})$.

Problem 4.1.1
Show that the plane wave satisfies the uncertainty principle.

Exercise 4.1.1
Calculate the de Broglie wavelength λ_e for the electron that has crossed a potential drop V_0 (see Figure 4.2).
The energy of a particle is equal to the potential drop V_0 times the elementary charge e, so we can obtain the momentum of the particle as a function of V_0:

$$\varepsilon = p^2/2m = eV_0.$$

Bearing in mind the de Broglie relation in Equation 4.3, we obtain

$$p = \hbar k = \frac{h}{2\pi}\cdot\frac{2\pi}{\lambda_e} = \frac{h}{\lambda_e}.$$

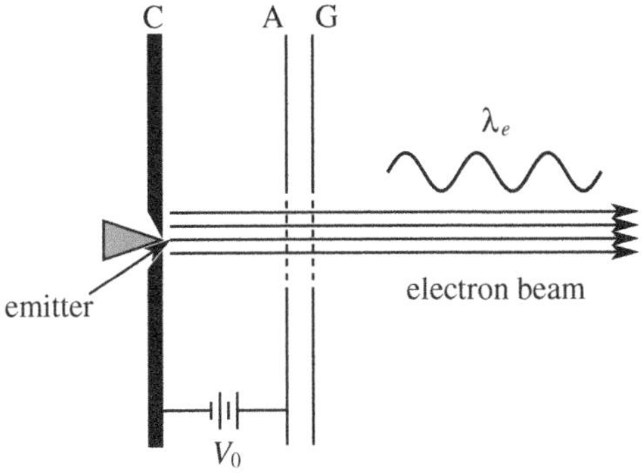

FIGURE 4.2 A stationary electron beam with mono-energetic energy eV_0 produced in a vacuum.

Finally, we have

$$\begin{aligned}\lambda_e &= \frac{h}{(2meV_0)^{1/2}} \\ &= \frac{6.63 \times 10^{-34}}{(2 \times 9.11 \times 10^{-31} \times 1.6 \times 10^{-19} \times V_0)^{1/2}} = \left(\frac{1.50}{V_0[\mathrm{V}]}\right)^{1/2} [\mathrm{nm}]. \quad (4.5)\end{aligned}$$

Exercise 4.1.2

Obtain the dispersion relation between the frequency and the wave number for a plane matter wave $\Phi(\boldsymbol{r}, t)$ in free space (see Equation 4.4).

If we start from the time- and space-dependent Schrodinger equation

$$\mathbf{H}\Phi(\boldsymbol{r}, t) = j\hbar\frac{\partial}{\partial t}\Phi(\boldsymbol{r}, t), \quad (4.6)$$

with the Hamiltonian for a free particle given by

$$\mathbf{H} = -\frac{\hbar^2}{2m}\nabla_r^2 + V(\boldsymbol{r}, t), \quad (4.7)$$

then the solution in free space ($V = 0$) to this system with the plane wave in Equation 4.4 gives us

$$\frac{\hbar^2 k^2}{2m}\Phi(\boldsymbol{r}, t) = \hbar\omega\Phi(\boldsymbol{r}, t).$$

Finally, we obtain the dispersion relation (between the wave number k and the angular frequency ω)

$$\omega(k) = \frac{\hbar k^2}{2m}. \quad (4.8)$$

Problem 4.1.2

The wave-front of a plane wave is defined as $\boldsymbol{k} \cdot \boldsymbol{r} = const$ at time t, and in that case the wave front is perpendicular to the wave vector $\boldsymbol{k}$. Show that a spherical wave with a circular wave front is expressed as

$$\Phi(\boldsymbol{r}, t) = \frac{A_0}{r}\exp[j(\boldsymbol{k} \cdot \boldsymbol{r} - \omega t)]. \quad (4.9)$$

4.1.2 Locally Isolated Particle Group and Wave Packets

An isolated group of particles moving in the z-direction is expressed as a wave packet. To obtain a wave packet, we can sum a group of plane waves in respect to $\boldsymbol{k}$ within the boundaries ($k_0 \pm \Delta k$). It is assumed that $(k - k_0) = \xi \ll k_0$. The summation may be converted to an integral as follows:

$$\begin{aligned}\Phi(z, t) &= \sum_{k_0 - \Delta k}^{k_0 + \Delta k} \Phi(z, t; k) \\ &= \int_{k_0 - \Delta k}^{k_0 + \Delta k} A(k)\exp[j(kz - \omega t)]\, dk. \quad (4.10)\end{aligned}$$

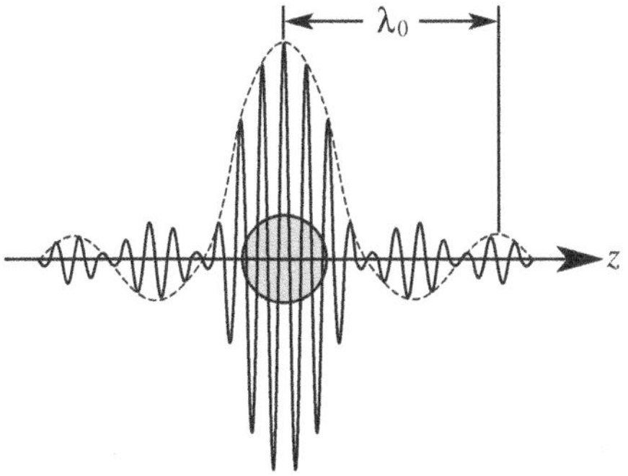

FIGURE 4.3 Wave packet of a group of particles moving in the z-direction.

Furthermore, we apply the Taylor expansion to the dispersion relation around $k = k_0$:

$$\begin{aligned} \omega &= \omega(k - k_0) \\ &= \omega_0 + \left.\frac{d\omega}{dk}\right|_0 (k - k_0) + \frac{1}{2}\left.\frac{d^2\omega}{dk^2}\right|_0 (k - k_0)^2 + \cdots, \end{aligned}$$

and, as a result, we obtain

$$\begin{aligned} \Phi(z,t) &= \int_{-\Delta k}^{\Delta k} A(k) \exp\left(j(k_0 + \xi)z - j\left(\omega_0 + \left.\frac{d\omega}{dk}\right|_0 \xi\right) t\right) d\xi \\ &\sim A(k_0) \exp[j(k_0 z - \omega_0 t)] \int_{-\Delta k}^{\Delta k} \exp\left(j\left(\xi z - \left.\frac{d\omega}{dk}\right|_0 \xi\right) t\right) d\xi. \end{aligned}$$

By combining the above equation with

$$\int_{-\Delta k}^{\Delta k} \exp(ja\theta) d\theta = \frac{e^{ja\Delta k} - e^{-ja\Delta k}}{ja} = \frac{2\sin(a\Delta k)}{a}$$

we obtain

$$\Phi(z,t) \sim 2A(k_0) \frac{\sin\left(\left(z - \left.\frac{d\omega}{dk}\right|_0 t\right) \Delta k\right)}{z - \left.\frac{d\omega}{dk}\right|_0 t} \exp[j(k_0 z - \omega_0 t)]. \tag{4.11}$$

In Figure 4.3, we show a plot of Equation 4.11. The locally isolated particle group is represented by a strongly modulated wave with the fundamental frequency $\omega_0(k_0)$. There is an envelope with a maximum in the position where a particle should be in a classical model (i.e., the center-of-mass) and the undulations extend approximately over the one wavelength λ_0. The exponential term in Equation 4.11 determines the phase of the wave.

The position of the maximum of the wave packet ($z = d\omega/dk|_0 t$) moves with a velocity

$$V_g = \frac{dz}{dt} = \left.\frac{d\omega}{dk}\right|_0, \tag{4.12}$$

also known as the group velocity. For the particular example given in Equation 4.8, the result is

$$V_g = \frac{d}{dk}\left(\frac{\hbar k^2}{2m}\right)\Big|_0 = \frac{\hbar k_0}{m}. \tag{4.13}$$

If we remember that $p(= mv) = \hbar k_0$, then V_g corresponds to the classical velocity v.

Exercise 4.1.3

Derive the phase velocity of the wave packet in Equation 4.11.
The phase in Equation 4.11 is

$$\Theta = k_0 z - \omega_0 t = const,$$

and the phase velocity V_p is equal to

$$V_p = \frac{dz}{dt} = \frac{d}{dt}\left(\frac{\omega_0 t}{k_0}\right) = \frac{\omega_0}{k_0}.$$

Using Equation 4.8 allows the phase velocity to be related to the group velocity

$$V_p = \frac{1}{k_0}\frac{\hbar k_0^2}{2m} = \frac{\hbar k_0}{2m} = \frac{V_g}{2}.$$

Problem 4.1.3

A pulsed electron beam is formed in a vacuum by using the electrical shutter between grids A and G in Figure 4.2. The half width of the group Δz is 1 mm, and V_0 is 150 V. Discuss the uncertainty principle using $\Delta v_z / v_z$.

4.2 COLLISIONS AND CROSS SECTIONS

An atomic collision is a stochastic process that takes place when two particles approach one another at a sufficiently short distance so that their interaction through different forces becomes appreciable. In weakly ionized plasmas, we deal mainly with two-body, short-range collisions, which are then regarded as occurring very quickly and over a very short distance as compared with the mean (free) time and mean (free) path between successive collisions. Because the Coulomb force has infinite range, collisions between charged particles are described as many-body, long-range collisions.

There is also a whole group of three-body processes where the third particle is required to satisfy the conservation laws. A three-body process occurs, for example, when one of the species is left in an excited state after the two-body

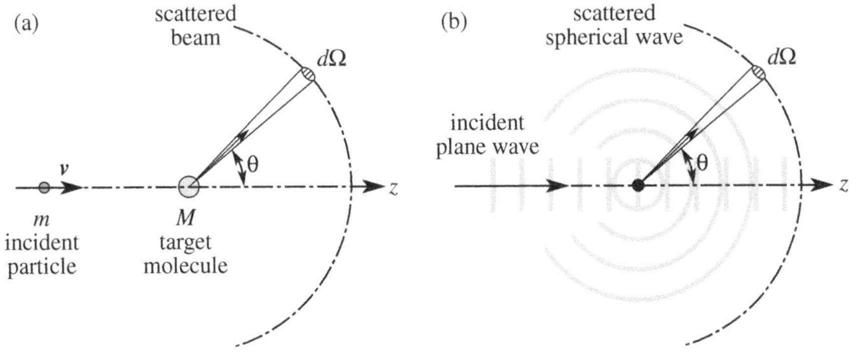

FIGURE 4.4 Classical (a) and quantum (b) representation of a collision.

collision and the excited species needs a subsequent collision with a third body to transfer its extra energy for stabilization. Three-body processes will occur only if the gas density is sufficiently high.

Here we deal primarily with two-body, short-range collisions. In classical mechanics, particles have definite identities, positions, and velocities (momenta). The classical collision of a light projectile of mass m and velocity $\boldsymbol{v}$ on a heavy target of mass M is depicted in Figure 4.4a. After the collision the light particle will be scattered at an angle θ with respect to its original direction z and into the solid angle $d\Omega(= \sin\theta d\theta d\phi)$. It will have a definitive velocity and path, and the differential cross section $\sigma(\theta, \phi, \varepsilon)d\Omega$ will give the probability of scattering into the solid angle $d\Omega$. In the quantum case, an incoming free particle will be represented as a plane wave described by Figure 4.4b and Equation 4.4. Then, as a result of spherical scattering, the outgoing wave will be given by the sum of the plane and the spherical waves.

4.2.1 Conservation Laws in Collisions

First we assume that the collision event occurs in a force-free space, in which the momentum and energy of the two particles are conserved. This is true, for example, for an electron collision with a neutral molecule, which is the most frequent type of collision occurring in collisional plasmas, where the interaction time between the two particles is on the order of 10^{-16}s, and the effect of the external force due to an imposed electric field is negligible compared with the huge internal force between the two particles. The conservation of momentum is then simply given by

$$m\boldsymbol{v}' + M\boldsymbol{V}' = m\boldsymbol{v} + M\boldsymbol{V}, \tag{4.14}$$

where (m, M) are, respectively, the masses of the charged and neutral particle, and $(\boldsymbol{v}', \boldsymbol{V}')$ are the velocities before and $(\boldsymbol{v}, \boldsymbol{V})$ the velocities after the collision. A similar conservation law is written for the energy. However, in this case internal excitation can occur and the inelastic process should also be accounted for. In the energy conservation law, it is convenient to change the frame of reference to a center-of-mass (CM) frame, and the kinetic energy is represented as

$$\varepsilon_{kin} = \frac{1}{2}mv^2 + \frac{1}{2}MV^2 = \frac{1}{2}(m+M)v_g^2 + \frac{1}{2}\mu_r v_r^2, \tag{4.15}$$

where the CM velocity $\boldsymbol{v}_g$ and the relative velocity $\boldsymbol{v}_r$ are given as

$$\boldsymbol{v}_g = \frac{m\boldsymbol{v} + M\boldsymbol{V}}{m+M} \tag{4.16}$$

and

$$\boldsymbol{v}_r = \boldsymbol{v} - \boldsymbol{V}. \tag{4.17}$$

The CM velocity remains constant before and after the collision under the conservation of momentum expressed in Equation 4.14. As a result, the transformation from the standard laboratory (LAB) system to the CM system effectively allows us to describe a two-particle process as a single-particle process. The reduced mass of the effective particle is therefore

$$\mu_r = \frac{mM}{m+M}. \tag{4.18}$$

Using these transformations to the CM frame, we can write the energy conservation law as

$$\frac{1}{2}\mu_r {v'_r}^2 = \frac{1}{2}\mu_r v_r^2 + \varepsilon_R, \tag{4.19}$$

where ε_R is the energy loss (reaction energy) due to the excitation of internal energy levels of one or both particles. We define three kinds of collisions, each defined by their value and sign of the reaction energy.

i. $\varepsilon_R > 0$: *Collisions of the first kind*, in which we have loss of the total kinetic energy to inelastic processes;

ii. $\varepsilon_R < 0$: *Collisions of the second kind*, in which we have gain of the total kinetic energy from the internal excitation energy; and

iii. $\varepsilon_R = 0$: *Elastic collisions*, in which the total kinetic energy is conserved.

Problem 4.2.1

Explain that the CM energy is conserved in the case of binary collision, and that in the case of inelastic scattering the energy to excite the molecule is provided from the relative kinetic energy.

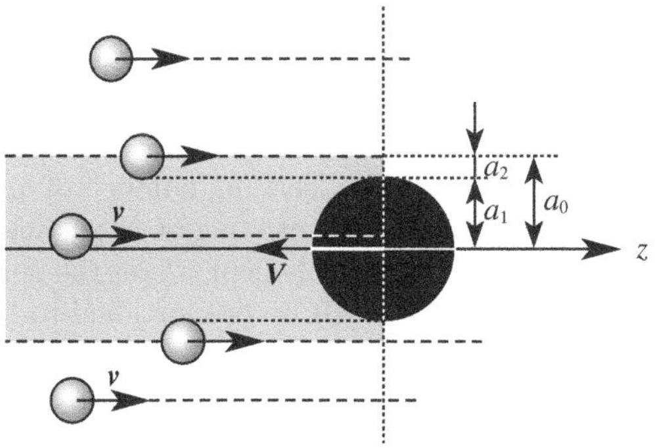

FIGURE 4.5 Binary encounter of solid spheres with radii of a_1 and a_2, and the concept of the collision cross section.

4.2.2 Definition of Collision Cross Sections

The concept of a collision cross section can be best understood when we consider a collision between two particles represented as hard spheres with radii of a_1 and a_2, approaching with a relative velocity $\boldsymbol{v}_r$ along the z-axis. A hard sphere is an object for which the interaction potential is zero for distance $r > a_0$ and infinitely repulsive for $r \leq a_0$. We now introduce the impact parameter b to the binary collision with b representing the perpendicular distance from the z-axis along the initial relative velocity to the center of the target. We then observe (Figure 4.5) that only those particles with an impact parameter b less than $a_1 + a_2$ will collide, whereas others will continue their motion without changing their direction. Thus the circle with radius $a_0 = a_1 + a_2$ is the effective size of the target for projectiles represented as points. The area of this circle $\pi a_0^2 = \pi(a_1 + a_2)^2$ is the effective collision cross section for the two hard spheres.

Problem 4.2.2
Derive the energy transfer as a function of scattering angle in an elastic collision of two hard spheres $(\boldsymbol{r}, m, \boldsymbol{v}')$ *and* $(\boldsymbol{R}, M, 0)$, *both defined by their radius, mass, and velocity before collision.*

The concept of an "effective area" associated with target particles for any pair of colliding particles (two-body collision) is useful even if the hard sphere approximation is not valid. There are, however, some conditions that must be met for it to be applicable.

i. The collisions between incident particles should be negligible; that is, the density of the incident particles should be sufficiently low;

ii. The density of the target particles should be low in order to avoid collisions of the incident particles with more than one target at the same time. That is, the de Broglie wavelength of the incident particle should be shorter than the mean distance between target molecules.

In defining the collision cross section we return to the differential cross section that was mentioned earlier. When a certain number of projectiles n with energy ε are directed toward a target consisting of N particles, the number of particles dn scattered into the surface element dS on the spherical surface at a distance r from the target center will be proportional to the density of the projectiles and the targets and also proportional to the area of the surface dS and r^{-2}, that is, the solid angle in the collision experiment:

$$dn \propto nN\, ds/r^2,$$

where $ds/r^2 = d\Omega(= \sin\theta d\theta d\phi)$.

It is known that dn depends on the incident energy of the projectile and the quantum property of the target molecule. If we define the constant of proportionality as $\sigma(\theta, \phi; \varepsilon)$, then the relation above becomes

$$dn = \sigma(\theta, \phi; \varepsilon) nN d\Omega. \tag{4.20}$$

In Equation 4.20, $\sigma(\theta, \phi; \varepsilon)d\Omega$ is the probability that one projectile is scattered into a small solid angle $d\Omega$ at scattering angle (θ, ϕ). $\sigma(\theta, \phi; \varepsilon)$ has the dimensions of the area and so is termed the "differential cross section."

The differential cross section may be expanded into the Legendre polynomials $P_n(\cos\theta)$ under the azimuthal symmetry

$$\sigma(\theta; \varepsilon) = \frac{1}{4\pi} \sum_0^{\infty} (2n+1) Q_n(\varepsilon) P_n(\cos\theta), \tag{4.21}$$

where $Q_n(\varepsilon)$ is the nth integral cross section. Multiplying the equation by $P_n(\cos\theta)$ and using the property of orthogonality of the Legendre polynomials (see Section 5.6.1), we obtain the following definition of the integral cross section $Q_n(\varepsilon)$:

$$Q_n(\varepsilon) = 2\pi \int_0^{\pi} \sigma(\theta; \varepsilon) P_n(\cos\theta) \sin\theta d\theta. \tag{4.22}$$

The coefficients $Q_n(\varepsilon)$ and the differential cross section $\sigma(\theta; \varepsilon)$ are defined for each of the many scattering processes that can occur. They are interpreted as the scattering probability averaged over all angles and weighted by some Legendre polynomial. For example, $Q_0(\varepsilon)$ is the total cross section for the process (i.e., the probability of scattering at any angle) and corresponds to the area of the target as seen by a point projectile. The most commonly used angular averages of the differential cross section are the following:

- Total cross section

$$Q_0(\varepsilon) = \int \sigma(\theta, \phi; \varepsilon) d\Omega = 2\pi \int_0^{\pi} \sigma(\theta; \varepsilon) \sin\theta d\theta; \tag{4.23}$$

- Momentum transfer cross section

$$\begin{aligned} Q_m(\varepsilon)(= Q_0 - Q_1) &= \int (1-\cos\theta)\sigma(\theta;\varepsilon)d\Omega \\ &= 2\pi \int_0^{\pi} (1-\cos\theta)\sigma(\theta;\varepsilon)\sin\theta d\theta; \text{ and} \qquad (4.24) \end{aligned}$$

- Viscosity cross section

$$\begin{aligned} Q_v(\varepsilon)\left(= \frac{2}{3}(Q_0 - Q_2)\right) &= \int (1-\cos^2\theta)\sigma(\theta;\varepsilon)d\Omega \\ &= 2\pi \int_0^{\pi} (1-\cos^2\theta)\sigma(\theta;\varepsilon)\sin\theta d\theta. \qquad (4.25) \end{aligned}$$

These integral forms of differential cross section play a special role in transport theory. For inelastic collisions with threshold energy ε_j, the definitions are the same except that the momentum transfer cross section is defined as

$$Q_m(\varepsilon) = 2\pi \int_0^{\pi} \left(1 - \sqrt{1 - \frac{\varepsilon_j}{\varepsilon}}\cos\theta\right)\sigma(\theta;\varepsilon)\sin\theta d\theta, \qquad (4.26)$$

Thus, at $\varepsilon \gtrsim \varepsilon_j$ when the scattering is nearly isotropic, $Q_m(\varepsilon)$ will be very small.

Exercise 4.2.1

Explain the physical meaning of the momentum transfer cross section for elastic scattering.

Assuming that the initial momentum is $\boldsymbol{p}$ and that particles are scattered at an angle θ with no loss of the magnitude of the velocity, the difference of momentum along the axis of the initial velocity is

$$\Delta p = p - p\cos\theta = p(1-\cos\theta).$$

In other words, the fractional change of the momentum is

$$\frac{\Delta p}{p} = (1-\cos\theta). \qquad (4.27)$$

Therefore, the cross section Q_m in Equation 4.24 is in some way a representation of the momentum transfer to the gas molecules. Q_m for elastic collision has a finite magnitude, and the elastic energy loss $(2m/M)\varepsilon$ of the electron with energy ε is negligibly small.

Problem 4.2.3

Discuss the physical meaning of the viscosity cross section.

Exercise 4.2.2

Gas molecules with density N and temperature T_g are present in a reactor.

Estimate the limitations of the two-body approximation for the collisions between molecules.
This criterion may be related through the relationship between the mean distance between molecules d and the de Broglie wavelength λ_{de} of the molecule,

$$d \gg \lambda_{de}.$$

The mean distance between the gas molecules may be obtained from the gas number density N as $d = N^{-1/3}$. On the other hand, the wavelength, which is used to approximate the range over which quantum effects are appreciable, will be calculated for the most probable speed of molecules, $\sqrt{2kT_g/M}$. Hence the de Broglie wavelength is given by

$$\lambda_{de} = \frac{h}{p} \approx \frac{h}{M\tilde{v}} = \frac{h}{\sqrt{2MkT_g}}.$$

The condition therefore becomes

$$\frac{\lambda_{de}}{d} \approx \frac{hN^{1/3}}{\sqrt{2MkT_g}} \ll 1. \tag{4.28}$$

Problem 4.2.4
Calculate Q_0, Q_m, and Q_v for the following models of differential scattering cross sections:

- *Isotropic ($\sigma(\theta)$ is independent of angle);*
- *Forward ($\sigma(\theta)$ is a delta function at zero angle);*
- *Backward ($\sigma(\theta)$ is a delta function at π);* and
- *Right angles ($\sigma(\theta)$ is a delta function at $\pi/2$).*

Discuss the relative magnitudes Q_m/Q_0 and Q_v/Q_0 in these cases.

The unit usually used in presenting a collision cross section between an electron and a molecule is 10^{-16} cm^2. This unit originates from the Angstrom (Å), an old unit of length equal to 10^{-8} cm (0.1 nm). Other units are often applied as well. For example, it is convenient to represent the cross section by using the Bohr radius (i.e., the first Bohr orbit of H), $a_0 = 0.529 \times 10^{-8}$cm as

$$[a_0^2] = 0.280 \times 10^{-16} \text{ cm}^2. \tag{4.29a}$$

The effective cross section of a Bohr radius is then given as

$$[\pi a_0^2] = 0.880 \times 10^{-16} \text{ cm}^2. \tag{4.29b}$$

In the case of photon scattering, it is common to use the "barn"

$$[barn] = 10^{-24} \text{ cm}^2. \tag{4.29c}$$

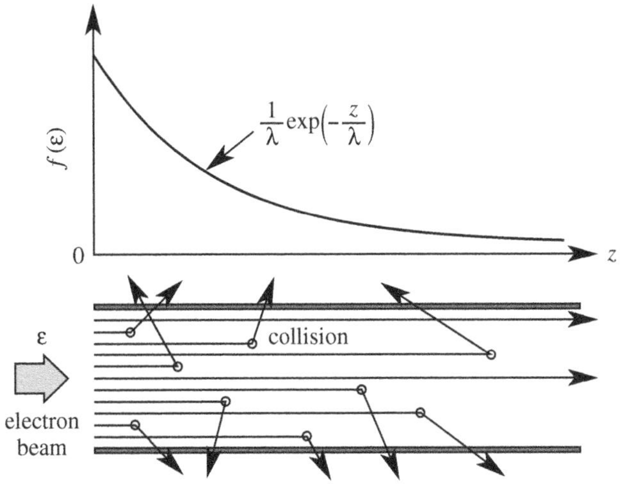

FIGURE 4.6 Distribution of the mean free path $f(z;\varepsilon)$ for a beam of particles with energy ε incident on a gas target with random distribution.

Finally, we note that the total collision probability P_c is often used to represent the number of collisions in a gas at some pressure p and at temperature T_g[K]. The total collision probability is equal to $P_c = (p/kT_g)Q$. Here, the standard value that is used is $p = 1$Torr at room temperature, where we have

$$P_c = 9.66 \times 10^{18} Q[\mathrm{cm}^2]/T_g[\mathrm{K}]\mathrm{cm}^{-1}\mathrm{Torr}^{-1} \tag{4.29d}$$

for the standard collision probability at 1 Torr $= 133Pa$.

4.2.3 The Distribution of Free Paths

In order to obtain the distribution of the free paths of a particle between collisions, we consider a beam of particles with density n_0 and energy ε entering gas at $z = 0$ (see Figure 4.6). It can be shown that the distribution function of the free path is given by

$$f(z;\varepsilon) = \frac{1}{<\lambda(\varepsilon)>} \exp\left(-\frac{z}{<\lambda(\varepsilon)>}\right), \tag{4.30}$$

where $<\lambda(\varepsilon)>$ is the mean free path of a particle with energy ε.

Exercise 4.2.3
Derive the distribution function of the free path (Equation 4.30).
When we define $F(z)$ as the integral distribution of collisions on the segment $(0, z)$, the number of collisions in the small segment $(z, z + \Delta z)$ is proportional

to the number of particles remaining in the beam $(1 - F(z))$ and to dz, so we have

$$\{F(z + dz) - F(z)\} = c\{1 - F(z)\}dz,$$

where c is a constant. Next we obtain the equation

$$\frac{-1}{1 - F(z)} \frac{F(z + dz) - F(z)}{dz} = -c,$$

which has a solution

$$\ln\{1 - F(z)\} = -cz \quad \text{or} \quad F(z) = 1 - \exp(-cz).$$

From the integral distribution of collisions $F(z)$ we determine the distribution of the free path between collisions:

$$f(z) = \left(\frac{dF(z)}{dz} = \right) c \exp(-cz).$$

Considering the definition of the mean free path, we have

$$<\lambda> = \frac{\int_0^\infty z f(z) dz}{\int_0^\infty f(z) dz} = \frac{1}{c},$$

and thus the formula (Equation 4.30) for the distribution function of the free path is obtained. The distribution $f(z)$ can be used to determine the probability of collision (or of the length of the free path) of a single particle as well. One should bear in mind that the mean free path depends on the energy of the incident particles, that is, $<\lambda(\varepsilon)>$. Thus if we have a distribution of incident energies we perform additional averaging and obtain the mean free path for the whole ensemble.

4.2.4 Representation of Collisions in Laboratory and CM Reference Frames

In Section 4.2.1, we have shown that the motion of two particles of masses m and M and velocities $\boldsymbol{v}$ and $\boldsymbol{V}$ can be treated as a single-particle motion with relative mass μ_r and relative velocity $\boldsymbol{v}_r$ by considering the total energy conservation

$$\varepsilon_{kin} = \frac{1}{2}mv^2 + \frac{1}{2}MV^2 = \frac{1}{2}(m + M)v_g^2 + \frac{1}{2}\mu_r v_r^2, \tag{4.31}$$

where $\boldsymbol{v}_g$ is the CM velocity. A convenient way to describe a binary collision is to make a transfer from the laboratory frame of reference to the CM frame of reference of the relative mass μ_r and the velocity $\boldsymbol{v}_r$ (see Figure 4.7).

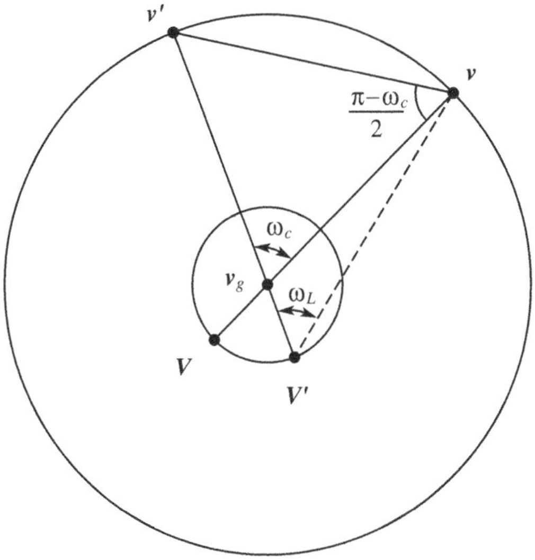

FIGURE 4.7 Velocities of two particles before and after collision and the corresponding angles in the CM frame.

Exercise 4.2.4

Obtain the relationship between differential cross sections for the laboratory and the CM reference frames in the elastic scattering.

Consider two particles, (m) and (M), moving toward one another with velocities $(\boldsymbol{v}', \boldsymbol{V}')$ before and $(\boldsymbol{v}, \boldsymbol{V})$ after the collision in the LAB frame. The magnitude of the relative velocity $\boldsymbol{v}_r$ is conserved before and after the elastic scattering in the CM frame, $(\mid \boldsymbol{v} - \boldsymbol{V} \mid = \mid \boldsymbol{v}' - \boldsymbol{V}' \mid = \mid \boldsymbol{v_r} \mid)$. We may thus represent the motion of particles before and after the scattering by two straight lines of equal lengths $(|\boldsymbol{v}_r|)$ intersecting at the point $\boldsymbol{v}_g$ (CM velocity) (see Figure 4.7). In the CM frame, the velocities of the two particles before and after the collision are located on spheres with radii of $\boldsymbol{v} = \boldsymbol{v}_r M/(m+M)$ and $\boldsymbol{V} = \boldsymbol{v}_r m/(m+M)$, respectively.

We can identify one equilateral triangle with the CM frame scattering angle ω_c, that is, the triangle $[\boldsymbol{v}_g, \boldsymbol{v}', \boldsymbol{v}]$. Thus we write

$$\frac{\overline{\boldsymbol{v}'\boldsymbol{v}}}{\sin \omega_c} = \frac{M v_r/(m+M)}{\sin[(\pi - \omega_c)/2]},$$

where the length between points $[\boldsymbol{v}, \boldsymbol{v}']$ is denoted $\overline{\boldsymbol{v}'\boldsymbol{v}}$. If, however, we take the triangle $[\boldsymbol{V}', \boldsymbol{v}', \boldsymbol{v}]$, the following relation is obtained by using the LAB frame scattering angle of particle m, ω_L:

$$\frac{\overline{\boldsymbol{v}'\boldsymbol{v}}}{\sin \omega_L} = \frac{v_r}{\sin[\pi - \omega_L - (\pi - \omega_c)/2]}.$$

Combining the results from the two triangles we obtain

$$\tan\omega_L = \frac{\sin\omega_c}{(m/M) + \cos\omega_c} \tag{4.32a}$$

and

$$\cos\omega_L = \frac{(m/M) + \cos\omega_c}{\{1 + 2(m/M)\cos\omega_c + (m/M)^2\}^{1/2}}. \tag{4.32b}$$

Because the same number of particles is scattered through the solid angle in both systems, we have

$$\sigma_L(\omega_L;\varepsilon)d\Omega_L \equiv \sigma_c(\omega_c;\varepsilon)d\Omega_c,$$

or

$$\begin{aligned}\sigma_L(\omega_L;\varepsilon) &= \sigma_c(\omega_c;\varepsilon)\frac{d\cos\omega_c}{d\cos\omega_L}\\ &= \sigma_c(\omega_c;\varepsilon)\frac{\{1 + 2(m/M)\cos\omega_c + (m/M)^2\}^{3/2}}{1 + (m/M)\cos\omega_c}.\end{aligned} \tag{4.33}$$

It is possible to show that the energy transfer from the projectile to the target (i.e., the energy loss to the first particle) is then equal to

$$\begin{aligned}\frac{\varepsilon_M}{\varepsilon'_m} = \frac{MV^2/2}{mv'^2/2} &= \frac{M}{mv'^2}\left(\frac{2mv'}{m+M}\cos\omega_L\right)^2\\ &= \frac{4mM}{(m+M)^2}\cos^2\omega_L.\end{aligned} \tag{4.34}$$

Problem 4.2.5
Determine the scattering angles ω_L and ω_{CM} when (a) $m = M$ and (b) $m \ll M$.

Problem 4.2.6
Show that the energy loss in elastic collisions is given by Equation 4.34, and that for the electron-molecule scattering it is simplified after averaging to $(2m/M)\varepsilon$.

Problem 4.2.7
Prove that for a head-on ($\omega_L = 0$) collision of a heavy particle with a light target $m \gg M$, the maximum velocity of the target will be $V = 2v'$.

It is also of interest to determine the velocities after the collision in the case of inelastic processes characterized by the inelastic energy transfer $\Delta\varepsilon$. After inelastic scattering, the velocity of the target in the LAB frame is equal to

$$v = \frac{1}{m+M}\left\{mv'\cos\theta_L \pm \sqrt{\frac{m}{M}[mMv'^2cos^2\theta_L - 2(m+M)\Delta\varepsilon]}\right\} \tag{4.35}$$

and the velocity of the projectile is

$$V = \frac{1}{m+M}\left[mv'\cos\omega_L \pm \sqrt{v'^2(M^2 - m^2 sin^2\omega_L) - 2(m+M)\frac{M}{m}\Delta\varepsilon}\right], \tag{4.36}$$

where θ_L is the scattering angle of the target (M) in the LAB frame. From Equations 4.35 and 4.36 it is possible to obtain the minimum velocity of the incident particle necessary to achieve inelastic excitation with a loss $\Delta\varepsilon$:

$$\frac{1}{2}mv'^2 = \frac{(m+M)M}{M^2 - m^2\sin^2\omega_L}\Delta\varepsilon, \tag{4.37}$$

which for *head-on* collisions reduces to

$$\varepsilon'_m = \frac{m+M}{M}\Delta\varepsilon.$$

For particles of the same mass ($m = M$) one needs at least $\varepsilon'_m = 2\Delta\varepsilon$ to achieve an inelastic process, whereas for light projectiles ($m \ll M$) the minimum energy is $\varepsilon'_m = \Delta\varepsilon$.

4.3 CLASSICAL COLLISION THEORY

In the classical theory of collisions it is possible to determine the differential cross section based on the interaction potential and conservation laws in a binary collision. It is always useful to reduce the problem of scattering of two particles to the scattering of one particle with a reduced mass. It is assumed that the target molecule has a centrifugal potential that permits spherical symmetry in the scattering event. With the exception of the *head-on* collision, the incoming particle will pass the target at some distance. The point of the closest approach is known as the impact parameter (b). Even in the case when there is an external force field to deflect the particle, we may define the instantaneous impact parameter for any point as a function of the spherical coordinates that describe the motion.

It is evident that impact parameter b will define the degree of interaction and final trajectory for a given interaction potential. It is also clear that the outcome of scattering is described by the angle of scattering θ. In Figure 4.8 we show several classical trajectories for a projectile scattered from a target molecule. There, we define the scattering angle θ and the corresponding annular solid angle. It is also obvious that, in general, the smaller the impact parameter, the greater the scattering angle, because the particle will approach close to the target and thus be subjected to a stronger interaction. There is a unique relation between the scattering angle and the impact parameter $b = b(\theta)$ (note this may not be true in some special circumstances, e.g., when orbiting occurs).

The number of particles scattered into the angle θ is proportional to the solid angle $d\Omega$ and the differential cross section $\sigma(\theta)$. The scattered particles

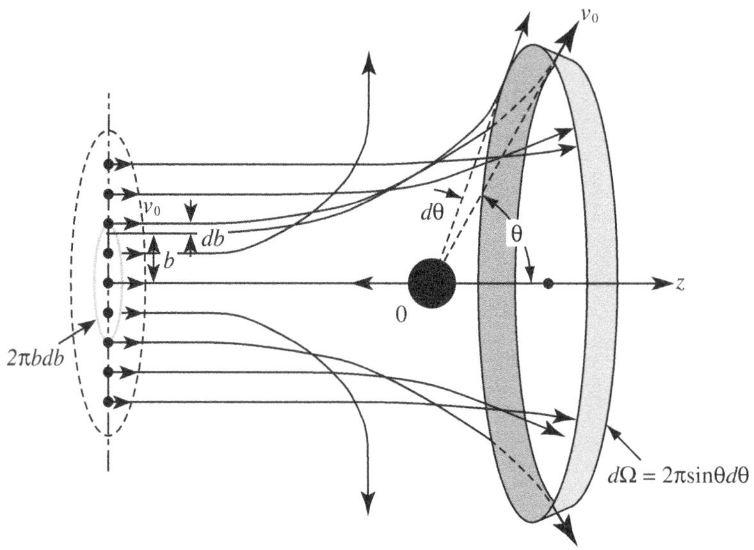

FIGURE 4.8 Definition of the scattering angles and the impact parameter in classical scattering phenomenology.

with probability $\sigma(\theta)d\Omega$ correspond to the particles incident on the area of the annular surface $2\pi bdb$. Therefore,

$$\begin{aligned}\sigma(\theta) &= 2\pi bdb/d\Omega \\ &= b(\theta)\left|\frac{db(\theta)}{d\theta}\right|\frac{d\theta}{\sin\theta d\theta} \\ &= \frac{1}{\sin\theta}b(\theta)\left|\frac{db(\theta)}{d\theta}\right|. \end{aligned} \quad (4.38)$$

Our goal is thus to determine the dependence $b(\theta)$ for any given potential interaction.

4.3.1 Scattering in Classical Mechanics

We now proceed to calculate the function $b(\theta)$ for scattering on a target placed at the scattering center $r = 0$. The projectile has velocity v and reduced mass μ. The interaction potential is $V(r)$, which is assumed to be spherically symmetric (i.e., a centrifugal potential). Additionally, the scattering is assumed to be purely elastic. Because we must satisfy the basic laws of conservation, the conservation of angular momentum L yields

$$L = \mu v_0 b = \mu(rd\xi/dt)r = \mu v_0' b'. \quad (4.39)$$

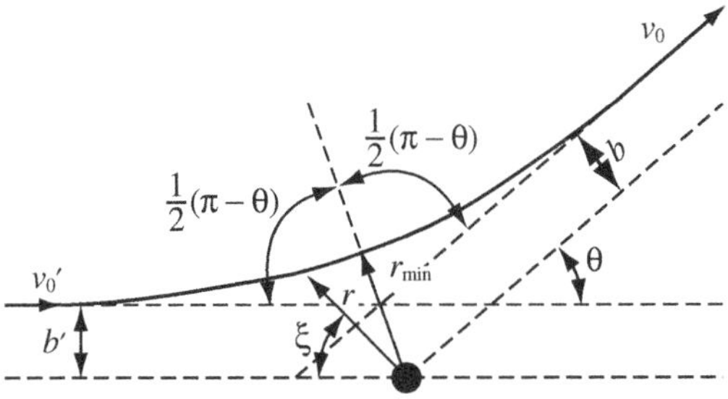

FIGURE 4.9 The geometry of classical scattering on a target with interaction potential $V(r)$.

The conservation of the total energy E yields

$$\begin{aligned} E &= \mu v_0^2/2 = \mu\{(dr/dt)^2 + r^2(d\xi/dt)^2\}/2 + V(r) \\ &= \mu {v'_0}^2/2, \end{aligned} \tag{4.40}$$

where ξ is defined as in Figure 4.9. At positions sufficiently far from the scattering center, Equations 4.39 and 4.40 give the relations of the velocities and the impact parameters before and after collision:

$$v_0 = v'_0, \quad b = b'. \tag{4.41}$$

In Figure 4.9 we define the coordinates for the scattering and basically the incoming particle with reduced mass μ follows the angle ξ. From Equation 4.39 it is shown that

$$d\xi/dt = L/\mu r^2, \quad dr/dt = (d\xi/dt)(dr/d\xi) = (L/\mu r^2)(dr/d\xi),$$

and from Equation 4.40 it follows that

$$\begin{aligned} E &= \mu\{(L/\mu r^2)^2(dr/d\xi) + r^2(L/\mu r^2)^2\}/2 + V(r) \\ &= \{L^2/2\mu r^4\}\{(dr/d\xi)^2 + r^2\} + V(r). \end{aligned}$$

The equation describing the particle trajectory is therefore

$$(dr/d\xi)^2 = \{2\mu r^4/L^2\}\{E - V(r)\} - r^2. \tag{4.42}$$

It is also worth noting that

$$\frac{L^2}{2\mu r^2} + V(r) = \frac{\mu v^2 b^2}{2r^2} + V(r) = V_{eff}(r), \tag{4.43}$$

and $V_{eff}(r)$ is the effective potential in the radial direction. The interaction potential usually has a minimum; as a result, the projectile is for a while accelerated to the target, but it will not be bound as it has a positive energy. The addition of the centrifugal term to the interaction potential leads to an increase in the potential in the region of the minimum as the centrifugal term increases. The potential may even become repulsive. It may also be possible that a maximum is formed at some distance r_c with a shallow minimum for smaller distances. Near the maximum the radial motion will be very small, but it will take several revolutions before the particle can leave after scattering as the angular momentum is large. This effect is known as "orbiting." Orbiting allows longer interaction times and therefore a greater probability for processes that involve transitions with finite lifetimes.

To calculate the trajectory from Equation 4.43, we must obtain an equation that will describe the trajectory through the angle ξ,

$$\frac{d\xi}{dr} = \pm \frac{1}{\left[\frac{2\mu r^4}{L^2}\{E - V(r)\} - r^2\right]^{1/2}}. \tag{4.44}$$

When $d\xi/dr < 0$, the incoming particle approaches the target and the interaction force increases. When $d\xi/dr > 0$, the projectile leaves the target. Because of the symmetry of the scattering (see Figure 4.9), we may perform the integration in two parts. First we perform it up to the point of the closest approach $r_{\min}$, from $\xi = 0$ (for $r \to \infty$) up to $\xi = (\pi - \theta)/2$. We obtain

$$\begin{aligned} \int_0^{(\pi-\theta)/2} d\xi &= \int_{\infty}^{r_{\min}} -[\{2\mu r^4/L^2\}\{E - V(r)\} - r^2]^{-1/2} dr \\ (\pi - \theta)/2 &= \int_{r_{\min}}^{\infty} [\{2\mu r^4/L^2\}\{E - V(r)\} - r^2]^{1/2} dr. \end{aligned} \tag{4.45}$$

The limit $r_{\min}$ (the point of the closest approach) may be obtained from $dr/d\xi = 0$, which in combination with Equation 4.45 leads to $\{2\mu r_{\min}^4/L^2\}\{E - V(r_{\min})\} - r_{\min}^2 = 0$. Introducing $u = 1/r$, we derive the following result from Equation 4.45:

$$\begin{aligned} (\pi - \theta)/2 = \int_0^{u_{\max}} [\{2\mu/L^2\}\{E - V(1/u) - u^2\}]^{-1/2} du; \\ (2\mu/L^2)\{E - V(1/u_{\max})\} - u_{\max}^2 = 0. \end{aligned} \tag{4.46}$$

An analogous equation may be obtained using the impact parameter b:

$$\begin{aligned} (\pi - \theta)/2 = \int_0^{u_{\max}} [(1/b^2)\{1 - V(1/u)/E\} - u^2]^{-1/2} du; \\ \{1 - V(1/u_{\max})/E - b^2 u_{\max}^2\} = 0. \end{aligned} \tag{4.47}$$

Exercise 4.3.1

Determine the dependence of the impact parameter $b(\theta)$ for classical hard sphere elastic scattering, that is, for the potential $V(r) = \begin{pmatrix} \infty (r \le r_0) \\ 0 (r > r_0) \end{pmatrix}$. Also determine the effective cross section for the scattering.

In Figure 4.10 we show that the hard sphere potential is a very good approximation for many classical potentials. The point of minimum approach is

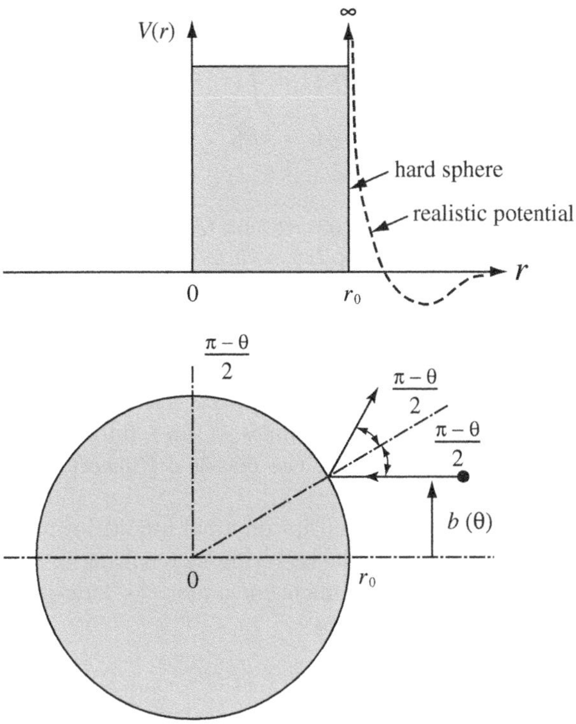

FIGURE 4.10 Comparison between the hard sphere interaction potential and a realistic potential (dashed line) and geometry of the hard sphere scattering for a given impact parameter.

$r_{\min} = r_0 = 1/u_{\max}$ and may be used in Equation 4.47 to obtain

$$\begin{aligned}(\pi - \theta)/2 &= \int_0^{1/r_0} [(1/b^2) - u^2]^{-1/2} du \\ &= \sin^{-1}[u/(1/b)]_0^{1/r_0} = \sin^{-1}[b/r_0]; (b \leq r_0) \\ &= \text{no collision}; (b > r_0).\end{aligned}$$

We obtain the solution, $\sin[(\pi - \theta)/2] = b/r_0$; that is,

$$\begin{aligned}\theta(b) &= 2\cos^{-1}(b/r_0) \quad (b < r_0) \\ &= 0 \quad (b > r_0).\end{aligned}$$

Substituting the above relation into Equation 4.38 we obtain the expression for the differential cross section:

$$\begin{aligned}\sigma(\theta) &= (1/\sin\theta)b(\theta)|db(\theta)/d\theta| \\ &= (1/\sin\theta)r_0\cos(\theta/2)| - (r_0/2)\sin(\theta/2)| \\ &= r_0^2 \sin\theta/(4\sin\theta) = r_0^2/4, \end{aligned} \tag{4.48}$$

and from the definition of the total cross section Equation 4.23 we obtain

$$\begin{aligned} Q_0 &= \int \sigma(\theta) d\Omega = \int (r_0^2/4) \sin\theta d\theta d\phi \\ &= (r_0^2/4) 4\pi = \pi r_0^2. \end{aligned} \tag{4.49}$$

Problem 4.3.1

Derive the momentum transfer cross section Q_m for hard sphere elastic scattering, and discuss the relationship between momentum transfer and the total cross section.

Exercise 4.3.2

Derive the impact parameter dependence of the scattering angle $\theta(b)$ and of the differential cross section $\sigma(\theta)$ for a light H^+ (with charge e and mass m) and massive ion (with charge Ze and mass M) in a fully ionized plasma (i.e., collisionless plasma) by considering the classical Rutherford scattering in a Coulomb field.

Classical Rutherford scattering and its physical quantities are shown in Figure 4.11. We may regard Coulomb scattering as resulting in small angle scattering or small changes of the momentum under the long-range interaction. Thus it is reasonable to assume that

$$\sin\theta \sim \theta = p_T/p.$$

Here, the Coulomb force will induce a velocity perpendicular to the original velocity and the resulting momentum is denoted by p_T. We may calculate the transverse momentum by integrating the effect of the force F_T in the region of interaction

$$\begin{aligned} p_T &= \int_{-\infty}^{\infty} F_T dt = \int_{-\infty}^{\infty} (Ze^2/4\pi\varepsilon_0 r^2) \cos\beta dt \\ &= \int_{-\infty}^{\infty} \{Ze^2/4\pi\varepsilon_0(b^2+v^2t^2)\}\{b/(b^2+v^2t^2)^{1/2}\} dt \\ &= 2Ze^2/4\pi\varepsilon_0 bv. \end{aligned}$$

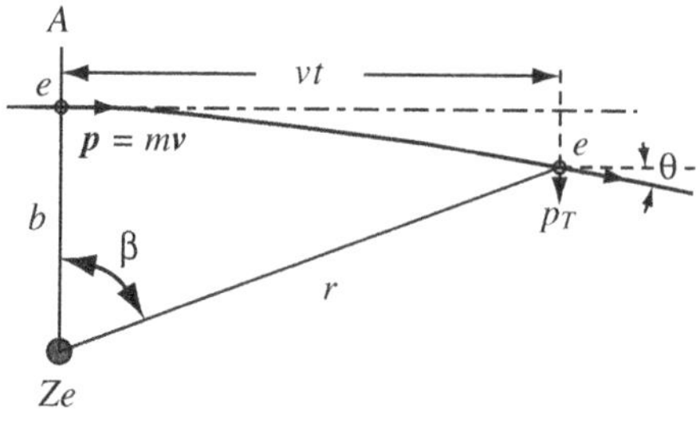

FIGURE 4.11 Classical Rutherford scattering between a light H^+ (with charge e and mass m) and a massive ion (with charge Ze and mass M) in a Coulomb field.

Therefore,

$$\begin{aligned}\theta(b) &= p_T/p \\ &= \frac{2Ze^2}{4\pi\varepsilon_0 bv}\cdot\frac{1}{mv} = \frac{2Ze^2}{4\pi\varepsilon_0 mv^2 b},\end{aligned} \tag{4.50}$$

and the differential cross section can be obtained from Equation 4.38 as

$$\begin{aligned}\sigma(\theta) &= (1/\sin\theta)b(\theta)|db(\theta)/d\theta| \\ &\sim \frac{1}{\theta}\frac{2Ze^2}{4\pi\varepsilon_0 mv^2\theta}\left|\frac{2Ze^2}{4\pi\varepsilon_0 mv^2}\frac{-1}{\theta^2}\right| \\ &\sim \left(\frac{Ze^2}{4\pi\varepsilon_0 mv^2}\right)^2\frac{4}{\theta^4}.\end{aligned} \tag{4.51}$$

A more detailed analysis gives $\theta(b)$ and $\sigma(\theta)$ for the Rutherford scattering:

$$\begin{aligned}\theta(b) &= 2\tan^{-1}\left(\frac{Ze^2}{4\pi\varepsilon_0 mv^2 b}\right); \\ \sigma(\theta) &= \frac{1}{4}\left(\frac{Ze^2}{4\pi\varepsilon_0 mv^2}\right)^2\frac{1}{\sin^4(\theta/2)}.\end{aligned} \tag{4.52}$$

It appears that it is not possible to integrate the differential cross section of the Rutherford scattering into total or momentum transfer cross sections as there is a singularity for $\theta \sim 0$. However, we should remember that the long-range effect of the Coulomb force will cancel out in ionized gas due to the effect of other charged particles. The effect of the Coulomb interaction is considered only within the so-called Debye radius λ_D; that is, we need to consider the interaction only at $b < \lambda_D$ with the minimum scattering angle $\theta_{\min}$ at $b = \lambda_D$. In other words, there is a limit on impact parameter $b \leq b_{\max}$. The momentum transfer cross section is therefore

$$\begin{aligned}Q_m &= \int_{\theta_{\min}}^{\pi}\frac{1}{4}\left(\frac{Ze^2}{4\pi\varepsilon_0 mv^2}\right)^2\frac{1}{\sin^4(\theta/2)}(1-\cos\theta)2\pi\sin\theta d\theta \\ &= 2\pi\left(\frac{Ze^2}{4\pi\varepsilon_0 mv^2}\right)^2\ln\left(\frac{2}{1-\cos\theta_{\min}}\right).\end{aligned}$$

Here, $1 - \cos\theta = 2/\{1 + [b/(Ze^2/4\pi\varepsilon_0 mv^2)]^2\}$ in Figure 4.11, and we may determine the cross section Q_m by using

$$1-\cos\theta_{\min} = 2[(Ze^2/4\pi\varepsilon_0 mv^2)/b_{\max}]^2, \quad \text{where } b_{\max} = \lambda_D.$$

We thus obtain

$$\begin{aligned}Q_m &= 4\pi(Ze^2/4\pi\varepsilon_0 mv^2)^2\ln[4\pi\varepsilon_0 mv^2\lambda_D/Ze^2] \\ &= \frac{Z^2e^4}{4\pi\varepsilon_0^2(mv^2)^2}\ln\Lambda.\end{aligned} \tag{4.53}$$

Here $\ln \Lambda$ is defined as

$$\ln \Lambda = \ln\left(\frac{4\pi\varepsilon_0 mv^2\lambda_D}{Ze^2}\right) \tag{4.54}$$

and is known as the Coulomb logarithm.

Problem 4.3.2
Show that the Coulomb logarithm corresponds to the ratio $b_{\max}/b_{\min}$.

4.3.2 Conditions for the Applicability of the Classical Scattering Theory

Two conditions must be met to allow us to use the classical scattering theory:

i. The trajectory of the particle must be clearly defined. In other words, the de Broglie wavelength λ_{de} must be much shorter than the effective range of the interaction potential d_0, or

$$\lambda_{de} = (h/\mu v_r) \ll d_0. \tag{4.55}$$

ii. The change of the momentum Δp must be much greater than the uncertainty $\Delta pd \gg h$. For an effective range of the potential V_0, the conservation of energy gives $v_r \Delta p \sim eV_0$. The condition is thus

$$eV_0 d/v_r \gg h. \tag{4.56}$$

4.4 QUANTUM THEORY OF SCATTERING

Collisions are considered in quantum theory under three major conditions:

(a) The incident particle is described by a plane wave.

(b) The total energy of the incident particle is arbitrarily positive.

(c) The scattered particle is analyzed at a point far from the scattering center, where the influence of the interaction potential is negligible.

We discuss the elastic scattering of a particle with energy E. A schematic diagram of the scattering is shown in Figure 4.12, where the region is divided into three points (a), (b), and (c).

(a) *Incident wave.* In this subsection, the incoming particle is represented as the wave function $\Psi(\boldsymbol{r}, t)$, which can be written as

$$\Psi(\boldsymbol{r}, t) = \exp[i(\boldsymbol{k} \cdot \boldsymbol{r} - \omega t)], \tag{4.57}$$

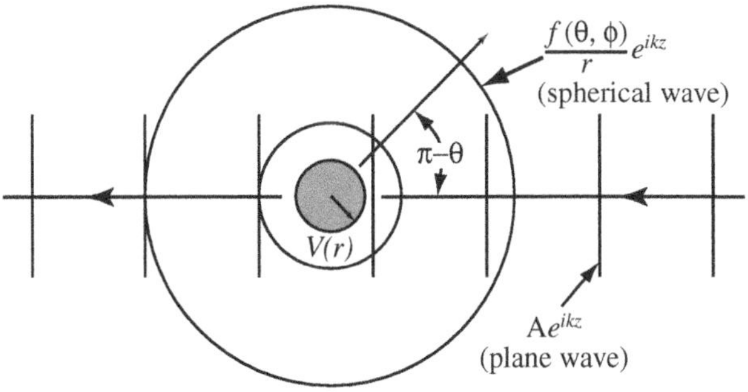

FIGURE 4.12 The planar wave of the incoming particle and the spherical wave of the scattered particle in quantum theory of elastic scattering.

where $\boldsymbol{p} = \hbar \boldsymbol{k}$ and $E = \hbar\omega$. The description in Equation 4.57 is valid only at a long distance where the interaction potential is negligible. When the velocity of the incoming particle is along the z-direction in the steady state, then

$$\Psi(z) = \exp(ikz). \tag{4.58}$$

(b) *Under scattering potential.* Here, the collision is characterized by the Schrodinger equation having a continuous positive energy E of the incoming particle and the interaction potential $V(r)$. Hence the wave equation reads as

$$-(\hbar^2/2\mu)\nabla^2\Psi(\boldsymbol{r}) + V(\boldsymbol{r})\Psi(\boldsymbol{r}) = E\Psi(\boldsymbol{r}). \tag{4.59}$$

(c) *After scattering.* From the condition that scattered particles eventually escape to far distances unaffected by $V(r)$ and distribute themselves spherically around the scattering center (target), we may conclude that the scattered particle can be described by a spherical wave. Thus, the total wave function is the sum of the scattered spherical wave and the unaffected plane wave:

$$\Psi(\boldsymbol{r}) = \exp(ikz) + \frac{f(k;\theta,\phi)}{r}\exp(ikr). \tag{4.60}$$

The spherical term is normalized by $f(k;\theta,\phi)$, which is known as the scattering amplitude. For example, if there is no interaction, the scattering amplitude is zero and the planar wave is unaffected. The number of particles Δn scattered into a small area dS normalized to the total number of the incident particles n represents the probability of the scattering in a given direction (θ, ϕ) and is therefore equal to

$$\begin{aligned}\Delta n/n &= \left|\frac{f(k;\theta,\phi)}{r}\exp(ikr)\right|^2 dS \cdot \frac{1}{|\exp(ikz)|^2}\\ &= |f(k;\theta,\phi)|^2\frac{dS}{r^2}.\end{aligned}$$

Here, the solid angle is $d\Omega = dS/r^2$, and by definition of the differential cross section σ, we have

$$\Delta n/n = \sigma(k;\theta,\phi)d\Omega.$$

We can then obtain the relation between the differential cross section and the scattering amplitude in quantum theory as

$$\sigma(k;\theta,\phi) = |f(k;\theta,\phi)|^2. \tag{4.61}$$

4.4.1 Differential Scattering Cross Section $\sigma(\theta)$

We can solve Equation 4.59 by using Equation 4.60, making a solution in the CM system for a stationary target and projectile with a reduced mass μ. The wave function can be separated into two terms, the radial (r) and angular (θ, ϕ) functions; that is,

$$\Psi(r) = R(r)Y(\theta,\phi).$$

In the case of azimuthal symmetry, it is useful to use Legendre polynomials (and take advantage of their orthogonality):

$$\Psi(\boldsymbol{r}) = R(r)P_l(\cos\theta). \tag{4.62}$$

This makes it possible to derive the differential equation for the radial part of the wave function $R(r)$:

$$\frac{1}{r^2}\frac{d}{dr}\left(r^2\frac{dR(r)}{dr}\right) + \left(\frac{2\mu E}{\hbar^2} - \frac{2\mu}{\hbar^2}\left[V(r) + \frac{\hbar^2}{2\mu}\frac{l(l+1)}{r^2}\right]\right)R(r) = 0. \tag{4.63}$$

Here, we have used quantization of the angular momentum. We define the effective radial potential as

$$V^l_{eff}(r) = V(r) + \frac{\hbar^2}{2\mu}\frac{l(l+1)}{r^2}. \tag{4.64}$$

As in the classical case, the effective radial potential is the result of the combined effect of the interaction potential and the centrifugal term. When we make transformation $R(r) = u(r)/r$, the differential equation can be written as

$$-\frac{\hbar^2}{2\mu}\frac{d^2u(r)}{dr^2} + V^l_{eff}(r)u(r) = Eu(r). \tag{4.65}$$

We now consider the asymptotic behavior of the solutions. From $V^l_{eff} \to 0$ when $r \to \infty$ it follows that

$$\begin{aligned} u(r) &\to A\sin(kr) + B\cos(kr) \\ &\to A\sin(kr + \eta_l - l\pi/2); \quad \text{at } r \to \infty. \end{aligned}$$

The radial component of the wave function has the effect of scattering through a phase-shift $(\eta_l - l\pi/2)$ at a long distance from the scattering center (target). It should be noted that without scattering $u(r)$ is exactly equal to the incoming plane wave, expressed below as $A\sin(kr - l\pi/2)$. Consequently, the solution for $R(r)$ is

$$R(r) = (1/kr)\sin(kr - l\pi/2 + \eta_l) \quad \text{at } r \to \infty, \tag{4.66}$$

and by combining Equations 4.59, 4.62, and 4.66, we obtain the general solution for the wave function as

$$\begin{aligned}
\Psi(\boldsymbol{r}) &= \sum_l A_l R(r) P_l(\cos\theta) \\
&\sim \sum_l A_l (1/kr)\sin(kr - l\pi/2 + \eta_l) P_l(\cos\theta) \quad (\text{at } r \to \infty) \\
&\sim \sum_l A_l (1/2ikr)\{\exp[i(kr - l\pi/2 + \eta_l)] \\
&\qquad - \exp[-i(kr - l\pi/2 + \eta_l)]\} P_l(\cos\theta) \quad (\text{at } r \to \infty).
\end{aligned}$$

We must now match the asymptotic forms of the above general solution and Equation 4.60. For that purpose we must expand the incoming plane wave, $\exp(ikz)$ in Equation 4.60 for $r \to \infty$ as

$$\begin{aligned}
\exp(ikz) &= \exp(ikr\cos\theta) \\
&= \sum_l P_l(\cos\theta)\frac{(2l+1)}{2}\int_{-1}^{1}\exp(ikrt)P_l(t)\,dt \quad t = \cos\theta \\
&\qquad\qquad (\text{Rayleigh formula}) \\
&= \sum_l P_l(\cos\theta)\frac{2l+1}{2}2i^l j_l(kr).
\end{aligned}$$

Recall some of the mathematical relations described above, including the orthogonal functions

$$f(x) = \sum_l a_l P_l(x) \quad \text{and} \quad a_l = \frac{(2l+1)}{2}\int_{-1}^{1} f(t)P_l(t)\,dt, \tag{4.67}$$

the integral expression of the spherical Bessel function

$$j_l(x) = (1/2i^l)\int_{-1}^{1}\exp(ixt)P_l(t)\,dt, \tag{4.68}$$

and the asymptotic form of the spherical Bessel function in the limit $x \to \infty$:

$$\begin{aligned}
j_l(x) &= (\pi/2x^l)^{1/2} J_{l+1/2}(x) \\
&\sim (1/x)\cos(x - (2l+1)\pi/2) \quad (\text{at } x \to \infty).
\end{aligned}$$

Thus, we may write the asymptotic expansion of the planar wave in the limit $r \to \infty$ as

$$\begin{aligned} &\exp(ikz)_{r\to\infty} \\ &\quad\sim \sum_l (2l+1)i^l(1/kr)\sin(kr - l\pi/2)P_l(\cos\theta) \\ &\quad\sim \sum_l \frac{(2l+1)i^l}{2ikr}\left(\exp\left[i\left(kr-\frac{l\pi}{2}\right)\right] - \exp\left[-i\left(kr-\frac{l\pi}{2}\right)\right]\right)P_l(\cos\theta) \end{aligned} \tag{4.69}$$

As a result, the scattering wave is obtained from

$$\begin{aligned} &\Psi(\boldsymbol{r}) - \exp(ikz) \\ &\quad= \sum_{l=0} A_l R(r) P_l(\cos\theta) - \sum_l P_l(\cos\theta)\frac{2l+1}{2}2i^l j_l(kr) \\ &\quad\sim \sum_{l=0} A_l \frac{1}{2ikr}\left(\exp\left[i\left(kr-\frac{l\pi}{2}+\eta_l\right)\right] - \exp\left[-i\left(kr-\frac{l\pi}{2}+\eta_l\right)\right]\right)P_l(\cos\theta) \\ &\quad -\sum_l \frac{(2l+1)i^l}{2ikr}\left(\exp\left[i\left(kr-\frac{l\pi}{2}\right)\right] - \exp\left[-i\left(kr-\frac{l\pi}{2}\right)\right]\right)P_l(\cos\theta) \\ &\quad\sim \sum_{l=0} P_l(\cos\theta)\left(\frac{1}{2ikr}\exp\left[i\left(kr-\frac{l\pi}{2}\right)\right]\left[A_l\exp(i\eta_l) - (2l+1)i^l\right]\right. \\ &\quad \left. -\frac{1}{2ikr}\exp\left[-i\left(kr-\frac{l\pi}{2}\right)\right]\left[A_l\exp(-i\eta_l) - (2l+1)i^l\right]\right) \quad (\text{at } r\to\infty). \end{aligned} \tag{4.70}$$

Equation 4.70 should be matched to the outgoing spherical wave $\exp(ikr)/r$. Therefore, the second term in Equation 4.70 must satisfy

$$\left[A_l\exp(-i\eta_l) - (2l+1)i^l\right] = 0, \quad \text{i.e.,} A_l = (2l+1)i^l\exp(i\eta_l).$$

The asymptotic form of the scattered spherical wave is

$$\begin{aligned} &\frac{f(k;\theta)}{r}\exp(ikr) \\ &\quad\sim \sum_{l=0}\frac{1}{2ikr}\exp\left[i\left(kr-\frac{l\pi}{2}\right)\right]\left[(2l+1)i^l\exp(2i\eta_l) - (2l+1)i^l\right]P_l(\cos\theta) \\ &\quad\sim \sum_{l=0}\frac{1}{2ik}\exp\left(-\frac{il\pi}{2}\right)i^l(2l+1)\{\exp(2i\eta_l)-1\}P_l(\cos\theta)\frac{\exp(ikr)}{r}, \end{aligned} \tag{4.71}$$

and from the expression for $f(k;\theta)$, it is possible to obtain the differential

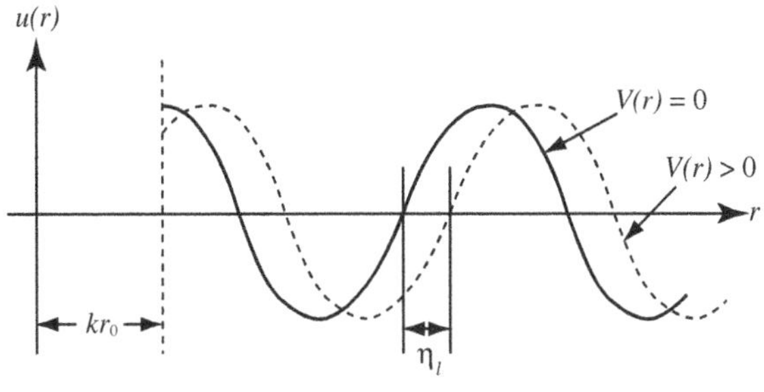

FIGURE 4.13 The effect of phase-shift on the wave function.

cross section $\sigma(\theta)$:

$$f(k;\theta) = \frac{1}{2ik}\sum_{l=0}(2l+1)\{\exp(2i\eta_l)-1\}P_l(\cos\theta) \tag{4.72}$$

$$\begin{aligned}\sigma(\theta) &= |f(k;\theta)|^2 \\ &= \frac{1}{k^2}\left|\sum_{l=0}\frac{2l+1}{2}\{\exp(2i\eta_l)-1\}P_l(\cos\theta)\right|^2.\end{aligned} \tag{4.73}$$

Hence the only quantities that describe the effect of the scattering are the phase-shifts η_l. The use of angular momentum quantum numbers l is analogous to the use of an impact parameter in classical scattering. The scattering cross section is obtained by summing up the different partial waves in respect to $l = 0, 1, 2, \ldots$, and the procedure shown here is known as the partial wave method. This technique was developed for light scattering by Rayleigh and applied to particle scattering by Faxen and Holtsmark.

Problem 4.4.1

Show that the total cross section Q_0 of the elastic scattering is given in partial wave expansion as

$$\begin{aligned}Q_0(\varepsilon) &= \int |f(k;\theta)|d\Omega \\ &= \frac{4\pi}{k^2}\sum_{l=0}(2l+1)\sin^2\eta_l.\end{aligned} \tag{4.74}$$

Exercise 4.4.1

Explain the phase difference η_l in Equation 4.66.

In Figure 4.13, we show two functions $u(r)$, one without any interaction with a target ($V(r) = 0$; solid line) and one with an interaction ($V(r) > 0$; dashed line). When the potential is repulsive, the outgoing wave function is pushed to the outside as compared with the wave at $V(r) = 0$, and then $\eta_l > 0$. On the other hand, if it is attractive, then $\eta_l < 0$.

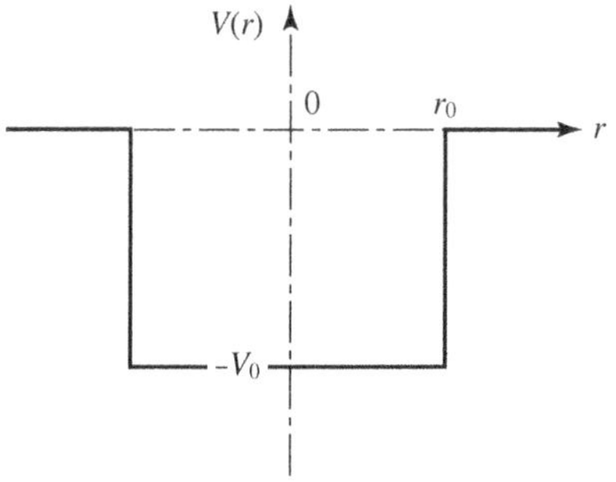

FIGURE 4.14 The spherical well potential (spherical attractive potential).

Exercise 4.4.2
In Figure 4.14 we show the spherical well potential:

$$\begin{aligned} V(r) &= -V_0 \quad r \leq r_0 \\ &= 0 \quad r > r_0. \end{aligned}$$

Calculate the total cross section and the phase-shifts for elastic scattering of a low-energy particle.
From Equation 4.65, we have

$$\frac{d^2u(r)}{dr^2} + \frac{2\mu}{\hbar^2}\{E - V(r)\}u(r) - \frac{l(l+1)u(r)}{r^2} = 0,$$

and for the potential it follows that

$$\{d^2/dr^2 + K^2 - l(l+1)/r^2\}u^{<}_{l,k}(r) = 0; \; K^2 = (2\mu/\hbar^2)(E + V_0); \quad (r \leq r_0),$$
$$\{d^2/dr^2 + k^2 - l(l+1)/r^2\}u^{>}_{l,k}(r) = 0; \; k^2 = (2\mu/\hbar^2)E; \quad (r > r_0).$$

The solution for $R(r) = u(r)/r$ must be finite at $r = 0$, and the $u(r)$ must be zero at $r = 0$. For low-energy scattering that satisfies $K^2 - l(l+1)/r^2 > 0$, the component in the partial wave expansion will mainly be $l = 0$ (s-wave). Therefore,

$$\begin{aligned} u_{0,k}(r) &= u^{<}_{0,k}(r) = A \sin Kr \\ &= u^{>}_{0,k}(r) = B \sin(kr + \eta_0). \end{aligned}$$

The wave function $u(r)$ and its derivative satisfying the Schrodinger equation must be continuous at $r = r_0$. Then,

$$A \sin Kr_0 = B \sin(kr_0 + \eta_0),$$
$$KA \cos Kr_0 = kB \cos(kr_0 + \eta_0),$$

and we obtain

$$\begin{aligned} (1/K)\tan(Kr_0) &= (1/k)\tan(kr_0 + \eta_0), \\ (\text{or})\, \eta_0 &= \tan^{-1}\left(\frac{k}{K}\tan(Kr_0)\right) - kr_0. \end{aligned}$$

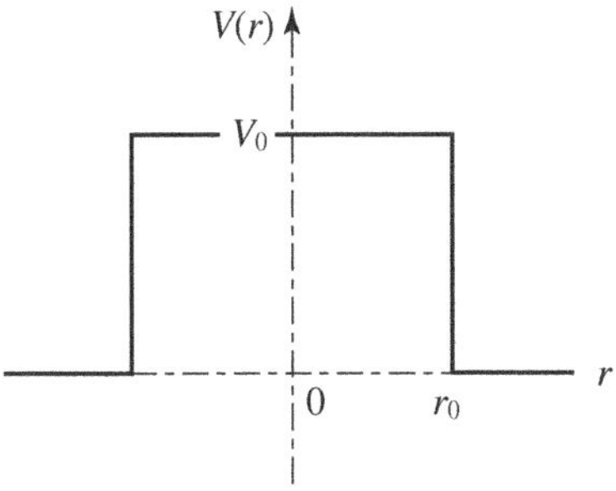

FIGURE 4.15 The profile of the spherical repulsive potential.

As a result, the scattering cross section for $l = 0$ (s-wave) is

$$Q_0(k) = \frac{4\pi}{k^2}\sin^2\eta_0 = \frac{4\pi}{k^2}\sin^2\left\{\tan^{-1}\left(\frac{k}{K}\tan Kr_0\right) - kr_0\right\}. \tag{4.75}$$

Problem 4.4.2

In Figure 4.15 we show the spherical repulsive potential:

$$\begin{aligned} V(r) &= +V_0 \quad r \le r_0 \\ &= 0 \quad r > r_0. \end{aligned}$$

Calculate the total cross section and the phase-shifts for elastic scattering of a low-energy particle. Show that the total cross section is equal to $Q_0 = 4\pi r_0^2$ in the limit $V_0(r) \to \infty$. Considering that the classical hard sphere cross section is equal to $Q_{0C} = \pi r_0^2$ (see Exercise 4.3.1), explain why the quantum result is so different ($Q_{0q} = 4\pi r_0^2$).

4.4.2 Modified Effective Range Theory in Electron Scattering

We have already established that the differential cross section in the elastic collision in basic quantum theory is expressed as $\sigma(\varepsilon) = |f(\theta;\varepsilon)|^2$, where the scattering amplitude $f(\theta)$ in the partial wave method is given by Equation 4.72

$$f(\theta) = \frac{1}{2ik}\sum_{l=0}(2l+1)\{(\exp(2i\eta_l) - 1\}P_l(\cos\theta)$$

in the partial wave method. Here k is the wave number of the incoming electron with energy ε. It follows that

$$Q_m(\varepsilon) = \frac{4\pi}{k^2}\sum_l(2l+1)\sin^2(\eta_l - \eta_{l+1}) \tag{4.76}$$

and also that

$$Q_0(\varepsilon) = \frac{4\pi}{k^2}\sum_l(2l+1)\sin^2\eta_l. \tag{4.77}$$

These formulae are sometimes used to analyze the differential cross section of the electron and perform analytic extrapolations or integrations of the experimental differential cross section.

A method for making an analytic representation of cross sections for targets without permanent dipole moments was proposed by O'Malley and coworkers under the name of the modified effective range theory (MERT) [6]. This method yields analytic expansion of phase-shifts:

$$\tan \eta_0 = -Ak\left[1 + \frac{4\alpha}{3a_0}k^2 \ln(a_0)\right] - \frac{\pi\alpha}{3a_0}k^2 + Dk^3 + Fk^4, \tag{4.78}$$

$$\tan \eta_1 = \frac{\pi\alpha}{15a_0}k^2 - A_1 k^3, \tag{4.79}$$

where α and a_0 are the polarizability and Bohr radius, respectively. For all higher-order phase-shifts ($l > 1$), the Born approximation is sufficiently accurate:

$$\tan \eta_l = \frac{\pi\alpha k^2}{(2l+3)(2l+1)(2l-1)}.$$

The parameters A (the scattering length), A_1, D, and F should be regarded as fitting parameters that may be obtained from transport coefficients of electrons and available cross section data. The value of parameter A may be determined using the dimension of the molecule at zero energy as $Q_0 = Q_m = 4\pi A^2$. For example, the best fit for argon is obtained for $A = -1.459a_0$, $A_1 = 8.69a_0^3$, $D = 68.93a_0^3$, and $F = -97a_0^4$. MERT was found to work well for scattering of electrons on atoms at low energies where the elastic collisions are the only scattering process. In other words, this approximation is valid for electron–atom scattering below 1 eV.

Problem 4.4.3

Calculate Q_0, Q_m for argon from 0 eV to 1 eV. Show that the Ramsauer–Townsend minimum (RTM) occurs when the contribution of both η_0 and η_1 are close to zero for the same energy. Show that the RTM occurs at different energies for Q_0 and Q_m and explain why.

4.5 COLLISIONS BETWEEN ELECTRONS AND NEUTRAL ATOMS/MOLECULES

Collisions of electrons with gas phase molecules are very frequent and are the most important in a majority of low-temperature nonequilibrium plasmas. Accordingly, these collisions are described in greater detail than other processes. A specific property of the collision between an electron and a molecule is the large difference in mass, which is more than four orders of magnitude. Thus the recoil of the molecule in collisions is small and the energy loss of electrons in elastic collisions is very small, whereas the energy gain for molecules is negligible. This is the basic property that allows formation of nonequilibrium plasmas. On the other hand, the transfer of energy to inelastic processes is close to 100% efficient. We can classify electron–molecule collisions by several criteria. The most important is division into *elastic* and *inelastic* processes. Another important distinction is between *conservative* and *nonconservative* processes of the electron, where, for example, ionization and attachment fall into the latter category.

A theoretical description of the electron–molecule collision is quite complex and an analytic representation of the cross sections is generally impossible. Classical and semiclassical treatments are only of very limited use in specific processes. Quantum theory is therefore almost always necessary and the Hamiltonian consists of the part describing the isolated molecule, the kinetic energy of the electron, and the electron–nuclei and the electron–molecule interactions. Inasmuch as the electrons in the outer orbit in a molecule and the projectile electron are fermions, the Pauli principle dictates that the corresponding potential is repulsive and nonlocal. There is no classical analogue for this interaction. This effect is of importance when the electron is close to the target, that is, in the *near* region. There it will compete with numerous electrostatic interactions.

If we move farther from the target the polarization effect becomes the dominant one. Polarization is induced by the projectile electron, and although the semiclassical picture is useful for understanding it, one needs quantum theory for practical calculations. At long distances r, the polarization potential is approximated by

$$V_{pol}(r) \rightarrow -\frac{1}{8\pi\varepsilon_0}\frac{\alpha(r)e^2}{r^4}, \qquad (4.80)$$

where α is the polarizability of the molecule. If there is a permanent dipole moment or if the target is charged, then the interaction may be dominated in the *far* region by electrostatic terms. Molecules with a permanent dipole moment, although neutral, have a very long-range potential, and thus the cross sections may be large, especially for low-incident energies. As a result of these interactions and of the internal degrees of freedom of molecules, there are numerous electron–molecule collisions that may occur; these are listed in Table 4.1.

4.5.1 Resonant Scattering

When an electron comes close to a molecule, it may be bound to the target molecule for a brief time (τ between 10^{-10} s and 10^{-15} s) depending on its initial energy and the structure of the molecule. When the time significantly exceeds the time required to cross the interaction region, we may talk about the formation of a compound state or temporary negative ion. The third term that may be used to describe this system is a *resonance*. Atoms in excited states may have positive electron affinity ε_a so that the energy of the bound electron is $\varepsilon_{ex} - \varepsilon_a$. However, this system is unstable as compared with the

TABLE 4.1 List of Electron–Molecule Collisions

Process	Reaction Scheme	Energy Loss	$\Delta\varepsilon$ (Typical)
Elastic	$e + M \xrightarrow{\varepsilon} e + M$	$\sim 2(m/M)\varepsilon$	~ 0
Electronic excitation	$e + M \xrightarrow{\varepsilon>\varepsilon_{ex}} e + M^j$	ε_{ex}	10 eV
Rotational excitation	$e + M(r) \xrightarrow{\varepsilon>\varepsilon_r} e + M(r')$	ε_r	kT
Vibrational excitation	$e + M(v) \xrightarrow{\varepsilon>\varepsilon_v} e + M(v')$	ε_v	0.1 eV
Dissociation	$e + AX \xrightarrow{\varepsilon>\varepsilon_d} e + A + X$	ε_d	10 eV
Ionization	$e + M \xrightarrow{\varepsilon>\varepsilon_i} e + e + M^+$	ε_i	15 eV
Attachment	$e + M \xrightarrow{\varepsilon>\varepsilon_a} M^-$	ε_a	$\sim$ kT
Dissociative attachment	$e + AX \xrightarrow{\varepsilon>\varepsilon_a} A + X^-$	ε_a	$\sim$ eV
Ion pair formation	$e + AX \xrightarrow{\varepsilon>\varepsilon_{ip}} e + A^+ + X^-$	ε_{ip}	20 eV
Superelastic collision	$e + M^j \xrightarrow{\varepsilon} e + M$	0.01$\sim$ –20eV	

ground state, so it may decay into a free electron with an energy close to $\varepsilon_{ex} - \varepsilon_a$ and the target molecule. In principle, if the state were long-lived it could survive long enough to suffer an additional collision with another gas molecule and it would be stabilized, but usually the system is unstable and it decays by autodetachment. Resonances are characterized by their energy width $\Delta\varepsilon$, which may be associated with the compound's lifetime through the uncertainty principle.

Problem 4.5.1

Calculate the energy width of the resonance that has a lifetime of 10^{-14} s. *Describe the elastic and the inelastic resonant scattering on a molecule.*

There are two types of resonances:

i. *Feschbach resonances.* Here a temporary potential well is induced when the target is excited by an electron and the electron is trapped. These lie from 0 eV to 0.5 eV below the parent excited state.

ii. *Shape resonances.* The resonance state is above the parent state and the scattering process strongly depends on the shape of the potential well.

It is evident that electrons must satisfy a very narrow energy condition in order to be able to excite the resonances. At the same time, enhanced interaction, due to a long-lasting bound state, will result in a much higher cross section, sometimes even by several orders of magnitude. It is difficult to distinguish whether scattering is resonant if the lifetime is very short (or the resonance is very broad). Yet resonant scattering is very useful in describing the shape of the cross sections.

The term *resonant* usually refers to a specific situation where two processes have to be matched in a very narrow range of conditions (in our case energy) in order to interact. We use the term in this broader meaning elsewhere in this book. A specific term *resonant scattering*, as described in this section, refers to electron molecule collisions.

4.6 ELECTRON–ATOM COLLISIONS

4.6.1 Energy Levels of Atoms

In the elastic collision between a light electron (mass m) and a heavier gas molecule (mass M), the loss of kinetic energy is on average equal to $2(m/M)\varepsilon$, where ε is the kinetic energy of the electron. Thus for low energies, the efficiency of energy transfer in one elastic collision is 10^{-4} as compared with the value for typical inelastic processes. In order to understand inelastic collisions between an electron and atom, one should first consider the energy levels of an atom. The energies between different levels determine the energy of the transition

$$\Delta\varepsilon = \varepsilon_{k+1} - \varepsilon_k. \tag{4.81}$$

The threshold energy ε_{ex} required for the upper excited state is equal to the excitation energy

$$\Delta\varepsilon = \varepsilon_{ex}, \tag{4.82}$$

and below that energy the cross section for the electron-induced excitation is equal to zero. In Figure 4.16 we show the energy level for helium. Only the allowed radiative transitions for bound electrons are possible (both absorption and emission). The lifetimes of excited states that have allowed radiative transitions to lower levels are on the order of 10^{-9} s.

Bound electrons can also be excited to triplet states, and these will eventually decay to the lowest excited triplet states that have no radiative transition to the ground state by dipole emission. Transitions by quadrupole and higher-order moments are weak, and the lifetimes of these levels are typically in the range of 10^2 s $\sim$ 100 s or even longer. These are the so-called metastable states. Of the lowest excitation levels of He (see Figure 4.16), two, 2^3S and 2^1S, are metastable states having lifetimes 7.9×10^3 s and 1.95×10^{-2} s, respectively. The levels such as 2^1P and 2^3P that have allowed transitions to the ground state are usually known as *resonant levels*, and their lifetimes are on the order of nanoseconds. If the electron collides with the excited metastable, then stepwise excitation or ionization is possible. This mode is more efficient, because the threshold for excitation of the higher level or the ionization is lower than that for the excitation from the ground state. On the other hand, the efficiency depends on the density of metastable atoms (or excited atoms in general). As the energy level increases, the energy gaps between excited states become smaller and smaller. The energy levels eventually converge to the ionization continuum where the excited electron is no longer bound to the atom forming an ion and free electron.

Exercise 4.6.1

Using the selection rules, explain the optically allowed and forbidden transitions in an atom.

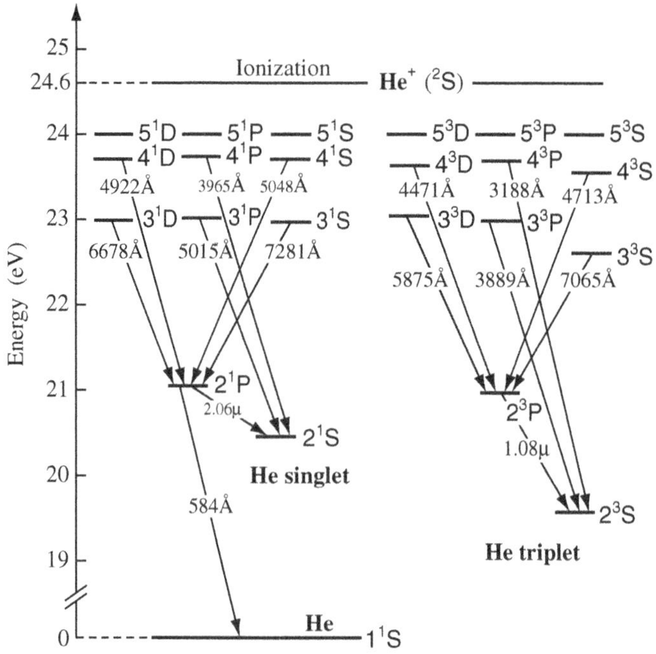

FIGURE 4.16 Energy levels of helium. Radiative transitions between singlet and triplet states are not allowed.

A typical term of an atom in a Russell–Saunders state is, for example, 4^3P_0. The first number is the principal quantum number n (in this case $n = 4$). Values of orbital angular momentum $\boldsymbol{L}$ are denoted by capital letters: $S(L = 0), P(L = 1), D(L = 2), \ldots$, (in this case $L = 1$). The spin angular momentum $\boldsymbol{S}$ is shown through multiplicity $(2S + 1)$ and written as a superscript in front of the letter annotating the value of L $(S = 1)$. The subscript to the right is equal to the total angular momentum $\boldsymbol{J} = \boldsymbol{L} + \boldsymbol{S}$ $(J = 0)$, which has integer values for odd multiplicity and takes values $1/2, 3/2, \ldots$ for even multiplicity. Parity is determined from the azimuthal quantum number for individual electrons l_i as $(-1)^{\sum l_i}$. The superscript to the right, if present, denotes parity (written for odd terms and not written for even terms). The magnetic quantum number is M. The selection rules are:

i. $\Delta J = 0, \pm 1$, except for $0 \to 0$;

ii. Parity must change;

iii. $\Delta L = 0, \pm 1$, except for $0 \to 0$;

iv. $\Delta S = 0$;

v. $\Delta l = \pm 1$; and

vi. $\Delta M = 0, \pm 1$.

Rules (iii) to (v) are only approximate, but even strict rules may be broken if there is, for example, strong spin-orbit coupling in strong magnetic fields, due to nuclear perturbations or for autoionizing states. Rule (iv) does not allow transitions among different multiplicities (e.g., triplet to singlet transitions). Transitions with $\Delta L = 0$ are very rare. For our purposes, it suffices to note that transitions to the ground state of helium 1^1S from 2^3S and 2^1S are forbidden because of the $\Delta S = 0$ rule in the former case and because of the $\Delta L = 0 \to 0$ rule in the latter case.

4.6.2 Electron–Atom Scattering Cross Sections

In Figure 4.17 we show a set of cross sections for electrons in helium. The elastic momentum transfer cross section at low energies is almost constant with a value close to 6×10^{-16} cm^2. At those energies helium is an excellent (and very rare) example of a constant cross section. Above 10 eV the elastic cross section falls. There are no inelastic losses until the incident energy exceeds 19.8 eV needed for the first electronic excitation. One should notice that the excitation to triplet states peaks close to the threshold and drops down rapidly due to the requirement of the spin exchange of the two bound electrons from $s(\frac{1}{2}, -\frac{1}{2})$ to $s(\frac{1}{2}, \frac{1}{2})$ by the aid of the incident electron. On the other hand, the

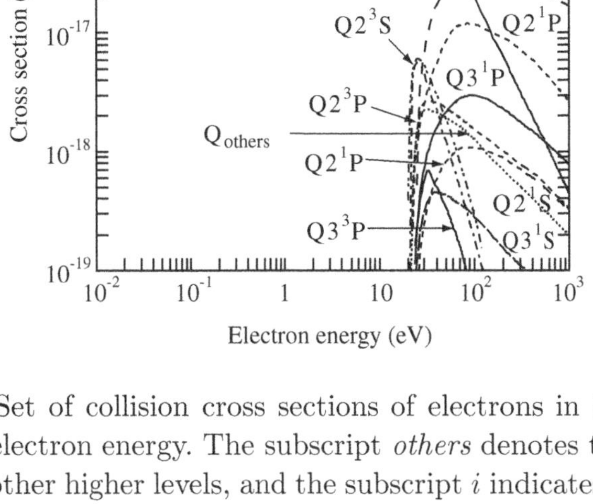

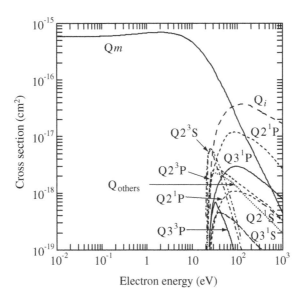

FIGURE 4.17 Set of collision cross sections of electrons in helium as a function of electron energy. The subscript *others* denotes the summed effect of all other higher levels, and the subscript i indicates ionization. All other levels are denoted by their terms.

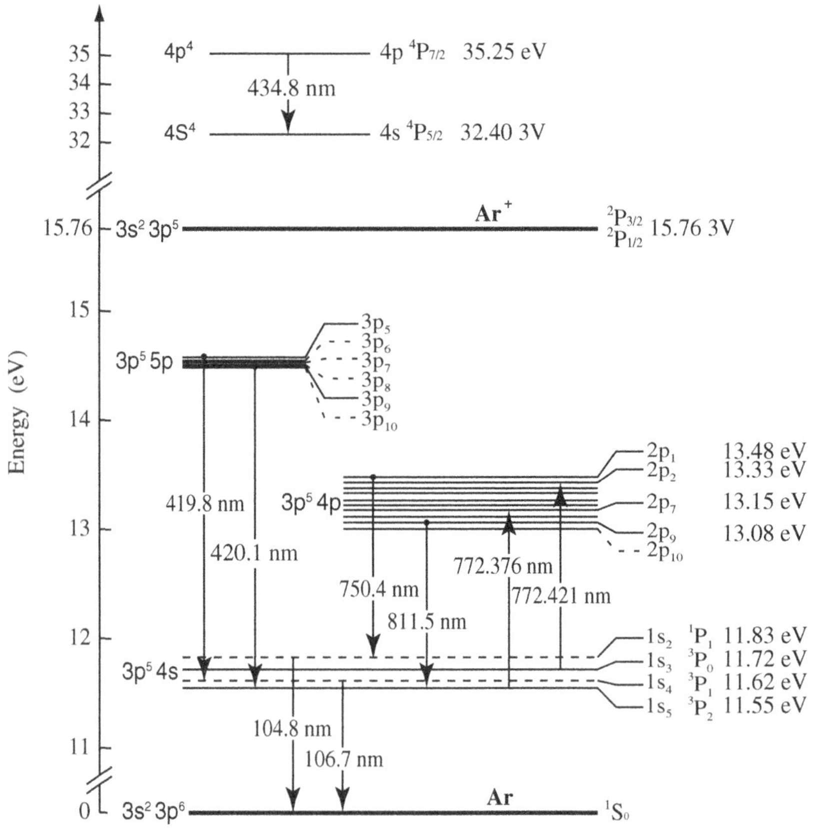

FIGURE 4.18 Energy levels of argon. The Paschen notation is also given.

cross section for excitation of singlet states is broad, peaking at around 100 eV, and resembles the ionization cross section in energy dependence.

Quite often Paschen notation is used to label the levels, especially for heavier rare gases (see Figure 4.18). This is really just an assignment of levels in their order rather than a notation associated with quantum numbers. Thus, in this notation the ground state would be $1s_1$, and the lowest metastable level would be $1s_5$. Two resonant lines from the four low-lying levels are in the ultraviolet region, but most lines from the 4p and 5p states are in the visible and near-infrared regions and are often used for diagnostics of argon plasmas. Properties of the ground state, of the low-lying levels, and of the ionization limit for the rare gases are given in Table 4.2.

Exercise 4.6.2

In the case of argon (see Figure 4.18), the ground state is ${}^1S_0$. Explain the allowed and forbidden transitions by using selection rules (and their possible

TABLE 4.2 Energies; Configurations; and Terms of Ground States, Low-Lying Levels, and Ionization Continuum for Rare Gases

			Excited	Levels				
Gas	**Configuration**	**Term**	**Resonant**	**(eV)**	**Metastable**	**(eV)**	**Ionization**	**(eV)**
He	$1s^2$	$1\,^1S_0$	$2\,^3P_1$	20.96	$2\,^3S_1$	19.82	$^2S_{3/2}$	24.59
			$2\,^1P_1$	21.12	$2\,^1S_0$	20.61		
Ne	$[He]2s^22p^6$	$2\,^1S_0$	$3\,^3P_1$	16.67	$3\,^3P_2$	16.62	$^2P_{3/2}$	21.56
			$3\,^1P_1$	16.85	$3\,^3P_0$	16.72		
Ar	$[Ne]3s^23p^6$	$3\,^1S_0$	$4\,^3P_1$	11.62	$4\,^3P_2$	11.55	$^2P_{3/2}$	15.76
			$4\,^1P_1$	11.83	$4\,^3P_0$	11.75		
Kr	$[Ar]4s^23d^{10}4p^6$	$4\,^1S_0$	$5\,^3P_1$	10.03	$5\,^3P_2$	9.92	$^2P_{3/2}$	14.00
			$5\,^1P_1$	10.64	$5\,^3P_0$	10.56		
Xe	$[Kr]5s^24d^{10}5p^6$	$5\,^1S_0$	$6\,^3P_1$	8.48	$6\,^3P_2$	8.32	$^2P_{1/2}$	12.13
			$6\,^1P_1$	9.54	$6\,^3P_0$	9.45		

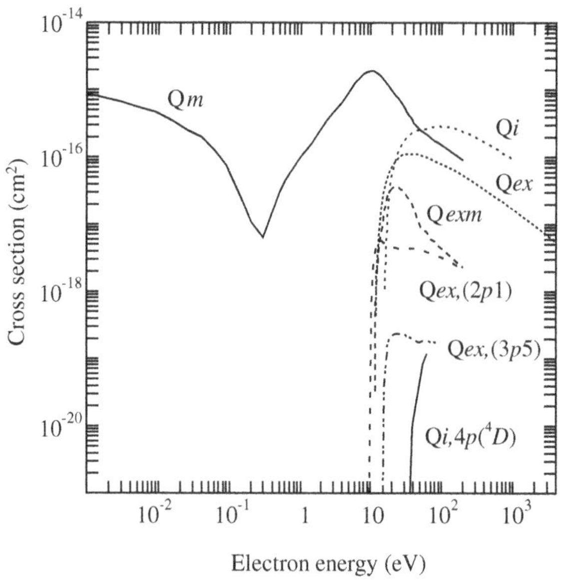

FIGURE 4.19 Cross sections for electron–argon scattering as a function of electron energy. Excitation is represented by effective summed cross sections and is separated into excitation of metastables and excitation of higher levels. Three cross sections for specific transitions are added, two for excitations into a neutral and the other for ionization into an excited ion from the ground state neutral. These are employed for comparison between modeling and the optical diagnostics of a low-temperature plasma in Ar.

breakdown). Discuss the selection rules in the case of other low-lying levels of the rare gases as given in Table 4.2.
In the case of argon we have four low-lying levels: two of these are metastable ($^3P_2$ and $^3P_0$), and the other two ($^3P_1$ and $^1P_1$) are resonant (with allowed transitions to the ground state $^1S_0$). Transitions from the two metastables are strongly forbidden by the rule that only $\Delta J = 0, \pm 1$ transitions, except for $0 \to 0$, are allowed. In this case, the $^3P_1$ level has a resonant transition to the ground state in spite of the weaker rule that only $\Delta S = 0$ transitions are allowed. All rare gases with the exception of helium have a similar configuration of metastable and resonant states.

Cross sections for electron–argon collisions are shown in Figure 4.19. One should observe that for energies below the threshold for the lowest metastable (11.55 eV in this case) there are no inelastic losses, as in helium. However, the elastic momentum transfer cross section has a broad minimum at around

0.23 eV. This is the RTM, which is the first quantum effect observed for particles with nonzero mass. When the RTM is present, the minimum value of the cross section is two orders of magnitude lower than the maximum (and the value that one would expect without the RTM). RTM is present in heavier rare gases (Ar, Kr, Xe) and in many molecules.

One should observe that above the four lowest levels there are a large number of excited levels having transitions to those four levels including the two metastables. If one wants to calculate the excitation rate to a particular level, one needs to consider the excitation to all higher states and include the radiative cascading by properly taking into account the branching ratios from higher levels. In addition to excitation by electrons and radiative de-excitation, there are several oher processes that provide nonradiative transitions to lower or nearby resonant levels. These represent collisional quenching with other gas molecules, electrons, or walls of the chamber and are discussed later. We note, however, that some of the measurements of cross sections for excitation suffer from the effect of cascading. All higher levels are also excited if the electron energy is sufficient. Short-lived radiative states decay and some of the transitions populate the excited state that is being studied. Thus the population of the excited state without collisional quenching will be

$$N_j = N_j^{direct} + \sum_{k>j} \frac{A_{kj}}{\sum_{i<k} A_{ki}} N_k,$$

where A_{kj} is the probability of the transition to level j from the higher-level k and $A_k = \sum_{i<k} A_{ki}$ is the inverse of the lifetime of the higher level. Because the direct population is proportional to the cross section (for a monoenergetic beam of electrons), the effective cross section consists of two parts: one due to direct excitation to the desired state j and the second due to cascading effects (where each of the higher levels is also subject to cascading). One should pay attention to the source of the data, the technique that was used, and other information in order to establish whether a cross section is subject to the contribution due to cascading.

4.7 ELECTRON–MOLECULE COLLISIONS

Elastic collisions between an electron and molecule have cross sections and energy dependencies similar to those for electron–atom scattering. There are examples both with and without the RTM. However, there are also two additional features important in electron–molecule scattering. For molecules with a permanent dipole moment, there is an enhanced elastic (momentum transfer) cross section peaking at zero energy. The second feature is the existence of either a sharp or broad structure due to resonant scattering for energies below the threshold for electronic excitation. Both are discussed later when complete sets for several molecules are presented. Let us begin by examining the most salient features of molecules from the viewpoint of electron–molecule collisions. Molecules having internal degrees of freedom bring something new to the kinetics of electron transport: the excitation to rotational and vibrational energy levels leads to inelastic energy losses at much lower energies. Hence with a molecular gas it may be possible to control the energy of electrons more precisely in a plasma.

4.7.1 Rotational, Vibrational, and Electronic Energy Levels of Molecules

When two atoms combine into a molecule, their dynamics will be determined by the forces that act between them. A combination of repulsive and attractive forces results in a potential energy for each internuclear position. A potential curve is plotted for each combination of atoms as a function of internuclear distance. The molecule will be stabilized at an equilibrium internuclear distance that presumably corresponds to the minimum potential. In Figure 4.20 we show the potential energy curves for a nitrogen molecule. The ground state is $X^1\Sigma_g^+$. Superimposed on the potential diagram are the vibrational (ε_v) and rotational (ε_r) energy levels. The energies of rotation and vibration of molecules are quantized; that is, the energy levels are discrete. The total energy of the bound electron in a molecule is thus

$$\varepsilon_T = \varepsilon_r + \varepsilon_v + \varepsilon_{ex}. \tag{4.83}$$

In Figure 4.20 we show only vibrational levels, inasmuch as rotational levels would need a much higher resolution to be observed. The excited state $A^3\Sigma_u^+$ also has vibrational and rotational levels. At room temperature molecules populate a large number of rotational levels and, consequently, the emission in transitions between two electronically excited states is broad and we talk about molecular bands rather than lines. The third potential curve, depicted in Figure 4.20, has no minimum. It is a repulsive curve, because the two atoms composing the molecule would separate with some kinetic energy if they found themselves on it.

The total energy of a diatomic molecule with internal potential energy $V(r)$ can be written using the relative velocity v_r and reduced mass μ in the CM frame in classical mechanics Equation 4.15 as

$$T = \frac{1}{2}\mu v_r^2 + V(r),$$

which can be transformed to the spherical coordinate system

$$T = \frac{1}{2}\mu r^2 \left(\frac{d\theta}{dt}\right)^2 + \frac{1}{2}\mu \left(\frac{dr}{dt}\right)^2 + V(r), \tag{4.84}$$

which corresponds to the Hamiltonian in quantum mechanics. In completing the Hamiltonian, we should remember that the angular momentum $\boldsymbol{L}$ is conserved and equal to $\boldsymbol{L} = \boldsymbol{r} \times \boldsymbol{P} = r\mu(rd\theta/dt)$. This introduces the Hamiltonian

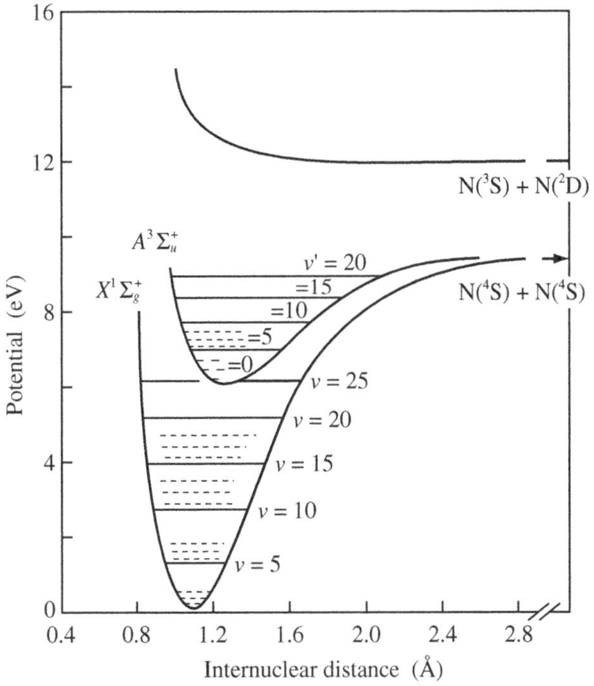

FIGURE 4.20 Potential energy curves for a nitrogen molecule.

along with the relation $\boldsymbol{P}(= -i\hbar d/d\boldsymbol{r})$ for quantization of momentum. The Hamiltonian as the operator of the total energy is given as

$$\mathbf{H} = \frac{L^2}{2\mu r^2} - \frac{\hbar^2}{2\mu}\frac{d^2}{dr^2} + V(r). \tag{4.85}$$

The Hamiltonian may be separated into two terms, one corresponding to the rotational θ-motion

$$\mathbf{H}_{rot} = \frac{L^2}{2\mu r^2} \tag{4.86}$$

and the other to the vibrational r-motion

$$\mathbf{H}_{vib} = -\frac{\hbar^2}{2\mu}\frac{d^2}{dr^2} + V(r). \tag{4.87}$$

4.7.2 Rotational Excitation

4.7.2.1 Rotational Energy Levels

Rotational energy levels are obtained as eigenvalues ε_r of the appropriately chosen eigenfunctions ϕ_r from the eigenvalue equation

$$\mathbf{H}_{rot}\phi_r = \varepsilon_r\phi_r, \tag{4.88}$$

where

$$\varepsilon_r = \frac{L^2}{2\mu r^2} = \frac{L^2}{2I}.$$

Here, I is the moment of inertia of the molecule, $I = \mu r^2$, where r is the equilibrium distance of the two nuclei. At the same time the angular momentum is quantized as $L^2 = l(l+1)\hbar^2$ with the quantum number $l = 0, 1, 2, \cdots$. As a result, we have the rotational energy levels

$$\varepsilon_r = \frac{\hbar^2 l(l+1)}{2I} = B_e l(l+1), \quad l = 0, 1, 2, \cdots. \tag{4.89}$$

Here, $B_e = \hbar^2/2I$ is a rotational constant, and for most molecules it is on the order of 10^{-4}eV$\sim 10^{-3}$ eV, that is, meV. The transition energy between two successive levels $\Delta\varepsilon_r$ with $\Delta l = 1$ is

$$\Delta\varepsilon_{rot} = \frac{\hbar^2}{2I}\{l(l+1) - (l-1)l\} = \frac{\hbar^2}{2I}2l = B_e 2l. \tag{4.90}$$

For molecules with a quadrupole moment, the selection rule is $\Delta l = \pm 2$ $(l, l+2)$, so in that case the energies of rotational transitions are $\Delta\varepsilon_{rot} = B_e(4l+6)$. (In both cases the transitions from the ground state are obtained by choosing $l = 0$.) So far it has been assumed that the molecule is rigid; that is, the distance between two atoms does not change as the molecule rotates. In reality, however, the distance between molecules changes and so the moment of inertia changes. This effect leads to the energy levels being described by

$$\varepsilon_r = \frac{\hbar^2 l(l+1)}{2I} = \left[B_e + \alpha\left(v + \frac{1}{2}\right)\right] l(l+1), \quad l = 0, 1, 2, \cdots,$$

where v is the quantum number of the vibrational level (discussed in the next section) and α is a constant tabulated in books on spectroscopy.

Exercise 4.7.1

Estimate the energy of rotational transitions of an HCl molecule, and determine the ratio of the population of $l = 1$ to that of $l = 0$ at 300 K based on this approximation. Prepare a table of exact values of B_e for different molecules and calculate exact energy losses for the first four levels of H_2, D_2, N_2, O_2, and CO_2.

For HCl, estimated values for the relevant quantities are

$$r \approx 1.27 \times 10^{-10}\ \text{m}, \quad \mu = \frac{M_H M_{Cl}}{M_H + M_{Cl}} \approx 1.63 \times 10^{-27}\ \text{kg},$$

which give the following estimate for the order of magnitude of the first rotational level:

$$\begin{aligned}\Delta\varepsilon_r(l=1) = \frac{\hbar^2}{I} = \frac{\hbar^2}{\mu r_0^2} \quad &\approx \quad \frac{(6.63 \times 10^{-34}/2\pi)^2}{1.63 \times 10^{-27} \times (1.27 \times 10^{-10})^2} \\ &\approx \quad 0.423 \times 10^{-21}\text{J} \approx 2.64 \times 10^{-3}\text{eV}.\end{aligned}$$

TABLE 4.3 Rotational Transition Energy of Several Molecules with a Quadrupole Moment

				$\Delta\varepsilon$	(eV)	
Molecule	q_4 (ea_0^2)	B_e **(eV)**	**0–2**	**1–3**	**2–4**	**3–5**
H_2	0.586	7.56E-3	0.0441	0.0735	0.1029	0.1323
D_2	0.2487	3.79E-3	0.0221	0.0368	0.0515	0.0662
N_2	1.04	2.48E-4	0.00146	0.00243	0.00340	0.00437
O_2	1.80	1.78E-4	0.00107	0.00178	0.00249	0.00320
CO_2	3.00	4.85E-5	0.00029	0.00048	0.00068	0.00087

The energy of rotational transition is on the order of meV. This should be compared with the thermal energy of molecules at room temperature, $3/2kT_g$ (where k is the Boltzmann constant). That is, the rotational state populations $N_r(l)$ obey the Boltzmann distribution with statistical weight G_l. The statistical weight (factor of degeneracy) is $(2l+1)$ and therefore from the estimate given above we get

$$\begin{aligned}\frac{N_r(l=1)}{N_r(l=0)} &= \frac{G_1}{G_0}\exp\left(-\frac{\varepsilon_r(l=1)-\varepsilon_r(l=0)}{kT_g}\right)\\ &= 3\times\exp\left(-\frac{2.64\times10^{-3}}{1.38\times10^{-23}\times300/1.60\times10^{-19}}\right)\approx 2.71.\end{aligned}$$

If we pursue this task more accurately we may assemble a table of values of B_e from the textbooks on spectroscopy or we may even calculate the values based on the data for stable configurations of molecules. Then one may determine the energy losses in the lowest rotational transitions. The results are shown in Table 4.3. It is noted that energies (and B_e) in spectroscopy are usually tabulated in units of cm^{-1} and that $1\mathrm{eV} = 8.0655\times10^5\ \mathrm{m}^{-1}$. In Table 4.4, we show typical wavelengths of radiation required to make rotational, vibrational, and electronic transitions and the corresponding energies.

TABLE 4.4 Typical Wavelengths and Transition Energies for Rotational, Vibrational, and Electronic Excitations

Transition	**Wave**	**Wavelength** (m)	**Energy**(eV)
Rotational	Microwave	$1\sim10^{-3}$	$1.24\times10^{-6}\sim1.24\times10^{-4}$
Vibrational	Infrared	$10^{-3}\sim0.8\times10^{-6}$	$1.24\times10^{-4}\sim1.55$
Electronic	Visible	$0.8\times10^{-6}\sim0.4\times10^{-6}$	$1.55\sim3.10$
Electronic	UV	$0.4\times10^{-6}\sim10^{-8}$	$3.10\sim30$

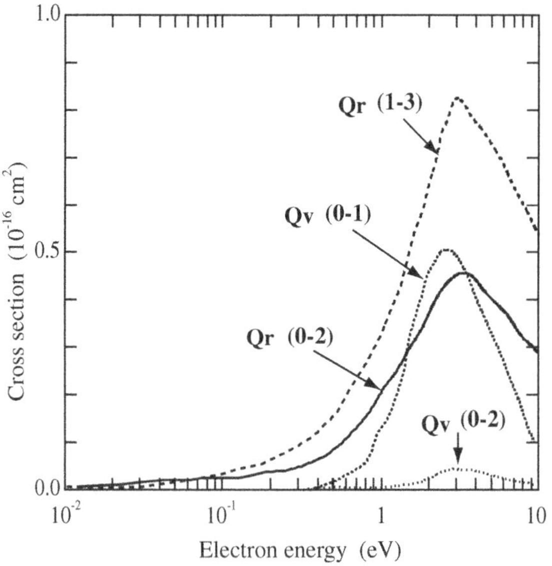

FIGURE 4.21 Rotational and vibrational excitation cross sections for hydrogen obtained from the swarm data.

4.7.2.2 Rotational Excitation Cross Sections

Rotational inelastic losses are typically small, but because they occur at small energies and there are no other losses in that range, it is important to include them properly. Rotational excitation is negligible for highly spherically symmetric molecules such as CH_4. As mentioned above, rotational excitation will have the selection rule $\Delta l = \pm 1$ for molecules with a permanent dipole and $\Delta l = \pm 2$ for molecules with a quadrupole moment. Gases such as hydrogen have very large rotational energies that are comparable with and even larger than their mean thermal energies at room temperature. Thus, we may expect that only a few rotational levels will have any significant population. For this reason, it should be possible to develop an accurate theory to represent rotational excitation and perhaps to use a normalizing factor for all the levels. For parahydrogen at 77 K, in which only the ground rotational level was populated, it was actually possible to determine the rotational excitation cross section (see Figure 4.21) for the $l = 0 \rightarrow 2$ transition from the transport data. It is then possible to extend the analysis to the $l = 1 \rightarrow 3$ transition from the data for normal hydrogen. The good agreement with available theories provided justification for the use of theory to determine cross sections for a few higher transitions that exist at room temperature.

TABLE 4.5 Rotational-Level Populations of H_2 and D_2

	$l = 0$	$l = 1$	$l = 2$	$l = 3$	$l = 4$	$l = 5$
p-$H_2$77 K	0.9946	0	0.0054	0	0	0
n-$H_2$77 K	0.2487	0.75	0.0013	0	0	0
n-$H_2$293 K	0.135	0.67	0.112	0.079	0.003	0
n-$D_2$77 K	0.5721	0.3307	0.0945	0.0026	0	0
n-$D_2$293 K	0.1832	0.2060	0.3858	0.1137	0.0924	0.0134

Exercise 4.7.2
What are para, the ortho, and the normal hydrogen? Calculate the population of rotationally excited states under thermal equilibrium at room temperature and at 77 K for hydrogen and deuterium. The nuclear spin for hydrogen is $I = 1/2$ and that for deuterium is $I = 1$.
The para and ortho states of diatomic molecules are those in which the nuclear spins of the two atoms are in the same direction and in the opposite directions, respectively. In normal hydrogen there is a 1:3 mixture of para to ortho states, and in deuterium the ratio is 2:1. The abundance of a rotational level of molecules in thermal equilibrium at temperature T is

$$\frac{N_l}{N} = \frac{p_l \exp(-\varepsilon_r/kT)}{\sum_{l=0}^{\infty} p_l \exp(-\varepsilon_r/kT)}. \tag{4.91}$$

Here, the statistical weight is equal to $p_l = (2I+1)(I+a)(2l+1)$. Here $a = 0$ for even l and $a = 1$ for odd l for molecules like H_2. The values of a are exactly the opposite for molecules like D_2 and N_2, which depend on the value of the nuclear moment. The results of the calculations are given in Table 4.5.

4.7.3 Vibrational Excitation

4.7.3.1 Vibrational Energy Levels

The energies of molecular vibrational states may be obtained as eigenvalues of the eigenvalue equation

$$\mathbf{H}_{vib}\phi_v = \varepsilon_v\varphi_v; \quad \boldsymbol{H}_{vib} = -\frac{\hbar^2}{2\mu}\frac{d^2}{dr^2} + V(r). \tag{4.92}$$

Furthermore, we may describe the interaction potential $V(r)$ as a function of departure from the equilibrium position r_0. The potential is approximated by a parabolic curve with minimum $V(r_0)$ (see Figure 4.20):

$$V(r) = A^2(r - r_0)^2 + V(r_0). \tag{4.93}$$

If we replace $x = r - r_0$ in Equation 4.93, we obtain

$$\frac{d^2\phi_v(x)}{dx^2} + \frac{2\mu}{\hbar^2}\left\{\varepsilon_v - V(r_0) - A^2x^2\right\}\phi_v = 0. \tag{4.94}$$

The solution to this equation is obtained by using Hermite polynomials H_n as

$$\phi_v(x) = \left(\frac{\sqrt{2\mu\nu/\hbar}}{2^v v!}\right)^{1/2} H_v\left(\sqrt{\frac{\mu\nu}{\hbar}}x\right)\exp\left(-\frac{\mu\nu x^2}{2\hbar}\right), \quad \nu^2 = 2A^2/\mu, \quad (4.95)$$

where v is the vibrational quantum number. The energy of the vibrational levels is then given by

$$\varepsilon_v = \left(v + \frac{1}{2}\right)\hbar\nu + V(r_0), \quad v = 0, 1, 2\cdots, \quad (4.96)$$

and the transition energy between successive vibrational levels is

$$\Delta\varepsilon_v = \varepsilon_v - \varepsilon_{v-1} = \hbar\nu. \quad (4.97)$$

Problem 4.7.1

A good approximation of the realistic potential of diatomic molecules is the Morse potential

$$V(r) = D\left[1 - e^{-a(r-r_0)}\right]^2.$$

Show that for small departures from the equilibrium position r_0, this potential may be represented by the harmonic potential 4.93. Perform the calculations and plot both potentials for the example of KCl where $D =$ Equation 4.42 eV and $a^2 = 0.602 \times 10^{-20}$ m^{-2}.

Problem 4.7.2

Derive the solution of the classical vibrational motion of a particle of reduced mass μ for force $-2A^2r$. Compare the angular frequency of the oscillation $\nu = \sqrt{2A^2/\mu}$ to that given in Equation 4.95. Relate the classical energy to that in Equation 4.96 and explain.

Problem 4.7.3

Estimate the magnitude of $\Delta\varepsilon_v$ for H_2.

Exercise 4.7.3

Assuming that the potential energy of a diatomic molecule at equilibrium distance is zero ($V(r_0) = 0$), the energy of the lowest vibrational state is $\varepsilon_{v=0} = \hbar\nu/2$. Discuss this zero point energy from the standpoint of quantum mechanics.

Quantum phenomena have a finite energy even at 0 K as fluctuation. It is well known that the nonzero energy arises from the uncertainty principle.

4.7.3.2 Vibrational Cross Sections

Vibrational excitation cross sections for H_2 are shown in Figure 4.21. For hydrogen the threshold for the $v = 0 - 1$ vibrational excitation is 0.5 eV, which is quite large. Other molecules have smaller vibrational energies on the order of 0.1 eV to 0.2 eV. In the case of hydrogen one also needs to consider the difference of losses for rotationally elastic $(0 - 0)$ and inelastic $(0 - 2)$ vibrational excitations. It is important to note that there is a systematic difference between the data for the vibrational excitation cross section obtained by fitting the transport coefficients and those from binary collision experiments or theory. The difference exceeds the combined uncertainties. On the other hand, the agreement for elastic scattering and rotational excitation between both methods is excellent. The "hydrogen problem" for vibrational excitation remains an open issue.

The smooth dependence of vibrational excitation cross sections for hydrogen as shown in Figure 4.22 is, however, not typical. In most molecules there exist resonances that induce a very pronounced structure in the cross section. Very strong resonances exist for N_2 and O_2. In Figure 4.23 we show a set of cross sections for NO, which includes resonant vibrational excitation peaking at around 1 eV. However, above 10 eV there is another peak, which may be obscured by competing with electronic excitation and dissociation, both of which are more efficient due to their larger energy losses. There are no useful analytic formulae for vibrational excitation. In most cases the $v = 0 - 1$ transition is the most important, but one may need to add data for higher-level

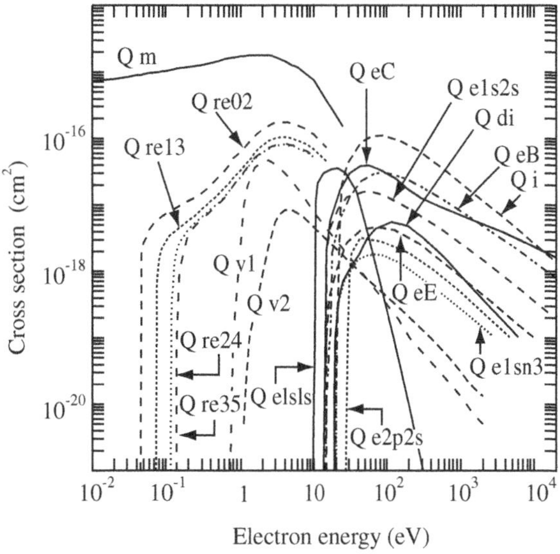

FIGURE 4.22 A set of cross sections for electron scattering of H_2.

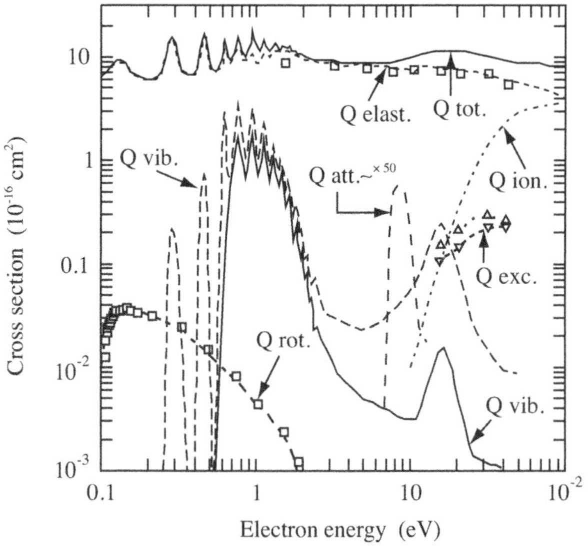

FIGURE 4.23 A set of cross sections for electron scattering on NO. Vibrational excitation shows a strong resonance that is also present in the total scattering cross section.

excitations for completeness. In such a case, relative magnitudes or complete cross sections derived from the theory or beam experiments are useful.

4.7.4 Electronic Excitation and Dissociation

4.7.4.1 Electronic States of Molecules

As shown in Figure 4.20, there are both bound and repulsive states of molecules above the ground state potential curve. Molecules are usually formed when electrons in the outer orbit are shared by both atoms. These electrons are in different states. Angular momentum in this case is denoted by the Greek letters $\Sigma, \Pi, \Delta, \cdots$, and the values ($\Lambda = 0, 1, \cdots$) correspond to the equivalent Latin letters S, P, D,$\cdots$.

Exercise 4.7.4
List the most important selection rules for transitions between excited states of molecules.
Some of the most important selection rules are:

i. $\Delta\Lambda = 0, \pm 1$, where Λ is the orbital angular momentum;

ii. $\Delta\Sigma = 0$, where Σ is the spin quantum number and $\Sigma = S, S-1, \cdots, -S$;

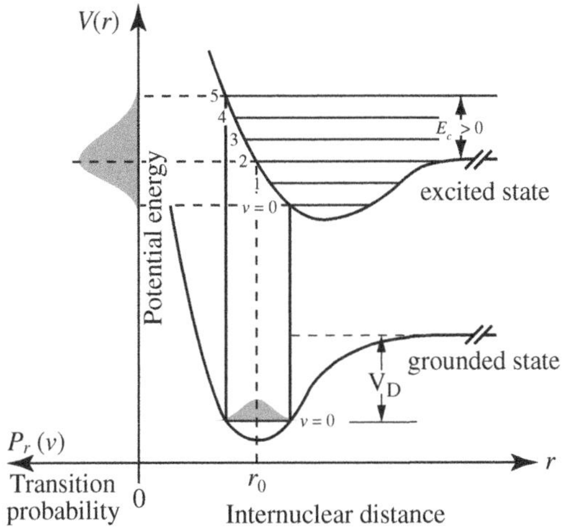

FIGURE 4.24 Dissociative electronic excitation and Franck–Condon principle of a molecule.

iii. $\Delta\Omega = 0, \pm 1$, where $\Omega = |\Lambda + \Sigma|$, and for $0 \to 0$ it follows that $\Delta J \neq 0$; and

iv. When the vibrational quantum number v changes, the $\Delta v = \pm 1$ transitions are the strongest but other transitions are not forbidden.

Problem 4.7.4
Interpret the molecular terms $X^1\Sigma_g^+$ and $A^3\Sigma_u^+$ of N_2 and discuss whether the upper state is metastable.

A very important rule in electronic excitation of molecules is that transitions take place over very short periods of time (on the order of 10^{-16} s), and therefore the intermolecular distance cannot change during these transitions. Thus, the so-called Franck–Condon principle states that the region of the upper potential curve that is directly below the lower vibrational state will be excited. In Figure 4.24 we show one such example. The ground state of the molecule at the $v = 0$ level may be excited to the $v' = 0 - 5$ levels of the excited state. The most probable transition (around r_0) will be to the $v' = 2$ level. One may also observe that the levels $v' = 0 - 2$ will be bound, although there is a possibility of dissociation with the levels $v' = 3 - 5$. In the case of dissociation, the potential curve may be reached above the dissociation limit and the additional energy converted to kinetic energy of the molecular fragments. The maximum kinetic energy will be E_c, as shown in the figure. If the upper state is repulsive, then all excited molecules will be dissociated with considerable kinetic energies of the fragments; these energies can be as

TABLE 4.6 Dissociation Energy (V_D) of Ground-State Molecules and Their Ions

Molecule	V_D (eV)
H_2	4.48
H_2^+	2.63
B_2	2.93
C_2	6.24
N_2	9.76
O_2	5.11
F_2	1.59
Cl_2	2.48
Br_2	1.97
I_2	1.54

large as 10 eV to 20 eV. Dissociative excitation is an efficient way of producing heavy particles of high kinetic energy.

Exercise 4.7.5

Describe a typical example of dissociation of molecules in thermal and nonequilibrium plasmas.

Dissociation will occur if energy sufficient to cross the dissociation barrier V_D in Figure 4.24 is available. Data for dissociation energies for several molecules are given in Table 4.6. In thermal plasmas, molecules will be dissociated through a number of rotational and vibrational transitions by electron–molecule, and even more so by molecule–molecule or ion–molecule collisions. However, in nonequilibrium low-temperature plasmas, only electron–molecule collisions to upper electronic excitation states will result in the dissociation due to the low kinetic energy of heavy particles.

4.7.4.2 Cross Sections for Electronic Excitation of Molecules

In addition to direct excitation of bound molecular states, it is also possible to have dissociative excitation of molecules. For example,

$$\begin{aligned} CH_4 + e &\rightarrow CH_2 + H(n = 3, 4, \ldots) + H \\ &\rightarrow CH(A^2\Delta) + H^* + H_2, \cdots, \end{aligned}$$

and the cross sections for excitation of Balmer α and β emission from a CH_4 molecule are shown in Figure 4.25.

4.7.5 Electron Collisions with Excited Atoms and Molecules

When an electron collides with an excited molecule or an excited atom, then there is a chance that the molecule will be de-excited with the electron gaining

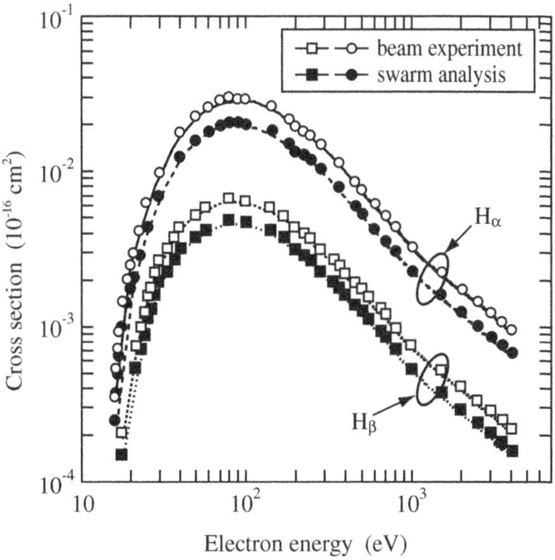

FIGURE 4.25 Cross section for electron-induced dissociative excitation of CH_4 into $H(n = 3, 4)$ levels.

energy equal to the excitation energy of the molecule. These are "superelastic collisions" in which the electron may gain a lot of energy and jump from small thermal energies to the high-energy tail of the energy distribution. The cross section for a superelastic collision Q_s may be calculated from the cross section Q_j for the excitation of state j (with the threshold energy of excitation ε_j and statistical weight g_j) by using the principle of detailed balance in a thermal equilibrium situation,

$$Q_s(\varepsilon) = \frac{g_0}{g_j}\frac{\varepsilon + \varepsilon_j}{\varepsilon} Q_j(\varepsilon + \varepsilon_j), \tag{4.98}$$

where g_0 is the statistical weight of the ground state. As an example we show data for the vibrational excitation of methane and the corresponding superelastic cross section (see Figure 4.26).

Exercise 4.7.6

Prove Equation 4.98 for a cross section of the superelastic collision based on the principle of detailed balance.

The principle of detailed balance states that, in thermal equilibrium (at a temperature T_g), the number of transitions of electrons from higher energies to lower energies due to inelastic collisions is equal to the opposite process of superelastic transition from low to high energies. In other words, for a combination of populations for ground state N_0 and excited state N_j separated

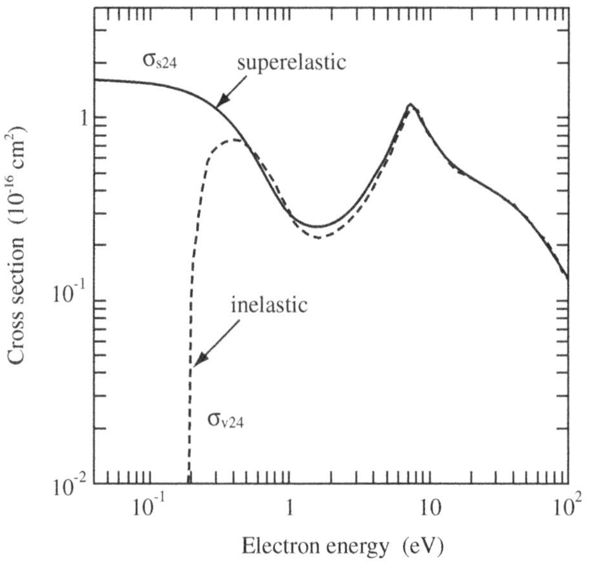

FIGURE 4.26 Cross section for electron-induced excitation and superelastic de-excitation of 2–4 mode of vibration of CH_4.

by excitation energy ε_j, we have a balance

$$N_0\sqrt{\varepsilon+\varepsilon_j}Q_{0j}^{exc}(\varepsilon+\varepsilon_j)f(\varepsilon+\varepsilon_j) = N_j\sqrt{\varepsilon}Q_{j0}^{sup}(\varepsilon)f(\varepsilon),$$

where the electron energy distribution is normalized to unity, $\int f(\varepsilon)d\varepsilon = 1$. Bearing in mind that for thermal equilibrium $f(\varepsilon) = \sqrt{\varepsilon}\exp(-\varepsilon/kT_g)$ and $N_j/N_0 = g_j/g_0\exp(-\varepsilon_j/kT_g)$, where g_j is the statistical weight of the excited state j, we obtain

$$N_0\sqrt{\varepsilon+\varepsilon_j}Q_{0j}^{exc}(\varepsilon+\varepsilon_j)\sqrt{\varepsilon+\varepsilon_j}\exp\left(-\frac{\varepsilon+\varepsilon_j}{kT_g}\right) = N_0\frac{g_j}{g_0}\exp\left(-\frac{\varepsilon_j}{kT_g}\right)$$
$$\times\sqrt{\varepsilon}Q_{j0}^{sup}(\varepsilon)\sqrt{\varepsilon}\exp\left(-\frac{\varepsilon}{kT_g}\right),$$

which eventually yields Equation 4.98.

In Figure 4.27 we show data for a process of excitation from vibrationally excited nitrogen ($N_2(X^1\Sigma_g; v = 1)$ to even higher levels. It is especially important to provide data for metastable levels because these are highly populated in nonequilibrium plasmas (depending on the electron density and the presence of quenching impurities and other losses). In Figure 4.28 we show the cross sections for collisional coupling between the metastable $Ar(^3P_2)$ and the resonant $Ar(^3P_1)$. This coupling results in a significant loss of metastables, that is, electron-induced quenching. Excitation to a higher state $Ar(2p)$ is

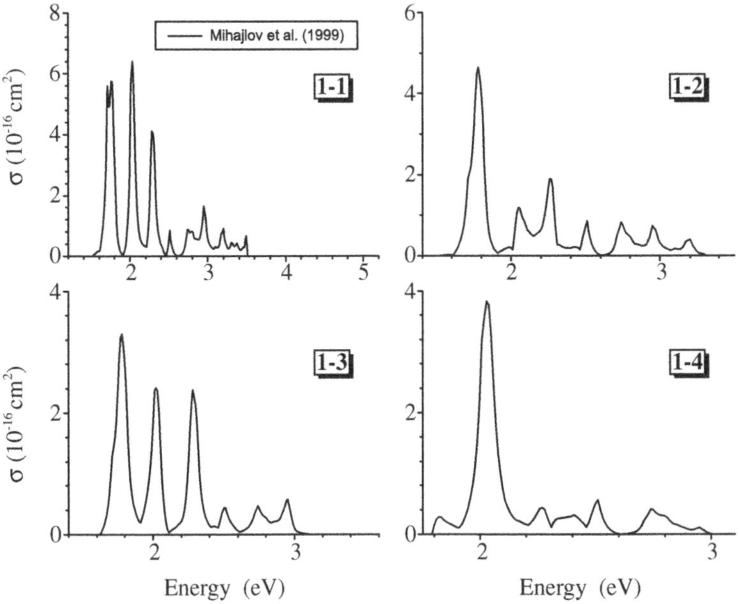

FIGURE 4.27 Cross section for the electron-induced vibrational excitation of a $N_2(X^1\Sigma_g; v = 1)$. 1–1 denotes elastic scattering, and the excitations to higher vibrational state are indicated by 1–v.

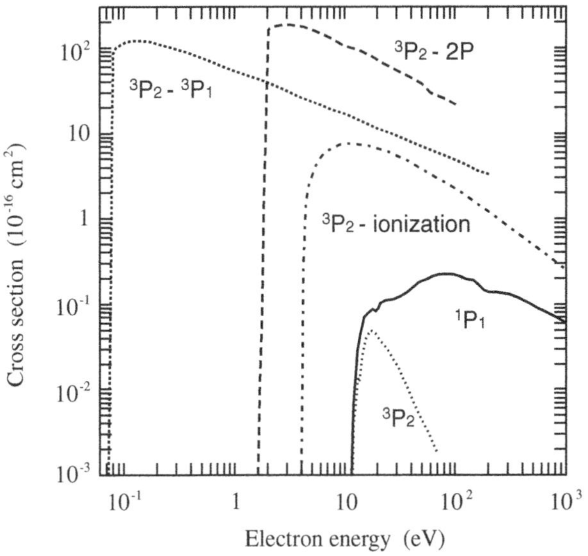

FIGURE 4.28 Cross section for the electron excitation of the metastable $Ar(^3P_2)$.

also shown together with ionization. The thresholds for these processes are 1.55 eV and 4.16 eV, respectively. Hence these processes may become very important once the metastable density is sufficiently high. For comparison we have also shown the cross sections to the lowest metastable state $Ar(^3P_2)$ and the resonant state $Ar(^1P_1)$ from the ground state $Ar(^1S_0)$.

4.8 NONCONSERVATIVE COLLISIONS OF ELECTRONS WITH ATOMS AND MOLECULES

Collisions in which the number of electrons changes either being produced or removed are regarded as nonconservative. Ionization, attachment, electron-induced detachment from negative ions, and electron–ion recombination all fall into this category. Although these processes may not differ significantly from the viewpoint of atomic and molecular physics, they may have a marked influence on the electron transport properties of any plasma (see Chapter 5).

4.8.1 Electron-Induced Ionization

If we excite a molecule (or atom) to the ground state of a singly charged ion, then ionization of the molecule (or atom) is said to have occurred and an additional electron is free to leave the target particle. One typical example of this process is written as:

$$e + \mathrm{A} \rightarrow e + e + \mathrm{A}^+. \tag{4.99}$$

The energy loss of the incoming (primary) electron with ε is ε_i (i.e., the ionization energy). The remaining energy $(\varepsilon - \varepsilon_i)$ is split between the two electrons, and usually the slower one is regarded as secondary. The double differential cross section, as a function of energy ε of the primary electron and ε_s of the secondary [7] is approximated by an analytic formula provided from beam measurements:

$$\sigma(\varepsilon, \varepsilon_s) = \frac{Q_i(\varepsilon)}{\bar{\varepsilon} \tan^{-1}\left(\frac{\varepsilon - \varepsilon_i}{2\bar{\varepsilon}}\right)} \frac{1}{1 + \left(\frac{\varepsilon_s}{\bar{\varepsilon}}\right)^{2.1}}.$$

In Table 4.7 we show the data required to determine the double differential cross section $\sigma(\varepsilon, \varepsilon_s)$ for atomic ionization.

The ionization cross section is usually determined very accurately in electron beam experiments and it is often used directly in cross section sets for plasma modeling. In Figure 4.29 we show data for ionization of an argon atom to several different charge states, Ar^+, Ar^{2+}, and so on. We should also note that there is a possibility that excited states of an ion or even multiply charged ions may be excited/ionized directly from the ground state of the neutral. For most nonequilibrium plasmas for which applications currently exist, only the ionization leading to a singly charged ion needs to be considered. The collisions of ions and electrons leading to further ionization may be neglected. The ionization cross section peaks at a few hundreds of eV, and at these energies it becomes the dominant collision process, even exceeding the elastic

TABLE 4.7 Double Differential Cross Section Parameters for Ionization of Several Atoms and Molecules

Molecule	ε_i (eV)	$\bar{\varepsilon}$ (eV)
He	24.6	15.8
Ne	21.6	24.2
Ar	15.8	10.0
Kr	14.0	9.6
Xe	12.1	8.7
H_2	15.4	8.3
O_2	12.2	17.4
N_2	15.6	13.0
CH_4	13.0	7.3

scattering. For molecules, ionization is often accompanied by dissociation. Simultaneous ionization and excitation often occur from the ground state of the neutral molecule, and emission from the excited ions may be used as a probe for high-energy electrons. Close to the ionization threshold, the cross section increases as $Q_i \sim (\varepsilon - \varepsilon_i)^{1.127}$, according to Wannier's classical theory. For very high energies the cross section decays according to the Bethe–Born

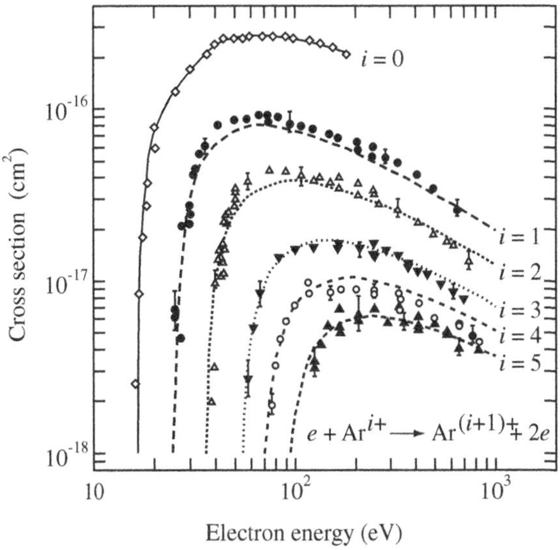

FIGURE 4.29 Ionization cross section for $e + A^{i+} \rightarrow A^{(i+1)+} + 2e$ ($i =$ 0–5). This cross section is different from one for multiple ionizations from the ground state of a neutral atom.

approximation as

$$Q_i \sim \frac{1}{\varepsilon}(A \ln \varepsilon + B),$$

where A and B are the terms for optically allowed and forbidden transitions, respectively.

4.8.2 Electron Attachment

For some molecules it is possible that an incoming electron may be attached and a stable negative ion formed. Electron attachment is also a nonconservative process. In this case, we need to consider a possibility that the electron will be subsequently released in some detachment processes. Cross sections for electron attachment are typically smaller than ionization cross sections. On the other hand, they may occur at lower energies between zero and the ionization energy. There are different processes that may lead to production of negative ions, including dissociative electron attachment (DEA), nondissociative electron attachment (NDEA), and ion pair formation. These processes may be standard in both two-body and three-body processes. They are shown through the following equations:

$$e + \mathrm{AX} - \xrightarrow{Q_0(\varepsilon)} [\mathrm{AX}^-]^* \quad \nearrow \quad \xrightarrow{P_d} A + X^- \qquad (4.100)$$

$$\searrow \quad (+\mathrm{M}) \rightarrow \mathrm{AX}^- + (\mathrm{energy}) \qquad (4.101)$$

and

$$e + \mathrm{AX} \rightarrow (\mathrm{AX}^* + e) \rightarrow \mathrm{A}^+ + \mathrm{B}^- + e. \qquad (4.102)$$

In the first step, the electron is attached to the excited state $[AB]^*$. The state may decay, through dissociation with probability p_d, to form two fragments, one being a negative ion (Equation 4.100). The other possible channel is for the excited state to be stabilized in collisions with the third particle M, preferably a heavy buffer gas molecule. It will then proceed to form a stable ground-state negative ion of the parent molecule (Equation 4.101). Excess energy will be taken away by the third particle. Another way to describe these processes is by following the dynamics on potential energy curves $V(r)$, as shown in Figure 4.30.

i. $(AX^-)^*$ is formed through excitation to the repulsive upper state. As the molecule breaks apart, while $r \leq r_c$ there is a chance for autodetachment to occur and the electron to be released again, leaving a neutral molecule. When $r > r_c$ the molecule will definitely break up into one negative ion and one neutral fragment.

ii. The excited state has a minimum. In this case, a group of molecules, excited to the potential curve at an energy higher than the dissociation limit, will undergo dissociative attachment. Other molecules that do not have the energy to dissociate will oscillate in the potential well until

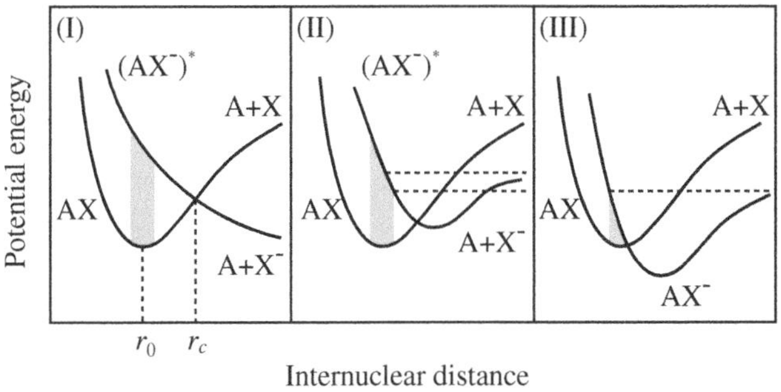

FIGURE 4.30 Potential energy curves for dissociative and nondissociative electron attachment.

autodetachment occurs. Even if the minimum of the excited state is lower than the ground state of the neutral molecule there is a threshold for this process to occur.

iii. If the potential bottom of the excited state $(AX^-)^*$ is below that of the lowest level of the neutral molecule, then there is a chance that the curves will cross close to the minimum and that even electrons with zero energy may be attached. The excited negative ion will spend time oscillating until it is stabilized by collision with the third body or until autodetachment releases the bound electron.

The overall rate of attachment will strongly depend on the ratio of the mean free time between collisions with gas molecules and the lifetime of the excited state. A good example of the potential curves of ground-state neutrals and negative ions is shown in Figure 4.30. An example of the potential curves is given in Figure 4.31 for oxygen. There, one may expect nondissociative attachment to occur due to crossing of the ground state of the molecule by the potential curve for the negative ion O_2^-. The electron affinity E_a is the energy released if a zero-energy electron is detached from a negative ion. The values of affinities for electrons for several atoms and molecules are shown in Table 4.8.

4.8.2.1 Dissociative Electron Attachment

Let us take molecular oxygen as an example of molecules having dissociative electron attachment. The efficiency of attachment or the effective cross section is given as $Q_0(\varepsilon) \times p_d$ (where Q_0 is the cross section to produce an excited unstable negative ion). The set of electron cross sections of O_2 is given as Q_a in Figure 4.32. One may observe resonant vibrational excitation, a number of electronic excitation cross sections, and the ionization cross section, which begins to dominate just below 100 eV. Finally, we can see that dissociative attachment is indicated as a small narrow cross section peaking at around

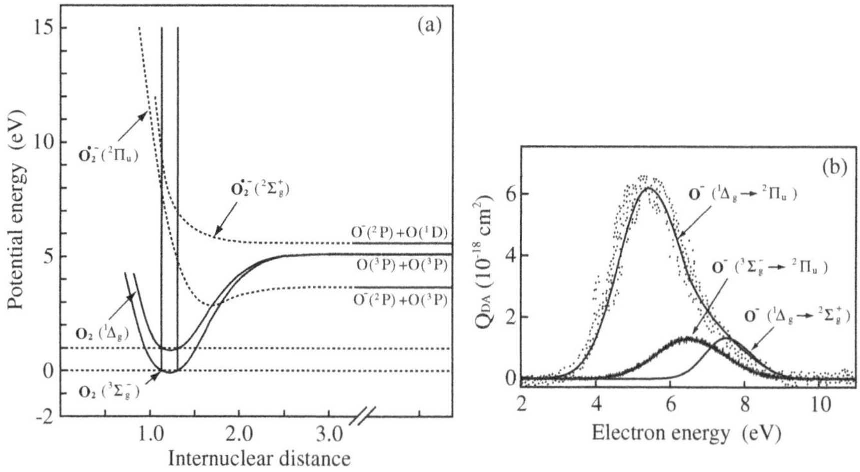

FIGURE 4.31 Potential energy curves for oxygen showing the ground state $O_2(^3\Sigma_g^-)$, metastable $O_2(^1\Delta_g)$, and higher excited states (a), cross sections for dissociative attachment of the $O_2(^3\Sigma_g^-)$ and $O_2(^1\Delta_g)$ states (b).

TABLE 4.8 Electron Affinity

Atom	E_a(eV)	Molecule	E_a (eV)
F	3.40	F_2	3.08
Cl	3.61	Cl_2	2.38
Br	3.36	Br_2	2.51
I	3.07	I_2	2.58
O	1.46	O_2	0.44
S	2.08	SF_6	0.60
C	1.26	CF_2	2.10
		CF_3	1.92

10^{-18}cm^2. The process that leads to production of O^- ions may be described as

$$e + O_2(^3\Sigma_g^-) \xrightarrow{\varepsilon>3.7eV} \left[O_2^-\ (^2\Pi_u)\right]^* \rightarrow O^-(^2P) + O(^3P). \tag{4.103}$$

The O^- ions produced may then undergo any of the following reactions:

$$O^- + O_2 \xrightarrow{\varepsilon>1.02eV} O + O_2^-\ \text{(charge transfer)}, \tag{4.104}$$

$$O^- + O_2 \xrightarrow{\varepsilon>0.39eV} O_3 + e\ \text{(associative detachment)}, \tag{4.105}$$

or

$$O^- + O_2 \xrightarrow{\varepsilon>1.46eV} O_2 + O + e\ \text{(electron detachment)}. \tag{4.106}$$

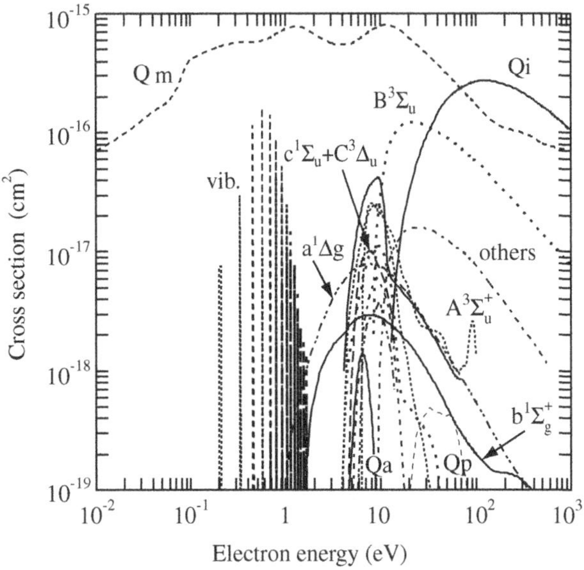

FIGURE 4.32 A set of cross sections for electrons in O_2.

4.8.2.2 Nondissociative Electron Attachment

The first example of nondissociative attachment is the electron-SF_6 attachment at low energies. The process may be described as

$$e + SF_6 \longrightarrow [SF_6^-]^* \xrightarrow{+M} SF_6^-, \quad \tau_{ad} \sim 25\,\mu s \searrow SF_5^- + F. \tag{4.107}$$

Here, the lifetime of the excited state is quite long and it is almost certain that the molecule will be stabilized in collisions with the third particle M. Effectively, this process will behave like a two-body process rather than a three-body process. It is also possible to produce SF_5^- in a dissociative attachment process. Low-energy attachment to SF_6 is close to the upper theoretical limit for attachment for s-wave scattering dominant at low energies. A set of cross sections for electrons in SF_6 is shown in Figure 4.33.

A second example for nondissociative attachment is that of O_2, in which a real three-body process will be observed at normal pressures due to the short lifetime of the excited state. The process may be described as

$$e + O_2 \xrightarrow{Q_0(\varepsilon)} \left[O_2^-\right]^* \begin{cases} \nearrow \xrightarrow{\nu_1} O_2 + e \\ \tau_{ad} \sim 0.1\ ns \searrow (+M) \xrightarrow{\nu} O_2^- + M, \end{cases} \tag{4.108}$$

where ν is the collision rate with a third-body M with pressure dependence, and ν_1 is the frequency to the autodetachment. Thus the effective cross section

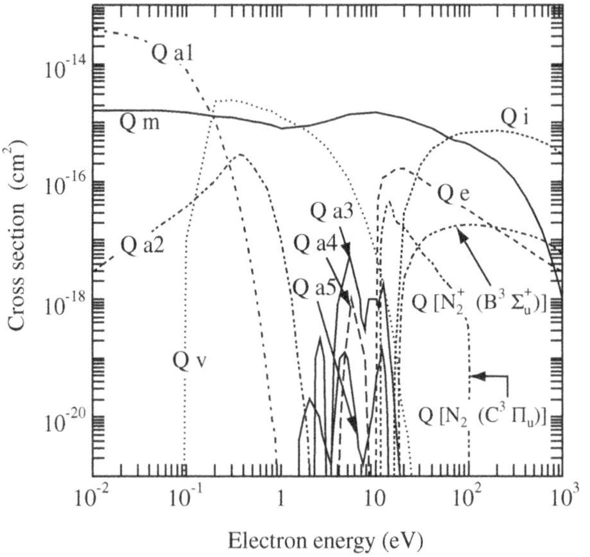

FIGURE 4.33 A set of cross sections for electrons in SF_6. There is one nondissociative (a_1) and several dissociative channels for attachment. Cross sections for excitation of two levels of N_2 are also shown.

of the electron attachment will depend on pressure by way of ν:

$$Q_{a3}(\varepsilon) = \frac{\nu}{\nu_1 + \nu} Q_0(\varepsilon). \tag{4.109}$$

4.8.2.3 Ion Pair Formation

The cross section for ion pair formation in electron–oxygen collisions

$$e + O_2 \xrightarrow{\varepsilon > 17.28\,\text{eV}} O^+ + O^- + e \tag{4.110}$$

is shown as Q_p in Figure 4.32. The threshold is at 17.28 eV and the maximum of 0.48×10^{-18} cm^2 is at 34 eV.

4.8.2.4 Electron Attachment to Excited Molecules

At elevated temperatures, populations of higher rotational and even vibrational levels increase, and one may expect that effective thresholds for inelastic processes and attachment decrease. However, in nonequilibrium plasma, populations of excited states usually greatly exceed the thermal populations for the gas temperature. Thus, effective rotational and vibrational temperatures on the order of 500 K to 3000 K are not uncommon. A classic example of the large enhancement of the cross section for dissociative attachment occurs in

TABLE 4.9 Cross Sections for Dissociative Attachment to $H_2(^1\Sigma_g^+)$

v	r	ε_a (eV)	Q_a (cm^2)
0	0	3.73	$1.6\ 10^{-21}$
0	10	3.13	$2.2\ 10^{-20}$
1	0	3.23	$5.5\ 10^{-20}$
2	0	2.73	$8.0\ 10^{-19}$
3	0	2.28	$6.3\ 10^{-18}$
4	0	1.85	$3.2\ 10^{-17}$
5	0	1.45	$1.1\ 10^{-16}$
9	0	0.13	$4.8\ 10^{-16}$

hydrogen (and even more so in deuterium). In Table 4.9 we show the dissociative attachment cross section to hydrogen as a function of rotational and vibrational excitation. v and r are the vibrational and rotational quantum numbers, respectively.

In nonequilibrium low-temperature plasmas, metastable molecules may also be populated with considerable concentrations and may provide the basis from which attachment and ionization can proceed. Here, we give as one example the metastable state, $O_2(a^1\Delta_g)$, which is 0.98 eV above the ground state (see Figure 4.31) with a lifetime of 2700 s. Electron impact excitation for this state may be described as

$$e + O_2(^3\Sigma_g^-) \xrightarrow{\varepsilon>0.98\,\text{eV}} e + O_2(a^1\Delta_g). \tag{4.111}$$

The excitation of the ground state $O_2(^3\Sigma_g^-)$ to the higher $O_2^{*-}(^2\Pi_u)$ state dissociates to two atoms. On the other hand, when the electron collides with the metastable $O_2(a^1\Delta_g)$, electron attachment proceeds with threshold energy 2.76 eV, which is 0.98 eV less than that of the ground state as

$$e + O_2(a^1\Delta_g) \xrightarrow{\varepsilon>2.76\,\text{eV}} [O_2^-(^2\Pi_u)]^* \rightarrow O^-(^2P) + O(^3P), \tag{4.112}$$

and it will result in dissociative attachment; however, the difference of 0.98 eV in threshold energy will result in a large increase in cross section (by a factor of almost four).

4.8.2.5 Rate Coefficients for Attachment

Many experiments on electron attachment to molecules have been conducted in such a way that the rates for such processes are defined by

$$R_a = \sqrt{\frac{2}{m}} \int_\varepsilon N Q_a(\varepsilon) \sqrt{\varepsilon} f(\varepsilon) d\varepsilon,$$

where the energy distribution of electrons, $f(\varepsilon)$, is normalized to unity. R_a describes directly the loss of electrons and the increase of negative ion density

TABLE 4.10 Thermal Attachment Rate Coefficients (300 K)

Molecule	$k_a = R_a/N$ ($\mathbf{cm^3 s^{-1}}$)
Cl_2	2.8×10^{-10}
F_2	3.1×10^{-9}
SF_6	2.27×10^{-7}
CCl_4	2.8×10^{-7}
CH_3I	7×10^{-8}
CH_3Br	6.7×10^{-12}
CF_2Cl_2	1.2×10^{-9}

in plasma kinetics. The data for nonequilibrium conditions as a function of E/N are provided by swarm experiments with drift tubes. There is a separate group of experiments including the Flowing afterglow, the Cavalleri diffusion experiment, and others that provide attachment rates for thermal equilibrium. In Table 4.10, we show the thermal attachment rate coefficients for several molecules. Measurements were performed for thermal electrons in equilibrium with background gas at room temperature (300 K).

4.8.3 Electron–Ion and Ion–Ion Recombination

Although attachment leads to loss of more mobile electrons, it still maintains the same number of free charges in plasma. On the other hand, the process of recombination leads to decay of the density of charges. For example, if electrons with density n_e and positive ions with density n_p recombine and the rate coefficient of the loss is k_{rm}, then the charged particle kinetics are described by

$$\frac{d}{dt} n_e(t) = -k_{rm} n_e(t) n_p(t), \tag{4.113}$$

where the rate coefficient is associated with the cross section through

$$k_{rm} = \langle Q_{rm}(\varepsilon_r) v_r \rangle.$$

Recombination processes are numerous, and some of the most important ones for nonequilibrium plasmas are listed in Table 4.11. Dissociative recombination is very fast but nondissociative recombination is very weak, and a three-body process is more efficient in high-pressure plasma conditions.

If there is a stable ion AB^+ then its potential curve must have a minimum that is considerably higher than that of the AB molecule. The AB^+ ion will require some finite dissociation energy V_{D+} to dissociate into A^+ and B. However, there will be a chance that this potential curve will be crossed in the region of the minimum by some of the repulsive potential curves of neutral molecule AB, which would allow dissociation when the electron approaches the ion. There will also be a finite probability that autoionization will occur before fragments separate very far. The fragments A' and B will be produced with a large kinetic energy (ε_k). In Figure 4.34 we show a diagram of dissociative

TABLE 4.11 Recombination Processes and Estimated Rate Coefficients (300 K)

Recombination Process	Reaction Scheme	Thermal Rate Coefficient
Radiative	$A^+ + e \to A^j + h\nu$	$\sim 10^{-12}$ cm^3 s^{-1}
Dissociative	$AX^+ + e \to (AX)^{**} \to A^j + X + (\varepsilon_{kin})$	$\sim 10^{-8}$ cm^3 s^{-1}
Dielectronic	$A^+ + e \to (A)^{**} \to A^j + h\nu$	—
Ion pair	$A^+ + X^- \to A^j + B$	$\sim 10^{-7}$ cm^3 s^{-1}
Three-body	$A^+ + e + X \to A + X + h\nu$	$\sim 10^{-26}$ cm^6 s^{-1}
	$A^+ + X^- + M \to A + X + M + (\varepsilon_{kin})$	$\sim 10^{-25}$ cm^6 s^{-1}
	$A^+ + e + e \to A^j + e$	$\sim 10^{-19}$ cm^6 s^{-1}

recombination. One of the most important processes in the ionosphere is the dissociative recombination of the NO^+ ion

$$e + NO^+ \to N + O + \varepsilon_k.$$

Exercise 4.8.1

Explain why radiative attachment and recombination are less effective than other nonradiative processes.

Radiative attachment is described by

$$A + e \to A^- + h\nu.$$

The lifetime of the excited state is typically on the order of 10^{-8} s, and the time that the electron spends in the neighborhood of the atom is on

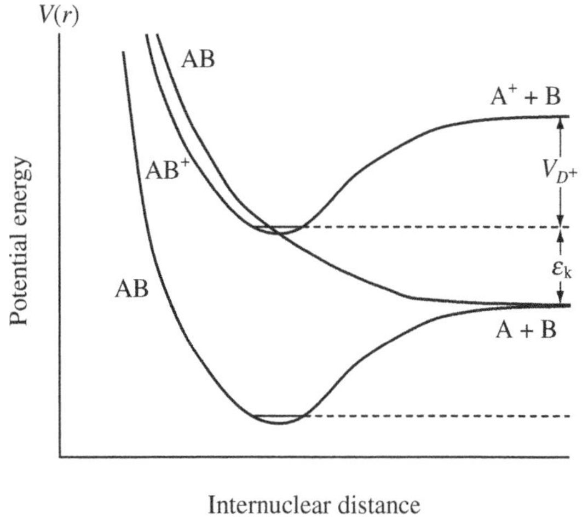

FIGURE 4.34 Potential diagram for dissociative recombination of the AB^+ molecular ion.

the order of 10^{-15} s (dimension 10^{-7} cm and velocity 10^8 cm s^{-1}). Thus, the probability that radiative stabilization of the transient negative ion will occur is 10^{-7}. Therefore, the effective cross section will be on the order of 10^{-21} cm^2. For slower electrons the cross section will be larger. For pairs of electrons and ions, the interaction will also occur at a greater distance, which will lead to a somewhat longer interaction time and a larger cross section. In the case of dissociative processes we do not need photon emission to reduce the energy of the electron from a free to a bound state. Thus, the efficiency will be defined by the collision cross section only. Radiative attachment and recombination will be roughly seven to five orders of magnitude less effective than the corresponding dissociative processes.

Exercise 4.8.2
Explain the inverse processes of radiative recombination and photoionization. Define and explain the dissociative and dielectronic recombination and their inverse processes.
It is obvious that the photo-ionization, that is,

$$h\nu + A \leftrightarrow A^+ + e,$$

is the opposite of radiative recombination. Dielectronic recombination follows a dielectronic capture, where the incoming electron excites the ionic target but, as a result, loses energy and is bound to some high (Rydberg) excited state. Doubly excited targets may relax without losing the incoming electron. If the excited electron returns to the ground state, by emission of a photon, the electron in the Rydberg state is left bound to the target (dielectronic recombination). The opposite process is autoionization, in which one electron is relaxed by photon emission, which is providing energy for the other (Rydberg) electron to leave the target. Dielectronic recombination is of great importance for astrophysical plasmas, but it is not significant in most nonequilibrium plasma applications.

4.8.4 Electron–Ion and Electron–Electron Collisions

In addition to electron–ion recombination, a number of different collisions may occur between two charged particles. Because inelastic electron–ion processes are similar to those for neutrals, we are primarily interested in momentum transfer collisions between two charged particles. We have already discussed the Rutherford scattering of two charged particles. In general, collisions of charged particles thermalize electrons and increase the temperature of ions and neutrals.

Exercise 4.8.3
The transport of electrons in liquid under an external field shows different characteristics from that in a gas. In particular, it is known that electrons injected into liquid Ar, Kr, and Xe exhibit a quasifree behavior in the conduction

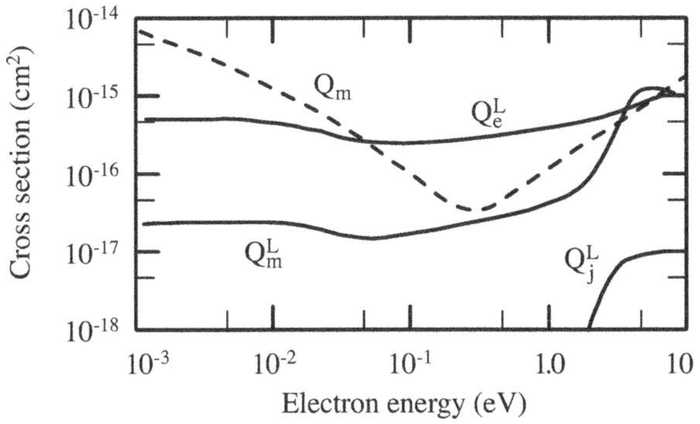

FIGURE 4.35 Collision cross sections, $Q^L(\epsilon)$ of electrons in Liquid-Ar [8]. For comparison, Q_m in gas-phase Ar is also shown. *From Sakai, Y, Nakamura, S. Tagashira, H. Drift Velocity of Hot Electrons in Liquid Ar, Kr, and Xe, Electrical Insulation, Volume: EI-20 Issue: 2. 1985. With permission from IEEE.*

band formed due to the Pauli exclusion principle. Compare the fundamental collision cross section of electrons in between gaseous Ar and liquid-Ar.
$\Rightarrow$ It means that the momentum transfer cross section Q^L_m in liquid-Ar is quite different from Q_m in gaseous Ar as shown in Figure 4.35. As expected, the inelastic cross section, Q^L_j in liquid-Ar has a very low threshold energy as compared with that in gas-phase Ar.

4.9 HEAVY PARTICLE COLLISIONS

The term "heavy particle collisions" covers a wide range of very different processes involving ions, fast neutrals, slow neutrals, excited species, and chemical reactions.

4.9.1 Ion–Molecule Collisions

Some types of ion–molecule collisions are listed in Table 4.12. The list could be extended, especially by a large number of additional possibilities for ion–molecule chemical reactions.

In gas discharges the mean energy of ions is quite low and normally ions cannot achieve the energies required to perform excitation or ionization. However, this is not always true, and recently some low-pressure discharges have been studied with the result that, at very high E/N and low pressures, conditions may be met for significant heavy particle excitation and sometimes even ionization. Such conditions may be found in a low-pressure magnetron sputtering, sheaths of DC and rf discharge plasmas, and plasma thrusters. A

TABLE 4.12 Ion–Molecule Collisions

Process	Reaction Scheme
Elastic Scattering	$\mathbf{X^{\pm}} + AB \rightarrow \mathbf{X^{\pm}} + \mathbf{AB}$
Positive Ions	
Nonresonant charge transfer	$X^{+} + AB \rightarrow AB^{+} + X$
Resonant charge transfer	$AB^{+} + AB \rightarrow AB + AB^{+}$
Dissociative charge transfer	$X^{+} + AB \rightarrow A^{+} + B + X$
Projectile excitation	$X^{+} + AB \rightarrow X^{+*} + AB$
Target excitation	$X^{+} + AB \rightarrow X^{+} + AB^{*}$
Target ionization	$X^{+} + AB \rightarrow X^{+} + AB^{+} + e$
Atom interchange	$X^{+} + AB \rightarrow XA^{+} + B$
Ion interchange	$XB^{+} + A \rightarrow X + AB^{+}$
Addition	$X^{+} + AB \rightarrow XAB^{+}$
Three-body association	$X^{+} + A + B \rightarrow XA^{+} + B$
Negative Ions	
Heavy atom transfer	$X^{-} + AB \rightarrow XA + B^{-}$
Associative detachment	$X^{-} + AB \rightarrow XAB + e$
Detachment	$X^{-} + AB \rightarrow X + AB + e$

set of cross sections for the elastic, charge transfer, and inelastic processes of the ion is shown for two ions in their parent gases. The first is an example of Ar^{+} ions in Ar (see Figure 4.36) and the second an example of N^{+} ions in N_2 (see Figure 4.37).

4.9.1.1 Charge Transfer, Elastic, and Inelastic Scattering of Ions

In principle, in the case of ions and molecules of the same mass, charge exchange (or charge transfer [CT]) could be regarded as elastic scattering in the backward direction; that is, it is highly anisotropic. The process of charge exchange is defined as

$$\underset{(fast)}{A^{+}} \quad + \quad \underset{(slow)}{A} \quad \rightarrow \quad \underset{(slow)}{A^{+}} \quad + \quad \underset{(fast)}{A}$$

whereby the ion that has gained energy from the electric field collides with a slow particle, and in the process it appears that an electron jumps from the neutral to the ion. Therefore, the fast neutral continues with almost the same kinetic energy as the incoming ion. The ion is left with a very small energy, equal to that of the target atom.

If we have ions in their parent gas, the charge transfer is said to be resonant (because of the matching energy levels) or symmetric. This process is extremely efficient and it appears to dominate the transport of ions in their parent gas at all energies, though it has a maximum cross section at zero energy. A good example of this process is the rare gas ions in their parent gas (see Figure 4.36). The symmetric (resonant) CT cross section may be as large

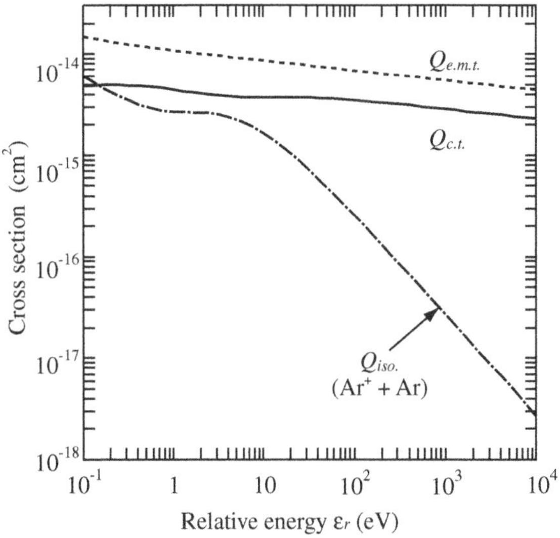

FIGURE 4.36 A set of cross sections for the Ar^+ ions in Ar.

or even larger than the gas kinetic cross section. The cross section decreases slowly with the energy and thus controls the energy of ions at all energies.

It is, however, important to be aware of one possible source of confusion if using data for the cross section of the CT Q_{CT} from the literature. Most

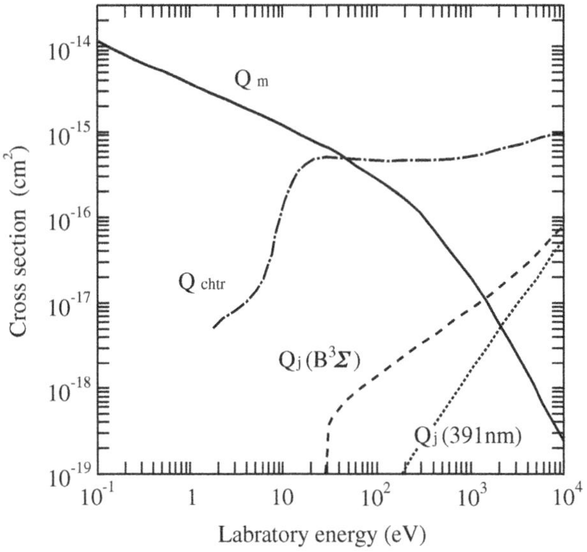

FIGURE 4.37 A set of cross sections for N^+ ions in N_2.

of these data originate from the analysis of ion mobilities. The result of that analysis is Q_m. These data were converted to the CT cross section under the assumption that the momentum transfer cross section corresponds only to the backward scattering, so that $Q_{CT} = Q_0 = Q_m/2$. It was thus implicitly assumed that all processes were negligible except for Q_{CT}. However, these data could also be used assuming isotropic elastic scattering so that $Q_0 = Q_m$. The momentum transfer data were applied in both approaches in the literature. Regardless of the choice, one would correctly predict the mobility of ions, and thus it was not possible to test the data. However, other transport coefficients could be incorrect. Most importantly, the transverse diffusion ND_T is calculated incorrectly. In the first approach $ND_T = 0$, and thus such an application of mobility data would effectively be equivalent to the one-dimensional model. A more complicated and more difficult approach would be to calculate the differential cross sections and use them to analyze the experimental data and for plasma models. However, at low energies we can split the cross section into two parts, isotropic Q_{iso} and backward scattering Q_b, so that $Q_m = Q_{iso} + Q_b$ [9].

Obviously, additional information is required, and the best way is either to use a measured differential cross section and fit it or to use available ND_T data. In the low-energy limit the isotropic part of the total cross section will reach the polarization limit of the cross section. In Figure 4.36 we can see a typical example of the ion–atom cross sections with resonant CT. Elastic isotropic scattering is very high at zero energy, but it diminishes and is negligible as compared with CT scattering beyond 100 eV. However, it is considerable for the largest range of nonequilibrium plasma conditions (i.e., for energies up to 10 eV). The resonant CT cross section is almost constant at all energies and it effectively controls the energy of ions. Still, at very high E/N ion energies may become high enough that inelastic processes may become observable.

When an ion collides with a neutral atom (or molecule) other than those of the parent gas, the CT is not resonant; such a transfer is called an asymmetric CT. This situation is typical for ions such as H^+ and H_3^+ in H_2. Isotropic elastic scattering, however, behaves in the same way. The asymmetric CT will be small, and there is the chance of an ion runaway (creation of a beamlike distribution where ions keep gaining energy with less and less chance to suffer collisions and dissipate the energy). One example of data with an asymmetric CT is given in Figure 4.37. Unfortunately, detailed data that are of interest for plasma processing exist only for a limited number of ion–background gas combinations. Some data exist for rare gases; basic diatomic molecules such as H_2, N_2, and O_2; and alkaline ions in numerous gases. Recently, data have begun to be made available for asymmetric CT of rare gas ions in other rare gases.

4.9.1.2 Ion–Molecule Reactions

Apart from elastic scattering and CT that occur at thermal energies, there are numerous ion–molecule processes that are of possible interest. A brief list is given in Table 4.12. It would require a separate book to give a detailed presentation of all possible low-energy ion molecule reactions. We just give several examples that have been shown to play a role in some discharges. First, we must separate these processes into those involving positive and those involving negative ions.

In addition to the processes given quite generally in Table 4.12, we add some more specific processes. A very important class of chemical processes involves the transfer of a hydrogen atom or its ions as follows:

$$\begin{aligned} \text{proton transfer: } H_2S^+ + H_2O &\rightarrow HS + H_3O^+; \\ \text{(hydrogen) atom transfer: } C_2H_4^+ + C_2H_4 &\rightarrow C_2H_5^+ + C_2H_3; \\ \text{hydride ion transfer: } R_a^+ + R_{bH} &\rightarrow R_{aH} + R_b^+. \end{aligned}$$

A particularly important reaction that determines the dominant ion in hydrogen discharges at low energies is

$$H_2^+ + H_2 \rightarrow H_3^+ + H.$$

Examples involving more general rearrangement processes with positive ions are

$$\begin{aligned} N_2^+ + O &\rightarrow NO^+ + N \quad \text{and} \\ O^+ + N_2 &\rightarrow NO^+ + N, \end{aligned}$$

which play a role in atmospheric chemistry. Cross sections (rates) for ion–molecule reactions are quite high and often are almost as large as the orbiting limit. Orbiting occurs for special combinations of the potential and centrifugal forces, when the radial component of velocity vanishes and the projectile makes many revolutions around the target before being released (see Section 4.5). The extended interaction time makes it possible to complete long-lived transitions.

As an example of reactions with negative ions, typical reactions involving oxygen ions are

$$\begin{aligned} O^- + H_2O &\rightarrow OH^- + OH, \\ O^- + 2O_2 &\rightarrow O_3^- + O_2 \quad \text{and} \\ O^- + O_2 &\rightarrow O_2^- + O, \end{aligned}$$

but there are also reactions involving thermal detachment (including associative detachment):

$$\begin{aligned} O^- + O &\rightarrow O + O + e, \\ O^- + O_2 &\rightarrow O + O_2 + e, \\ O_2^- + O_2 &\rightarrow O_2 + O_2 + e, \\ O_2^- + O &\rightarrow O_3 + e. \end{aligned}$$

Detachment is an extremely important process, and in addition to heavy particles it may be induced by electrons or photons. Finally, we should include processes forming clusters.

Exercise 4.9.1

Determine the rate for orbiting collisions between a slow ion and neutral with an attractive interaction potential given by $V(r) = -\alpha e^2/8\pi\varepsilon_0 r^4$ for $r \geq a$ and $V(r) = \infty$ for $r < a$, where the interaction is due to the polarization determined by polarizability α.

The effective radial interaction energy is equal to

$$V_{eff}(r) = V(r) + \frac{b^2 E}{r^2},$$

where the second term is due to the centrifugal force, E is the energy of the incoming particle at $r \to \infty$, and b is the impact parameter (see Equation 4.43). The radial velocity is equal to

$$v_{radial} = \sqrt{2(E - V_{eff}(r))/M},$$

and for orbiting motion, the radial velocity must be equal to zero. Combining these two conditions one can derive the value of the critical impact parameter b_c, which corresponds to the orbiting motion as

$$b_c = \left(\frac{\alpha e^2}{2\pi\varepsilon_0 E}\right)^{1/4}.$$

The orbiting cross section is thus equal to

$$Q_{or} = \pi b_c^2 = \pi \left(\frac{\alpha e^2}{2\pi\varepsilon_0 E}\right)^{1/2}.$$

For a finite probability of transition, the cross section should be multiplied by probability ζ for the transition during orbiting.

4.9.2 Collisions of Fast Neutrals

Fast neutrals produced in a CT collision continue with very high energy. These fast neutrals will have quite different cross sections from those for ions. Studies of Townsend discharges at very high E/N revealed that fast neutrals are actually more effective in producing excitation (and possibly ionization) than are ions. The energy distribution of fast neutrals upon formation is the same as that of ions, but they do not gain energy. In Figure 4.38 we show a set of data for cross sections for fast neutral Ar(*fast*)-Ar collisions [10].

It was discovered that the production of fast neutrals may also take place on a surface. Collisions of fast ions with a surface often result in reflection and neutralization, sometimes with small or negligible energy loss. Furthermore, for molecular ions, the breakup of molecules occurs and molecular neutral fragments are reflected.

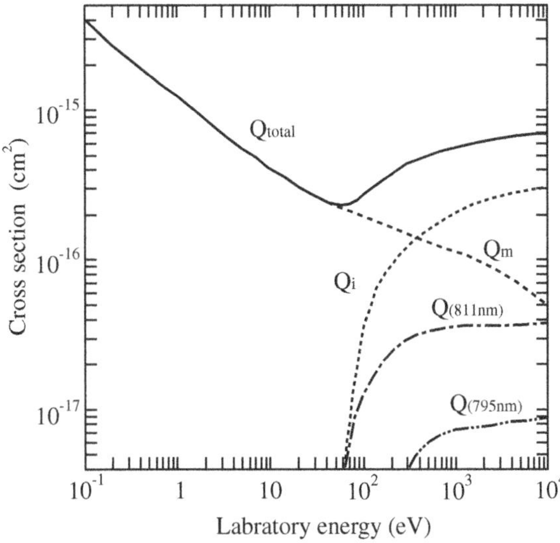

FIGURE 4.38 A set of cross sections for fast Ar atoms in argon. Ionization and ultraviolet excitation cross sections are identical.

4.9.3 Collisions of Excited Particles

Excited atoms or molecules play a considerable role in the kinetics of plasmas. Metastables constitute a large store of energy that may critically affect processes that sustain plasmas. On the other hand, they may induce chemical processes and participate in processes at surfaces. Thus, we are particularly interested in ionizing collisions in the gas phase and at surfaces. In this section, we concentrate on electronically excited atoms (and molecules). Some of the interesting processes are listed in Table 4.13. A very special process is the so-called energy-pooling process: $A^* + B^* = A^+ + B +$ e. For argon metastable–argon metastable collisions, the rate coefficient is 6×10^{-10} cm^3 s^{-1}. It may be one of the dominant sources of ionization in high-density plasmas containing high metastable densities.

4.9.3.1 Chemi-Ionization and Penning Ionization

The processes shown in the first half of Table 4.13 lead to some form of ionization. The necessary condition for these processes to occur is that the energy of the excited state is greater than the ionization potential. Metastables (and all higher excited states) of Ne and He have enough energy to ionize most molecules, so we use Ne as an example ($\varepsilon_m(\text{Ne}) > \varepsilon_i(\text{AB})$). If excited neon atoms collide with molecule AB, there are several different processes that may

TABLE 4.13 Collisions of Excited Species

Process	Reaction Scheme	Energy Condition
Leading to Ionization		
Penning ionization	$X^* + AB \rightarrow X + AB^+ + e$	$\varepsilon_j(X) \geq \varepsilon_i(AB)$
Dissociative ionization	$X^* + AB \rightarrow X + A + B^+ + e$	$\varepsilon_j(X) \geq$ $\varepsilon_d(AB)+\varepsilon_i(B)$
Associative ionization	$X^* + AB \rightarrow (XAB)^+ + e$	
Heavy atom ionization transfer	$X^* + AB \rightarrow XA^+ + B + e$	$\varepsilon_j(X) \geq \varepsilon_i(XA)$ $+\varepsilon_d(AB) - \varepsilon_d(XA)$
Energy pooling	$X^* + X^* \rightarrow X_2^+ + e$	
Ion pair formation	$X^* + AB \rightarrow X + A^- + B^+$	
Leading to De-excitation or Excitation Transfer		
Excimer formation	$X^* + X + M \rightarrow X_2^* + M$	
Transition to resonance levels	$X^M + X \rightarrow X^{res} + X \rightarrow 2\,X + h\nu$	
Collisional-radiative decay	$X^j + X \rightarrow X^k + X \rightarrow 2\,X + h\nu$	
Collision-induced emission	$X^* + X \rightarrow 2\,X + h\nu$	
Transfer of electronic excitation	$X^* + AB \rightarrow X + AB^*$	
Dissociative excitation transfer	$X^* + AB \rightarrow X + A^* + B + (\varepsilon_{kin})$	
Exciplex formation	$X^* + AB \rightarrow XA^* + B$	

occur:

$$\begin{aligned} Ne^* + AB &\rightarrow Ne + AB^+ + e, \\ &\rightarrow Ne + A^+ + B + e, \\ &\rightarrow NeAB^+ + e, \\ &\rightarrow NeA^+ + B + e. \end{aligned}$$

The first channel is known as Penning ionization, and the other three are examples of the general term "chemi-ionization." The second process is dissociative ionization, the third is associative ionization, and the fourth is one of the possible rearrangement channels. The metastables $Ne(^3P_{2,0})$ have energy of 16.62 eV and 16.72 eV, respectively, which is above the ionization limit of the $Ar(^1S_0)$ ground state, 15.76 eV. Thus the combination of Ne and Ar is an excellent example of a mixture that demonstrates Penning and chemi-ionization, and the data for this combination are shown in Figure 4.39. In addition to Penning and chemi-ionization, we show elastic scattering

$$Ne(^3P_2) + Ne(^1S_0) \rightarrow Ne(^3P_2) + Ne(^1S_0)$$

for comparison. One can see that Penning ionization is 10 times more efficient at relatively low energies (which is relevant for applications, because standard collisions occur at room temperature, which is equivalent to 24 meV). At higher energies the cross section for the associative process becomes equal to and even larger than that for the Penning ionization. The cross section for the elastic process is another 10 times larger. Thus a small mixture of Ne in Ar will have a very efficient ionization, and as such it is used in many applications.

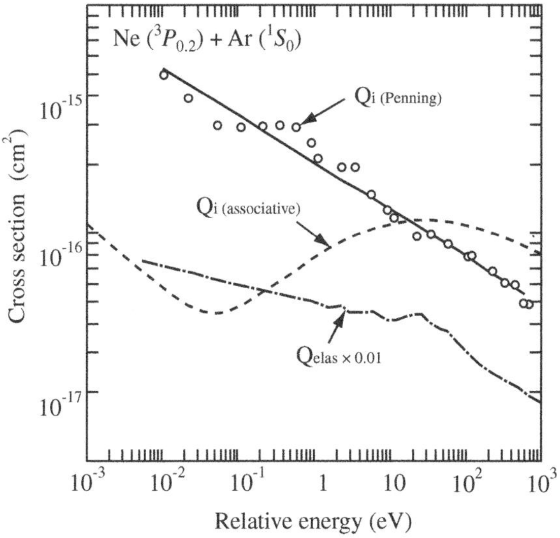

FIGURE 4.39 Cross sections for the Penning ionization, chemi-ionization, and elastic scattering.

Problem 4.9.1

The rate constant for Penning ionization in a mixture of a very small amount of Ar in Ne is expressed as,

$$k_{PI} = \int\int Q_{PI}(\epsilon_r)|\boldsymbol{V}_{Ar} - \boldsymbol{V}_{Ne}|G_{Ar}(\boldsymbol{V}_{Ar})G^*_{Ne}(\boldsymbol{V}_{Ne})d\boldsymbol{V}_{Ar}d\boldsymbol{V}_{Ne}, \quad (4.114)$$

where $Q_{PI}(\epsilon)$ is the cross section of Penning ionization. The velocity distributions $G_{Ar}(\boldsymbol{V}_{Ar})$ and $G^*_{Ne}(\boldsymbol{V}_{Ne})$ are normalized to unity, respectively. When the gas temperature is kept at T_g, both particles have the Maxwellian distribution with temperature T_g. Find the simple form,

$$k_{PI}(T_g) = \int\int Q_{PI}(\epsilon_r)V_r\left(\frac{\mu}{2\pi kT_g}\right)^{3/2} \exp\left(-\frac{\mu V_r^2}{2kT_g}\right) d\boldsymbol{V}_r. \quad (4.115)$$

Here, μ and $V_r = |\boldsymbol{V}_{Ar} - \boldsymbol{V}_{Ne}| = (2\epsilon_r/\mu)^{1/2}$ are the reduced mass and the relative speed between Ar and Ne, respectively. Calculate $k_{PI}(T_g)$ at T_g of 300 K by using the cross section in Figure 4.38.

Exercise 4.9.2

Explain how the ionization rate in the mixture of Ar and Ne exceeds that of either Ar or Ne as shown in Figure 4.40.
For the purpose of this evaluation, it is better to represent ionization through the number of ionizations per unit time, $k_i = v_d(\alpha/E) \cdot (E/N)$, where v_d is

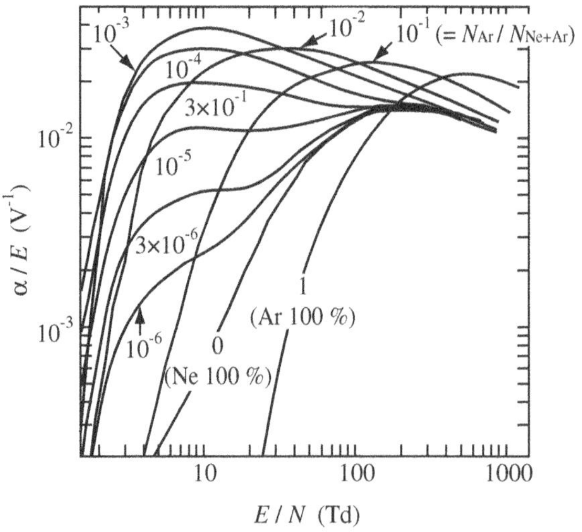

FIGURE 4.40 Ionization coefficients normalized by the field ($\alpha/E[V^{-1}]$ number of ionizations per unit of potential difference) for neon–argon mixtures.

the drift velocity. The production of electrons is given by

$$\frac{dn_e}{dt} = k_{i-Ar}n_eN_{Ar} + k_{i-Ne}n_eN_{Ne} + k_{Penn}N_{Ar}N^*_{Ne}.$$

Here, k_i denotes the ionization rate coefficients for two gases and k_{Penn} the rate coefficient for Penning ionization. At the same time, the population of neon metastables is

$$\frac{dN^*_{Ne}}{dt} = k_{Ne^*}n_eN_{Ne} - k_{Penn}N_{Ar}N^*_{Ne} - k_{Q-Ne}N_{Ne}N^*_{Ne}.$$

Here, k_{Ne^*} is the electron excitation coefficient, k_Q is the two-body quenching coefficient of the neon metastable by the ground-state neon (quenching is discussed in the next section), and we have safely assumed that Penning ionization dominates the quenching of neon metastables by argon. Under stationary conditions this leads to

$$N^*_{Ne} = \frac{k_{Ne^*}n_eN_{Ne}}{k_{Penn}N_{Ar} + k_{Q-Ne}N_{Ne}}.$$

The effective growth of ionization is therefore

$$\begin{aligned} \frac{dn_e}{dt}(= R_in_e) &= k_{i-Ar}n_eN_{Ar} + k_{i-Ne}n_eN_{Ne} \\ &+ k_{Penn}N_{Ar}\frac{k_{Ne^*}n_eN_{Ne}}{k_{Penn}N_{Ar} + k_{Q-Ne}N_{Ne}}. \end{aligned} \quad (4.116)$$

We can see that the effective ionization will be modified by a term that depends strongly on the abundance of the constituents in the mixture. As Ar density increases, the third term will increase substantially. Subsequently, Ar will affect the energy distribution through its inelastic processes with lower thresholds than those for Ne. However, when the argon density becomes large enough that the Penning term exceeds the quenching by Ne, the result will simply be

$$\frac{dn_e}{dt}(= R_in_e) = n_e(k_{i-Ar}N_{Ar} + k_{i-Ne}N_{Ne} + k_{Ne^*}N_{Ne}). \quad (4.117)$$

For small E/N the ionization rate R_i will be equal to the excitation coefficient for metastable Ne. Thus, a way to measure the excitation rate could be to determine the ionization rate for the Penning mixture.

Exercise 4.9.3

Explain the principle of operation of the He–Cd laser (its energy diagram is shown in Figure 4.41).

In collisions of metastable He($^3S_1$,$^1S_0$) with the ground state of Cd($^1S_0$), it is possible to ionize and excite Cd to high-energy levels Cd$^+$($^2D_{3/2,5/2}$), which are then more strongly populated than the lower-energy levels Cd$^+$($^2P_{3/2,1/2}$). Because inverse distribution of the population is achieved, it is possible to achieve lasing on the two lines that originate from the transitions ($\lambda = 325$ nm and 441.6 nm).

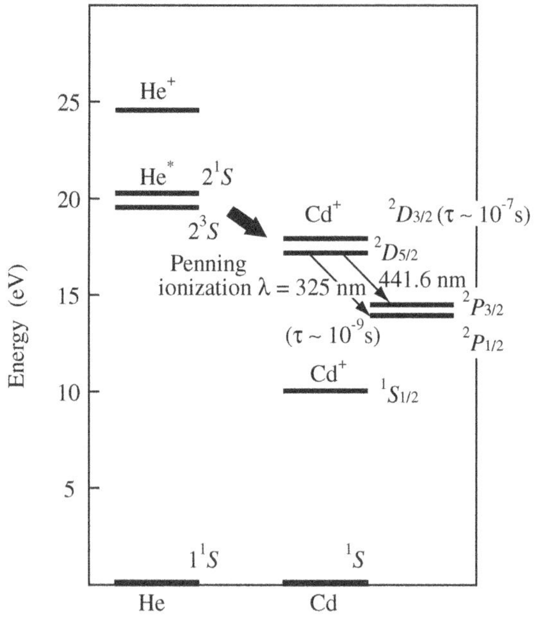

FIGURE 4.41 Energy diagram of He and Cd.

4.9.4 Collisions of Slow Neutrals and Rate Coefficients

4.9.4.1 *Quenching and Transport of Excited States*

There are a large number of processes in addition to radiative transitions that may depopulate excited states. We have considered several processes leading to ionization. However, there are numerous processes that lead to different excited states of the same or other molecules, some of which are listed in Table 4.13. Therefore, if one wishes to study particular channels and their products, data for each channel are required. However, often we just want to follow the kinetics of the excited state and thus need to evaluate total losses of population. In such cases we can define the total quenching coefficient k_q^m (of the excited state m) of N_m with a radiative lifetime $\tau_m = 1/A_m$. The kinetics of one excited state m for steady-state conditions is then described by

$$\frac{dN_m}{dt} = 0 = k_{exc}^m n_e N - A_m N_m - k_q^m N_m N,$$

which can be solved as

$$N_m = \frac{k_{exc}^m n_e N}{A_m + k_q^m N},$$

TABLE 4.14 Quenching of Excited States Often Used in Plasma Diagnostics

Molecule (Excited State)	Feed Gas	Excitation Energy (eV)	Rad-Lifetime τ_{rad} (ns)	Emission $h\nu$ (nm)	Quenching Coefficient k_q (cm^3 s^{-1})
$Ar(2p_1)$	Ar	13.57	21	750.4	0.16 10^{-10}
$Ar(3p_9)$	Ar	14.57	90	419.8	2.7 10^{-10}
$N_2(C^3\Pi_u; v=0)$	N_2	11.03	36	337.1	1.5 10^{-11}
$N_2^+(B^2\Sigma_u; v=0)$	N_2	18.75	63	391.4	4.5 10^{-10}
$H(n=3)$	H_2	12.1	15	656.3	3.8 10^{-9}
$H(n=3)$	Ar	12.1	15	656.3	4.6 10^{-10}
$H(n=3)$	CH_4	12.1	15	656.3	3.5 10^{-9}
$H(n=2)$	H_2	10.2	15	656.3	2.5 10^{-9}
$H_2(d^3\Pi_u; v=0)$	H_2	13.8	39	601.8	3.3 10^{-9}
$CH(A^2\Delta)$	CH_4	12.08	565	431.1	2.0 10^{-11}
$CH(B^2\Sigma)$	CH_4	12.43	357	387.1	6.8 10^{-11}
$O(3p^3P)$	O_2	10.98	34	844.6	6.3 10^{-10}
$O(3p^5P)$	O_2	10.73	34	777.4	10.8 10^{-10}

and therefore

$$N_m A_m = I_m = \frac{k_{exc}^m n_e N}{1 + \frac{k_q^m N}{A_m}}.$$

If one observes the intensity of emission ($I_m = A_m N_m$) for a fixed source of excitation, the intensity will decrease with pressure. The inverse of the intensity (I_m) is plotted as a function of N (the so-called Stern–Volmer diagram) and the intersection of the extrapolated line with zero pressure gives the value of intensity without the effect of quenching (which may be used to determine the excitation rate k_m^{exc} for the state m from the intensity of radiations).

Several quenching processes are listed in Table 4.14. Quenching of excited molecules may be induced by collisions (*collision–radiative decay*) that transfer molecules to a different excited state that subsequently decays radiatively. This can happen both for short- and long-lived states. In the latter case the most important transitions are made to nearby resonant states. Thermal collisions with molecules may provide enough energy to make the transition to nearby excited levels (which are within several times of kT from the original level). This may happen within one multiplet. Emission of a photon, even from a metastable state, may be induced by collisions (*collision-induced emission*). During collisions a temporary dipole moment is induced (or a transient molecule is formed), which provides a mechanism for radiative decay.

Special attention should be paid to the quenching of metastable levels. Two-body quenching consists mainly of collision-radiative and collision-induced emission processes, as outlined above. Three-body quenching mainly

TABLE 4.15 Radiative Lifetimes, Two- and Three-Body Quenching, and Diffusion Coefficients of Rare Gas Metastable States in Their Parent Gases

Metastable State	Natural Lifetime τ_{rad} (s)	Two-Body k_q^{2b} (cm^3 s^{-1})	Three-Body k_q^{3b}(cm^6 s^{-1})	Diffusion ND(10^{18}cm^{-1} s^{-1})
He(2^1S_0)	0.02	6 10^{-15}	—	15
He(2^3S_1)	9000	—	2 10^{-34}	16
Ne(3^3P_2)	24	1.6 10^{-15}	5 10^{-34}	5
Ar(4^3P_2)	56	1.2 10^{-15}	9 10^{-33}	1.8
Kr(5^3P_2)	85	2.3 10^{-15}	4.2 10^{-32}	1.03
Xe(6^3P_2)	56	5.8 10^{-15}	4.9 10^{-32}	0.55

leads to the formation of eximers [A_2^*]:

$$A^* + A + A = A_2^* + A(\rightarrow A + A + h\nu).$$

Metastable atoms and molecules, regardless of whether they have a short effective lifetime due to quenching, may still diffuse to the walls of the discharge vessel, where they will have plenty of opportunity to be de-excited. Thus we can calculate the losses to the walls by using diffusion coefficients. However, at moderate pressures (0.5–5 Torr and higher), gas phase collisions will dominate metastable loss. Typically, diffusion coefficients for rare gas metastables are approximately a factor of two smaller than those of the ground-state atoms. In Table 4.15 we show data for the quenching and diffusion of rare gas metastables within their parent gases.

Problem 4.9.2

Determine the effective lifetimes of the metastable levels of rare gases at 1 Torr and at 10 Torr. Compare them with natural radiative lifetimes.

In addition to parent gas quenching, even the smallest amount of impurities leads to quenching of the metastables. Two types of processes are responsible, depending on the relative positions of the energy levels: Penning ionization and all possible channels of chemi-ionization and excitation transfer (sensitized fluorescence), including chemi-luminescence. These processes are very effective, and even small amounts of impurities may change the effective lifetimes significantly. In Table 4.16 we show a survey of total quenching coefficients for combinations of rare gas metastables with impurities. It should be remembered that excitation transfer from metastables (with or without additional ionization) may be used as a way to detect the rare gas metastables. For example, the $N_2(C^3\Pi_u)$ state almost coincides with the Ar metastable and has been used as a probe of Ar metastable population densities. Similarly, $N_2^+(B^2\Sigma_u)$ may be used to probe the concentration of He metastables [11].

TABLE 4.16 Rate Coefficients for Quenching of Rare Gas Metastables by Molecular Gases (10^{-10} cm^3 s^{-1})

Metastable	N_2	H_2	O_2	NO	CH_4	CF_4	SF_6	Cl_2
He(2^1S_0)	1.7	0.49	5.8	4.3	3.7	—	6.7	—
He(2^3S_1)	0.7	0.32	2.1	2.4	1.4	—	2.6	—
Ne(3^3P_2)	0.84	0.5	—	—	—	—	4.1	—
Ar(4^3P_2)	0.36	0.66	2.1	2.2	3.3	0.4	1.6	7.1
Ar(4^3P_0)	0.16	0.78	2.4	2.5	5.5	0.4	1.7	7.2
Kr(5^3P_2)	0.04	0.3	1.6	1.9	3.7	—	1.8	7.3
Xe(6^3P_2)	0.19	0.16	2.2	—	3.3	—	2.8	6.5

Problem 4.9.3

Calculate the values of the cross sections for quenching of rare gas metastables by their parent molecules from the rate coefficients given in Table 4.15.

4.9.4.2 Kinetics of Rotational and Vibrational Levels

When molecules that are excited to some rotational or vibrational levels collide, several processes may occur. First, it is possible that rotational and vibrational excitation may be relaxed to translational energy (R-T, V-T). It is also possible to have a transfer of rotational or vibrational quantum to the other molecule (R–R, V–V). For example,

$$AB(v_1) + AB(v_2) = AB(v_1 + 1) + AB(v_2 - 1).$$

Thus, if we have a relatively large population of vibrationally excited molecules, then higher-order vibrational levels may be populated step by step to achieve nonthermal populations of very high levels. Due to a harmonicity, transfer of energy will be effective between nearby levels v_1 and v_2. If the V–V processes are effective, the distribution of populations of vibrationally excited levels will be different from the Boltzmann distribution.

If the effective temperature of the first vibrational level is T_v, which may be quite different from the gas temperature T_g, in the absence of V–T processes one obtains Treanor's distribution:

$$\frac{N_v}{N_{v+1}} = \exp\left(\frac{\varepsilon_{01}}{kT_v} - \frac{2\varepsilon_{01}\delta v}{kT_g}\right).$$

Here, δ is the factor of unharmonicity and ε_{01} is the energy of the first vibrational level. This distribution will have a plateau and even a maximum at moderate vibrational quantum numbers. The actual vibrational distributions should be calculated from kinetic equations, and they will be affected by the electron excitation from the ground and from vibrationally excited levels and by the V–T and V–V processes. However, excitation to higher levels will be mainly due to V–V transfers. On the other hand, highly excited molecules may dissociate to neutral fragments or have a very large cross section for dissociative attachment.

4.10 PHOTONS IN IONIZED GASES

Several processes involving photons may be of interest for nonequilibrium plasmas. These include emission, absorption, photo-ionization, and photo-detachment in the gas phase and photo-emission from surfaces. Photo-detachment may be important in the kinetics of negative ions in the majority of low-pressure nonequilibrium plasmas in electronegative gases. It can also be used as a diagnostic technique to establish the identity of negative ions, because the threshold for detachment is a unique characteristic of each ion. In Figure 4.42 we show the cross section for photo-detachment of negative ions in oxygen (O^- and O_2^-). Photo-detachment may lead to nonlinear phenomena in plasmas. Photo-ionization is not important under most conditions; photo-emission from surfaces are discussed in the next section. Absorption, followed by re-emission, are now discussed. Radiation originating from plasmas consists of lines from well-defined transitions between discrete levels or molecular bands, and there are also broad continua that may originate either from transitions to repulsive lower states of molecules or from free–free

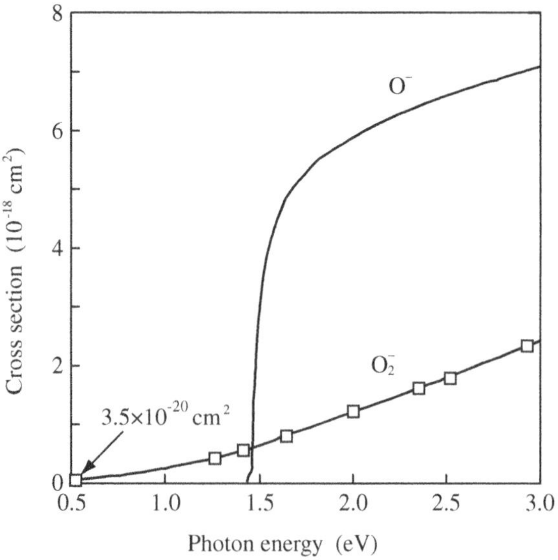

FIGURE 4.42 Cross sections for photon-induced detachment for negative ions in oxygen.

transitions. The relevant spectral range starts in the ultraviolet and extends to the far infrared. In any case, line radiation and molecular bands are the dominant features in low-temperature plasmas.

4.10.1 Emission and Absorption of Line Radiation

The kinetics of emission and absorption of line radiation can be represented by Einstein's probability coefficients. If we take two discrete levels with an energy difference $\varepsilon_{21} = \varepsilon_2 - \varepsilon_1$, the lower with density N_1 and the upper with density N_2, then the probability of spontaneous emission per unit time is defined by coefficient A_{21}, and thus the total number of emitted photons is equal to $A_{21}N_2$. If the transition $2 \to 1$ is the only radiative decay of the level 2, then the lifetime of the excited level is equal to $\tau_2 = A_{21}^{-1}$. In the presence of radiation with a density (per unit frequency) of line radiation of transition $2 \to 1$ equal to $\rho(\nu_{21})$, it is possible that either the radiation will be absorbed by the atoms in the lower level with probability $B_{12}\rho(\nu_{21})$ or a stimulated emission by the photons of exactly the same frequency will occur from the atoms in the upper levels, with probability $B_{21}\rho(\nu_{21})$.

Exercise 4.10.1
Assuming thermal equilibrium for a system with two excited levels 2 and 1, find the relationships between Einstein's coefficients.
In thermal equilibrium, the number of emissions (spontaneous plus stimulated) must equal the number of absorptions, so we have

$$A_{21}N_2 + B_{21}\rho(\nu_{21})N_2 = B_{12}\rho(\nu_{21})N_1.$$

At the same time, the populations of two excited states can be determined from the Boltzmann distribution

$$\frac{N_2}{N_1} = \frac{g_2}{g_1}e^{h\nu_{21}/kT}.$$

Therefore, we can solve the equation for radiation density as

$$\rho(\nu_{21}) = \frac{AN_2}{B_{12}N_1 - B_{21}N_2} = \frac{A_{21}}{B_{12}\frac{N_1}{N_2} - B_{21}} = \frac{A_{21}}{B_{12}\frac{g_2}{g_1}e^{h\nu_{21}/kT} - B_{21}}.$$

Using Planck's law

$$\rho(\nu_{21}) = \frac{8\pi h\nu_{21}^3/c^3}{e^{h\nu_{21}/kT} - 1}, \tag{4.118}$$

we obtain the following relations:

$$g_1B_{12} = g_2B_{21}, \tag{4.119}$$

$$A_{21} = \frac{8\pi h\nu_{21}^3}{c^3}B_{21}. \tag{4.120}$$

Stimulated emission reduces absorption, and if we have population inversion, that is, if the density of the excited state exceeds the density of the lower state, we will have overall amplification of emission. This is a possibility in nonequilibrium discharges and is the fundamental principle behind gas lasers. As discussed above, the kinetics of excited states will be more complicated due to possible nonradiative transitions such as quenching.

The probability of spontaneous emission can be related to the oscillator strength f_{21} through the formula

$$A_{21} = \frac{g_1}{g_2} \frac{2\pi e^2 \nu^2}{\varepsilon_0 mc^3} f_{21}. \tag{4.121}$$

The absorption of radiation can be represented by an absorption coefficient, and it will be associated with the attenuation of a beam of radiation of intensity $I_\nu = c\rho_\nu$ as it passes through a gas of absorbers

$$-\frac{1}{I_\nu}\frac{dI_\nu}{dx} d\nu = \frac{h\nu}{c}\left[B_{=12}dN_1 - B_{21}dN_2\right] k_\nu d\nu, \tag{4.122}$$

where dN refers to a subgroup of atoms that are absorbed within the frequency interval $d\nu$, and k_ν is the absorption coefficient. If we integrate over all frequencies, we have

$$\int k_\nu d\nu = \frac{h\nu}{c}\left[B_{12}N_1 - B_{21}N_2\right] = \frac{h\nu}{c}\frac{c^3 A_{21}}{8\pi h\nu_{21}^3}\frac{g_2}{g_1}N_1\left[1 - \frac{g_1}{g_2}\frac{N_2}{N_1}\right]. \tag{4.123}$$

In addition to the integration over all frequencies covered by the line profile, one can independently perform integration along the length of the absorbing medium (where one can expect variation in the density of absorbers). Thus the solution to Equation 4.123 for a uniform absorbing medium of length l is

$$I(l) = I(0)e^{-k_\nu l}, \tag{4.124}$$

and therefore one can determine the absorption coefficient k_ν from the experimental measurement of absorption profiles. One can then define the optical depth as $\tau_\nu = \int_0^l k_\nu dx = k_\nu l$. The frequency profiles of lines can be determined by Doppler broadening with the profile given by

$$k_\nu = k_0 \exp\left[-\frac{mc^2}{2kT}\left(\frac{\nu - \nu_0}{\nu_0}\right)^2\right], \tag{4.125}$$

with the half width equal to

$$\delta\nu_D = \frac{2\nu_0}{c}\sqrt{\frac{2RT\ln 2}{M}} = 7.18 \times 10^{-7}\sqrt{\frac{T}{M}}\nu_0. \tag{4.126}$$

The line profile may also have a Lorentzian shape

$$k_\nu = \frac{e^2}{(4\pi\varepsilon_0)mc} N_1 f_{12} \frac{\gamma/4\pi}{(\nu_{21} - \nu)^2 + (\gamma/4\pi)^2}, \tag{4.127}$$

where γ is the width of the line and ($\nu_{21} \gg \gamma$). If we integrate the absorption coefficient over all frequencies (i.e., over the line profile) we obtain [12]

$$\int_0^{\infty} k_\nu d\nu = \frac{e^2}{4\varepsilon_0 mc} N_1 f_{21}.$$

Problem 4.10.1
Evaluate the profile due to Doppler broadening, and determine k_0 and the width of the line by assuming a thermal distribution of absorbers at temperature T_g.

Problem 4.10.2
Plot the Lorentzian and Doppler profiles (normalized to the same integral value) with the same half width and compare the center of the line and the width.

4.10.2 Resonant Radiation Trapping

Absorption of photons is most probable between the same levels that participated in the emission (i.e., if there is a resonance). Therefore, if the gas density is sufficient, resonant levels (those with strong transitions to the ground level) will have their emission easily reabsorbed. As a result the resonant radiation may be trapped. In other words, the resonant radiation does not leave the discharge vessel immediately but rather diffuses out through many absorptions and re-emissions, hence the effective lifetime, and consequently the populations of the excited levels, may increase. Typically for few Torr, the populations of rare gas resonant levels are the same order of magnitude as the populations of metastables. The effective transition probability in the case of radiation trapping is denoted gA_{jk}, where g is the trapping factor ($g \leq 1$). There are approximate theories and Monte Carlo simulations that estimate these factors, and these may even be important for pressure ranges of interest in many nonequilibrium plasmas. However, these effects are unimportant for the transitions between two excited states that are frequently used in diagnostics.

Calculations of the resonant radiation trapping are performed, either by Monte Carlo simulations or by using some form of the propagator method where an equation similar to the Boltzmann equation is solved. In these studies a typical approximation is that of the complete redistribution of frequency, whereby the frequencies of the initial and re-emitted photons are not correlated. Approximate analytic solutions were found for g by Holstein. For Doppler broadening we have

- $g = \frac{1.875}{k_0 l \sqrt{\pi ln(k_0 l)}}$, for parallel plane geometry, and
- $g = \frac{1.6}{k_0 R \sqrt{\pi ln(k_0 R)}}$, for cylindrical geometry,

and for collisional broadening we have

- $g = \frac{1.15}{\sqrt{\pi k_\rho l}}$, for parallel plane geometry, and
- $g = \frac{1.115}{\sqrt{\pi k_\rho R}}$, for spherical geometry.

These formulae have been updated in numerous studies. However, it is important to state that g may be quite small even at pressures below 0.1 Torr. This may reduce the transition probability so that the lifetime and, consequently, the densities of the resonant levels are almost equal (less than an order of magnitude smaller) to those of the metastable levels. Under these circumstances the resonant states may behave in a similar fashion as metastables and may even suffer similar quenching and reaction processes.

4.11 ELEMENTARY PROCESSES AT SURFACES

With the exception of most astrophysical and atmospheric conditions, gas discharge plasmas are always maintained and enclosed inside a vessel. There are a large number of processes that may occur between active particles of the plasma and the reactor wall and these may strongly affect the plasma itself [13]. For example, the main sustaining mechanism of low-pressure gas discharges in DC and low-frequency power sources is the surface production of electrons at the cathode by ions and other active particles from the discharge.

In Figure 4.43 we give a schematic representation of a few of the many possible interaction processes between the plasma and the wall of a reactor or other surfaces such as electrodes. As can be seen, it is possible to have photoemission of electrons from the surface; to have a secondary electron production

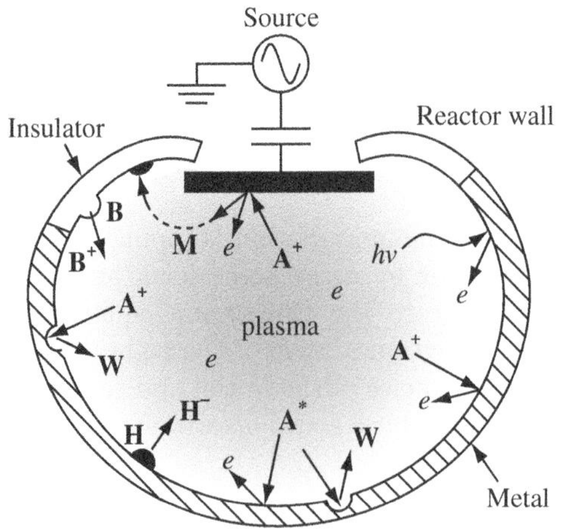

FIGURE 4.43 A schematic representation of plasma surface interactions in a typical plasma chamber.

by an ion or a metastable molecule impact; to have sputtering of surface material by an ion or fast atom bombardment; to have ion neutralization or charging at the surface; to have chemical reactions at the surface; to have adsorption or desorption; and finally, if the surface is heated or subjected to very high fields, to have emission of charged particles from the surface.

Because production and loss of charged particles are of key importance for the maintenance of plasmas, we primarily consider the production and the loss process of charged particles on a surface. On the other hand, for plasma technologies we require additional processes such as chemical reactions at the surface, physical sputtering, and implantation. A special form of interaction between a plasma and a surface is the formation of the sheath that protects plasma from losing the charged particles to the conductive surfaces. Thus, the properties of charged particles arriving at the surface and reaching the plasma from the surface will strongly depend on the sheath potential. We should note that inasmuch as we mainly have surface modification in mind when we discuss plasma processing, the understanding of plasma surface interaction is critical.

4.11.1 Energy Levels of Electrons in Solids

Interaction of electrons that are bound to a nucleus forming an atom is represented by a bound potential with a negative value and discrete levels. Due to the laws of quantum physics, each of the electrons must possess a different state, and each must fill a different energy level. For example, for the sodium atom Na, the electron configuration is

$$\mathrm{Na}: (1s)^2(2s)^2(2p)^6(3s).$$

The potential energy of these bound electrons is negative, as shown in Figure 4.44a and approximated by the Coulomb potential

$$V(r) = -\frac{11e}{4\pi\varepsilon_0 r}. \tag{4.128}$$

Consider a sodium lattice consisting of N atoms in one-dimensional space. The bound potential between two adjacent atoms with a distance a is

$$V(r) = -\frac{11e}{4\pi\varepsilon_0 r} - \frac{11e}{4\pi\varepsilon_0 (a-r)} \tag{4.129}$$

$$= -\frac{11ea}{4\pi\varepsilon_0 r(a-r)}. \tag{4.130}$$

When the electrons of the outermost shell $(3s)$ with energy level ε_k keep the following relation at the maximum value of $V(r)$,

$$|V(a/2)| > (\varepsilon_i - \varepsilon_k), \tag{4.131}$$

then each of the $(3s)$ electrons is not potentially restricted in one atom but free between N atoms (see Figure 4.44b). Here, ε_i is the ionization energy of

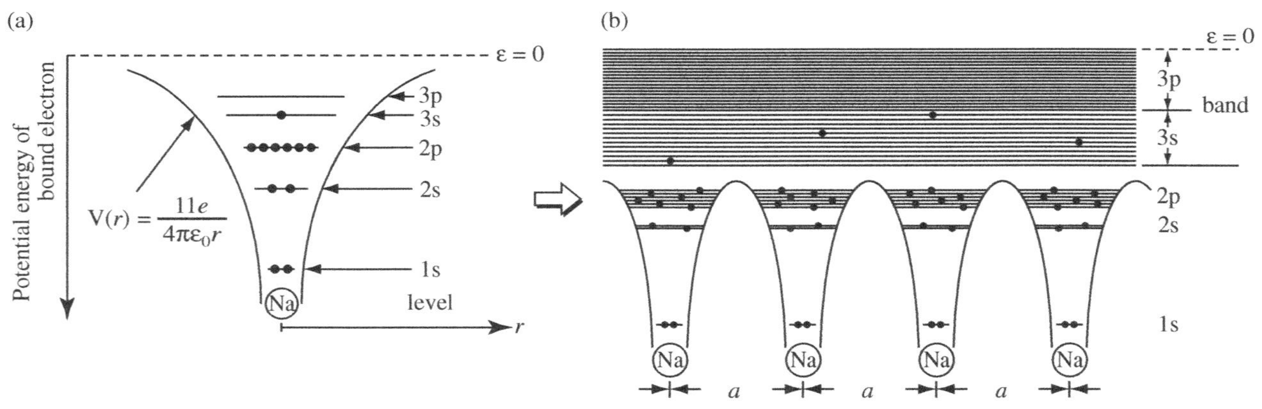

FIGURE 4.44 Energy levels of a single atom of Na (a) and of atoms in a solid (a one-dimensional array of Na atoms) (b).

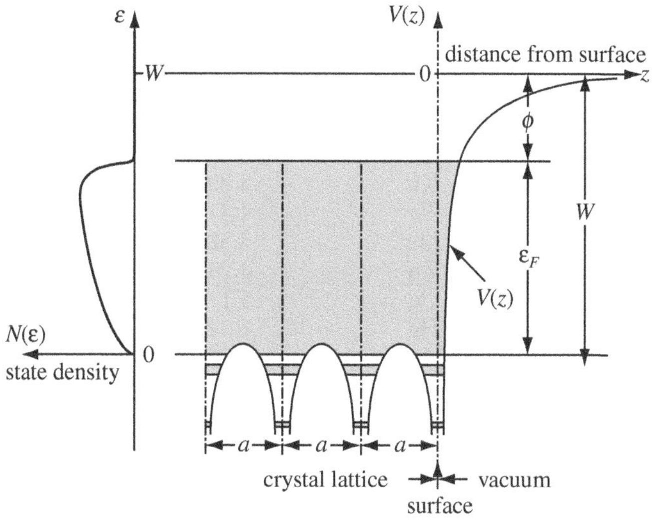

FIGURE 4.45 Potential energy and state density of bound electrons in a metal.

Na. When the external field is applied from both sides of the one-dimensional lattice, (3s) electrons can move to the neighboring atoms as free electrons.

The electrons that may freely leave atoms will combine their levels into a band because electrons are fermions. The band is known as the conduction band, and the electron as the conduction electron (or valence electron). If no levels populated by electrons satisfy the criterion 4.131 then the material is nonconducting.

Now we should consider a lattice of atoms and the potential at the edge of the solid, that is, at the surface (see Figure 4.45). The last atom facing the surface will have a potential, which increases toward the vacuum level unperturbed by the neighbors, and conduction electrons cannot leave the surface potential barrier. To determine this potential more accurately we observe the force acting on one electron that is leaving the surface. The attraction of one charged particle at a distance z from the surface may be described by using the image charge whereby the whole surface is replaced by the opposite charge at a distance $-z$.

Within the conduction band in the metal of temperature T, the electrons will fill the lowest states according to the Fermi–Dirac distribution,

$$f(\varepsilon) = \frac{1}{\exp\left(\frac{\varepsilon-\varepsilon_F}{kT}\right) + 1}, \tag{4.132}$$

where ε is the energy level of the conduction band, and ε_F is the Fermi level. At zero temperature ($T = 0$), the maximum energy level of electrons is equal to the Fermi energy ε_F. Electrons at the top of the conduction band require

TABLE 4.17 Work-Function of Typical Metals

Metal	$\varepsilon_F(eV)$	$\phi(eV)$	Metal	$\varepsilon_F(eV)$	$\phi(eV)$	Metal	$\varepsilon_F(eV)$	$\phi(eV)$
Li	4.7	2.38	Cu	4.3	2.80	In	—	3.8
Na	3.1	2.35	Sr	—	2.35	Ga	—	3.96
K	2.1	2.22	Ba	—	2.49	Tl	—	3.7
Rb	1.8	2.16	Nb	—	3.99	Sn	—	4.38
Cs	1.5	1.81	Fe	—	4.31	Pb	—	4.0
Cu	7.0	4.4	Mn	—	3.83	Bi	—	4.4
Ag	5.5	4.3	Zn	—	4.24	Pt	—	5.3
Au	5.5	4.3	Cd	—	4.1	W	—	4.5
Be	14.3	3.92	Hg	—	4.52	Ta	—	4.2
Mg	—	3.64	Al	5.6	4.25	Ni	—	4.6

energy equal to ϕ to leave the surface, and this energy is known as the work-function. In Table 4.17 we show the work-functions and the Fermi energies of several metals. Alkaline metals have the lowest work-function, typically on the order of 2 eV.

4.11.2 Emission of Electrons from Surfaces

4.11.2.1 Photo-Emission

Photons may provide sufficient energy to bound electrons to allow them to leave the surface, as proposed by Einstein in his explanation of the photo-electric effect 100 years ago (1905). In photo-electric emission, the energy of the photon $h\nu$ is spent on electrons that overcome the potential barrier ϕ to leave the metal, and the remaining energy is converted to the kinetic energy ε_k of the free electron:

$$\varepsilon_{kin_{\max}} = \frac{1}{2}mv_{\max}^2 = h\nu - \phi. \tag{4.133}$$

As a result, there is a cutoff frequency (wavelength) of a photon that can produce an electron regardless of the intensity (i.e., number of photons) of the light ($h\nu - \phi = 0$). That is, the kinetic energy of electrons does not depend on the photon flux but on the frequency. The photo-electron yield is the probability that one photon will induce emission of one electron, and it depends on the material and the surface condition. Data for several metals are shown in Figure 4.46. Photo-emission was recently recognized as one of the major mechanisms of secondary electron production in the maintenance of a high-pressure DC discharge.

Problem 4.11.1

Calculate the cutoff wavelengths for the materials in Table 4.17. Compare them with the well-known wavelengths of light sources, such as a mercury lamp, nitrogen lamp, and rare gas lamps.

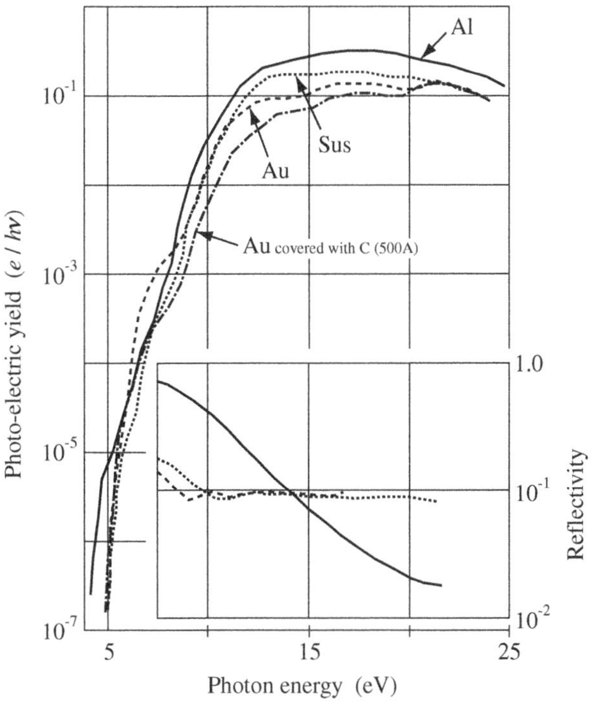

FIGURE 4.46 Photo-emission yield for some metals as a function of photon energy.

4.11.2.2 Thermionic Emission

When a metal is heated, there is a chance that some of the bound electrons will gain sufficient energy to leave the metal surface. Integrating the Fermi–Dirac distribution, one obtains the Richardson–Dushman equation

$$J = AT^2 \exp\left(-\frac{\phi}{kT}\right), \tag{4.134}$$

where J is the current density, $A = 4\pi emk^2/h^3 = 1.2 \times 10^6 Am^{-2}K^{-2}$, and k is the Boltzmann constant. In practice, either tungsten (coated by thorium) or oxide (e.g., BaO) cathodes are used as thermionic sources of electrons. A similar law may be defined for the thermal emission of positive ions that may also be released from the surface. Heating may also lead to the evaporation of neutral atoms.

4.11.2.3 Field-Induced Emission

If we apply an electric field to the surface of the metal, the effective potential of the bound electrons is a combination of the attractive potential due to the

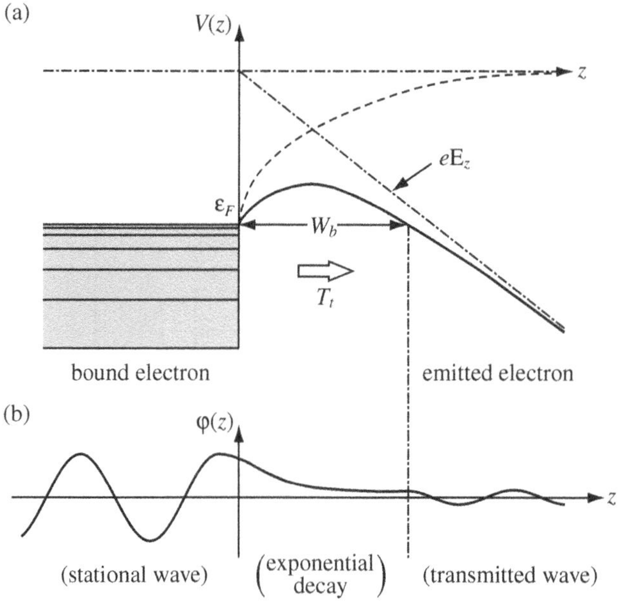

FIGURE 4.47 Field-induced emission is a result of the changes in potential due to external field: (a) formation of the potential barrier, (b) wave function of electrons.

image charge (at distance $2z$) and the potential due to external field E:

$$V(z) = -\frac{e^2}{16\pi\varepsilon_0 z} - eEz. \tag{4.135}$$

The bound potential has a negative minimum at $z_m = \sqrt{e/16\pi\varepsilon_0 E}$ that is equal to

$$V_{\max} = -\frac{e}{2}\sqrt{\frac{eE}{\pi\varepsilon_0}} \tag{4.136}$$

and is thus considerably smaller than the work-function ϕ. This potential is shown in Figure 4.47. Effectively, a barrier with the width w_b is formed. There is therefore a chance that electrons on the left-hand side with energy ε will pass through the barrier by tunneling, even though their energy is smaller than the height of the barrier. The probability of tunneling through a square barrier of the width w_b and height V_0 for an electron of mass m is equal to

$$T \sim 16\frac{\varepsilon}{V_0}\left(1 - \frac{\varepsilon}{V_0}\right)\exp\left[-\frac{2\left\{2m(V_0 - \varepsilon)\right\}^{1/2}}{\hbar}w_b\right]. \tag{4.137}$$

The current density of the field-induced emission is given by the Fowler–Nordheim equation

$$J = \alpha_e E^2 \exp(-\beta_e/E), \tag{4.138}$$

where α_e and β_e are constants that are best determined by fitting the experiments and both depend on the work-function and Fermi energy of the metal. Fields of the order of $E \sim 10^6$ Vcm^{-1} should be required for significant emission. However, even smaller fields are sufficient to produce the emission. For example, the microroughness of the surface produces local fields that are very high. In addition, very thin layers of surface oxide may have deposited charges on them, which produce very high fields and lead to enhanced emission.

Problem 4.11.2

Prove that the effective reduction of the work-function under the influence of the external field E is given by Equation 4.136.

Exercise 4.11.1

Explain the principle of operation of the scanning tunneling microscope (STM) as shown in Figure 4.48.

The STM consists of a very sharp probe made of W or Pt that is brought close to the surface at a distance of 0.1 to 0.5 nm. We label the quantities related to the sample substrate (metal) with subscript s and those related to the probe with the subscript p. A potential is applied between the probe and the sample, and the current is measured; therefore, STM may be applied only to conductive materials. The direction of current depends on the relative work-functions of the probe tip and of the substrate. The principle of the STM is described in Figure 4.48.

There are two possible modes of operation of an STM. It is possible to operate in a constant height mode, where the tip is held at a fixed position while variations in the current are recorded. Consider what happens as the probe tip gradually approaches the sample. At a distance of ~1 nm (1) under the condition, $\phi_p < \phi_s$, the conduction electrons tunnel from the tip to the sample substrate until both Fermi levels become equal, and the electric dipole with potential $V_b = \phi_s - \phi_p$ is formed between the two surfaces. When the external potential V is applied at the probe with respect to the sample, the tunneling current is detected at the probe. By scanning the substrate surface in the constant height mode, the quantum state close to the Fermi level is observed in the two-dimensional image.

The other mode in which STMs operate is the fixed current mode, where the position of the tip is modified to keep the current constant. As current due to tunneling decays exponentially within the barrier, the measured current is extremely sensitive to the distance between the tip and the substrate. That also means that most of the current flows through the atom at the top of the tip, and thus resolution may be very high (subatomic) provided that the movements of the tip are controlled to within one hundredth of an atomic diameter.

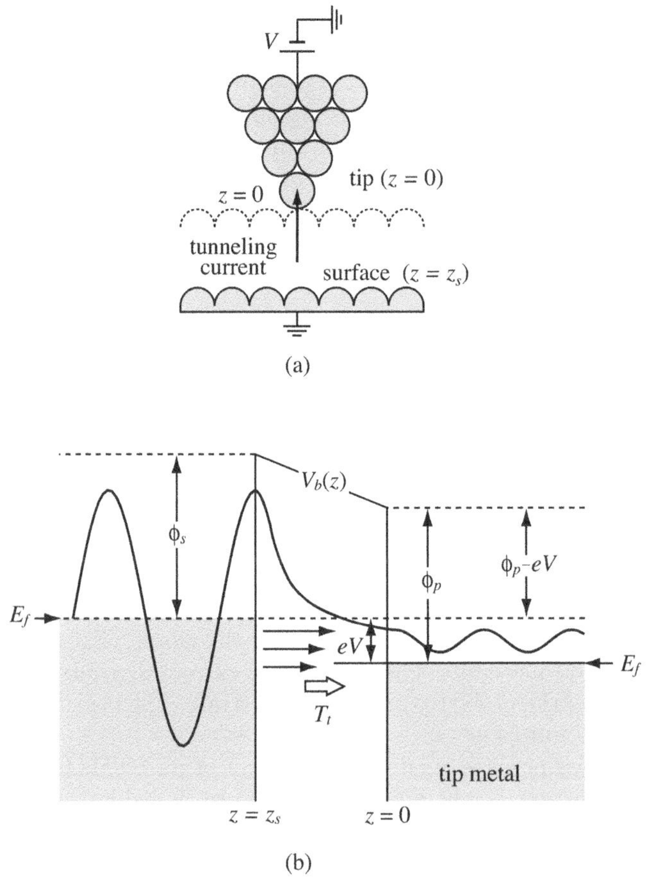

FIGURE 4.48 Schematic diagram of the scanning tunneling microscope (STM) and the relevant potentials.

4.11.2.4 Potential Ejection of Electrons from Surfaces by Ions and Excited Atoms

When an ion or long-lived excited molecule approaches a surface, there is a chance that an electron will be ejected from the surface (potential ejection), depending on the energy levels of the incident particle and of the surface. This phenomenon is different from the emission of electrons by the kinetic energy of an incoming particle that deposits sufficient energy onto the surface. We consider more closely the potential ejection mechanism. The schematic of such a particle–surface encounter (ion arriving at the surface) is given in

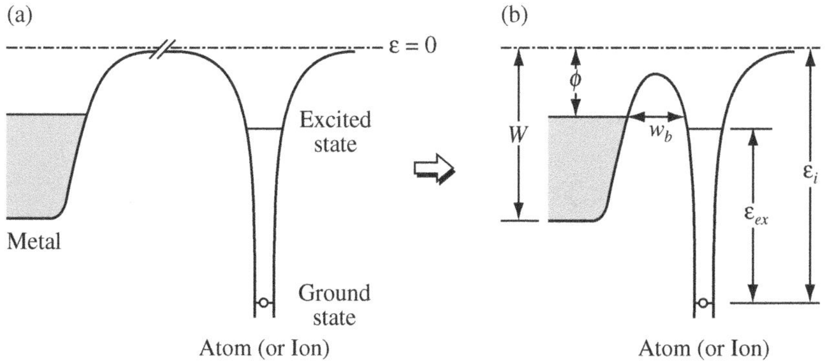

FIGURE 4.49 Schematics of a potential energy diagram for interaction of a particle with the surface (a) at a distance and (b) in closest contact.

Figure 4.49. There are four possible processes:

(1.) Resonance neutralization

(2.) Resonance ionization

(3.) Auger de-excitation

(4.) Auger neutralization

These four processes are discussed in detail in this section. We first explain the Auger processes. As a particle with excitation and ionization energy ε_{ex} and ε_i approaches the surface with the work-function ϕ, the potential barrier between the particle and the surface is reduced in height and especially in the width w_b, becoming so small that it assumes atomic dimensions. There are two interesting situations for the relative magnitude of potentials as follows:

i $(\varepsilon_i - \varepsilon_{ex}) > \phi$. In this case, the interaction during approach is as shown in Figure 4.50. As an ion A^+ approaches the surface, the first step (1) will be a transfer of one electron from the metal to the ion, and a long-lived excited state (metastable) A^* will be formed as the intermediate state:

(1) $$A^+ + Ne_m \to A^* + (N-1)e_m,$$

where N is the number of conduction electrons e_m in the metal. De-excitation of the metastable will lead to the emission of excitation energy ε_{ex}, which may be absorbed by one electron in the conduction band, which will then be free to leave the surface $e(\uparrow)$ as

(3) $$A^* + Ne_m \to A + e(\uparrow) + (N-1)e_m.$$

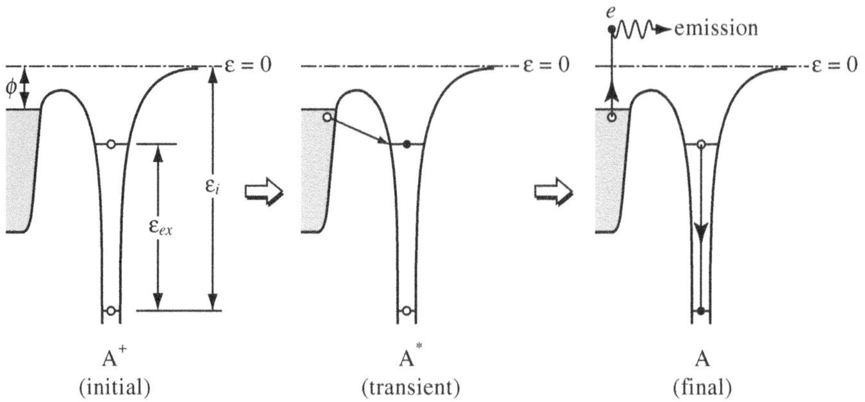

FIGURE 4.50 Electron ejection in the case of $(\varepsilon_i - \varepsilon_{ex}) \geq \phi$.

The maximum kinetic energy of the released electron $e(\uparrow)$ will occur if it originates from the top of the conduction band, that is,

$$\varepsilon_{\max} = \varepsilon_{ex} - \phi,$$

whereas the minimum kinetic energy of the secondary electron occurs when electrons from the bottom of the conduction band are released:

$$\varepsilon_{\min} = \varepsilon_{ex} - W.$$

The production of electrons by incoming ions on the electrodes is one of the major mechanisms to sustain low-temperature plasmas. A similar process of electron ejection (only step (3)) will occur if a metastable atom approaches the surface. Metastable atoms are perturbed close to the surface and may also experience radiative transitions to the ground state.

ii $(\varepsilon_i - \varepsilon_{ex}) < \phi$. In this case, the most energetic electrons in the conduction band cannot tunnel to the excited state of the incoming ion at the surface. However, if an excited atom approaches the surface, then it is possible that the electron will tunnel through the atom to the metal and into an unoccupied state in the conduction band (resonance ionization) as shown in Figure 4.51. The ion is formed as the intermediate state:

(2) $$\mathrm{A}^* + \mathrm{N}e_m \rightarrow \mathrm{A}^+ + (N+1)e_m.$$

This process is followed by a transition of the conduction electron to the ground state of the ion A^+ (Auger neutralization):

(4) $$\mathrm{A}^+ + \mathrm{N}e_m \rightarrow \mathrm{A} + e(\uparrow) + (N-2)e_m,$$

with a release of energy equal to $\varepsilon_i - \varphi$ and the subsequent release of an electron fromthe conduction band through absorption of the

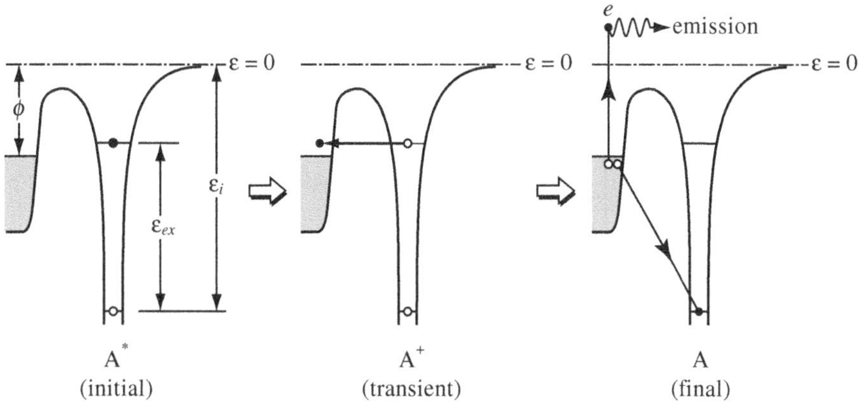

FIGURE 4.51 Electron ejection in the case of $(\varepsilon_i - \varepsilon_{ex}) < \phi$.

irradiated energy. In this case, the maximum and the minimum energies of secondary electrons are:

$$\varepsilon_{\text{max}} = \varepsilon_i - 2\phi \quad \text{and,} \quad \varepsilon_{\text{min}} = \varepsilon_i - 2W.$$

Ejected electrons typically have broad distributions with a mean energy of several eV.

Problem 4.11.3
When He**($^3S_1$) is approaching the W-surface, estimate the maximum kinetic energy of electrons ejected from W by Auger potential ejection.*

The efficiency of absorbing the released energy may be limited and there is always a chance that the electron may dissipate its energy on the way to the surface. Ejected electrons may be trapped at the surface by adsorbed molecules or may be reflected from gas molecules back to the metal surface. Therefore, there is a limited efficiency of secondary electron release per incident ion, which is described by the secondary electron yield. Examples of secondary electron yields for rare gas ions colliding with a tungsten surface are given in Figure 4.52.

4.11.3 Emission of Ions and Neutrals from Surfaces

From the viewpoint of the incident particles, the two processes in Figures 4.50 and 4.51 are important. In the latter case there is a chance that the excited atom will be released on a surface after the transfer of the electron and before the subsequent Auger neutralization. In that case the process would be an ionization process. In addition, there is an efficient resonant neutralization and ionization.

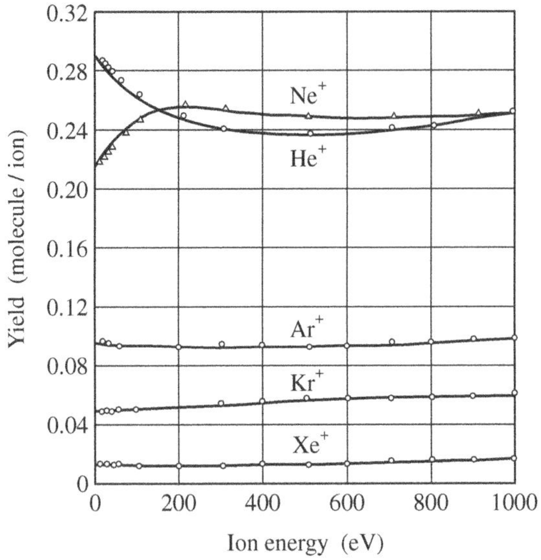

FIGURE 4.52 Secondary electron yields for rare gas ions colliding with a clean tungsten surface.

One should pay attention to the fact that processes shown in Figure 4.50, where the bound electron makes a transition to the excited state of an ion and then de-excites, and the process where the transition is made directly to the ground state of an ion (see Figure 4.51), are not the same and may differ in the energy distribution of ejected electrons. These processes may be adiabatic or nonadiabatic. In the former case, there is no energy exchange between the nuclear motion and the electrons. Adiabatic processes dominate at low energies.

4.11.3.1 Surface Neutralization

For those modeling a plasma–surface interaction, it may be interesting to calculate the probability that a beam of ions will be neutralized upon surface collisions. Neutralization may proceed through three processes: the radiative, the resonant, and the Auger neutralization. The first is highly unlikely as the transitions require 10^{-8} s, which is many orders of magnitude longer than the time of interaction. Both the resonant and Auger processes are probable. Because the transitions of electrons required to neutralize the ion take place through barriers, the tunneling effect will determine the probability, and the basic parameters will be associated with the time spent at the place closest to the surface. This, in turn, will lead to three principal parameters: the perpendicular velocity $V_{\perp}$ and two quantities, $A(\mathrm{s}^{-1})$ and $a(\mathrm{cm}^{-1})$, which are associated with the barrier height and width. Thus the probability of neutralization of a beam is given in a form typical for tunneling processes:

$$P_{neut}(V_\perp) = 1 - \exp\left(-\frac{A}{aV_\perp}\right). \tag{4.139}$$

One should pay attention to the fact that ions may be neutralized while approaching the surface and also while leaving it. Thus the probability will depend on both the initial and final velocities (V_i and V_f), which may be quite different:

$$P_{neut}(V_\perp) = 1 - \exp\left(-\frac{A}{a}\left(\frac{1}{V_{i\perp}} + \frac{1}{V_{f\perp}}\right)\right). \tag{4.140}$$

If there is a penetration of the solid, the formula becomes more complex and will include probabilities for CT in binary encounters inside the solid.

4.11.3.2 Surface Ionization

Neutral atoms on a surface may be ionized depending on their molecular properties and the work-function of the material. Regarding the relative magnitude of the ionization potential ε_i, the electron affinity ε_a of the molecule, and the work-function ϕ, there are three distinct situations:

i. $\varepsilon_i < \phi$: The production of positive ions.

   This is the situation in which ions with small ionization potential, such as alkaline metal atoms (Li, Na, K, Rb, Cs), interact with surfaces with relatively high work-function, and therefore a large percentage of the neutral atom density (n_a) is converted into positive ions (n_p). The degree of ionization at the surface η_i is given by

$$\eta_i = \frac{n_p}{n_p + n_a} = \left\{1 + w_0 \exp\left[\frac{e(\varepsilon_i - \phi)}{kT}\right]\right\}^{-1}, \tag{4.141}$$

   where T is the surface temperature, and w_0 is a parameter depending on the nature of the surface.

ii. $\varepsilon_a > \phi$: The production of negative ions.

   If, however, we combine atoms with very high electron affinity (F_2, Cl_2, Br_2, I_2) with surfaces that have low work-function, negative ions of the molecule will emerge from the surface.

iii. $\varepsilon_i > \phi > \varepsilon_a$.

   If the work-function of the surface is between the electron affinity and ionization energy of the molecule, then the surface will leave the molecule as neutrals without exchange of electrons. For example, hydrogen molecules H_2 ($\varepsilon_i = 13.6$ eV, $\varepsilon_a = 0.7$ eV) will scatter as neutrals off the tungsten surface ($\phi = 4.5$ eV).

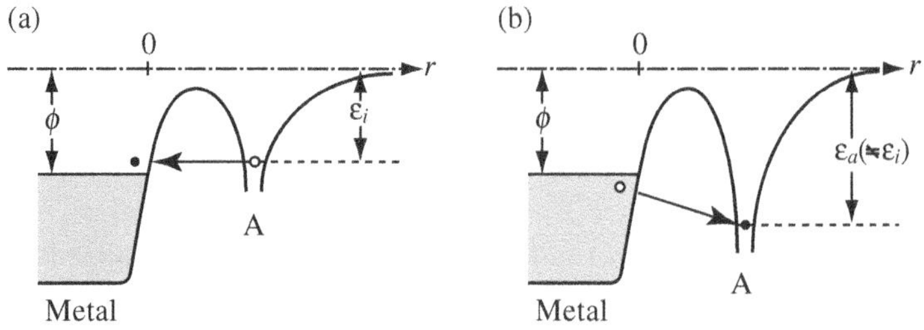

FIGURE 4.53 Potentials for surface ion production: (a) $\varepsilon_i < \phi$: production of positive ions, (b) $\varepsilon_a > \phi$: production of negative ions.

Problem 4.11.4

Explain the processes of (i) surface ionization and (ii) negative ion formation. The potential energy curves and the corresponding transitions are shown in Figure 4.53.

Even in case (iii) it is possible to have surface ionization if a strong electric field is applied externally. Typically, fields on the order of 10^8 V cm^{-1} are required to achieve significant field ionization at the surface. One should bear in mind that this process is different from field emission from the solid and is also different from possible ionization of a free atom or molecule. In this case the field allows the process of surface ionization by affecting the potential of particles close to the surface.

Exercise 4.11.2

A hydrogen atom resides at the surface of tungsten (W; $\phi = 4.5$ eV) in an external field 2×10^8 V cm^{-1}. Determine the critical distance z_c required to achieve field ionization.

Because the work-function of tungsten is $\phi = 4.5$ eV and the ionization energy of the ground-state hydrogen atom H(^{2}S$_{1/2}$;1s) has $\varepsilon_i = 13.6$ eV, we have $\varepsilon_i - \phi > 0$. Hence the surface ionization without an external field is not possible (see the dashed line in Figure 4.54). If we add the electric field, the potential of the bound electrons will be modified and the condition for ionization will be given by

$$(eEz_c + \phi - \varepsilon_i) > 0. \tag{4.142}$$

At that point the electron from the neutral H may tunnel through the barrier and a positive ion H^+ may emerge from the surface. Therefore the condition is achieved for $z_c > (\varepsilon_i - \phi)/eE$. The probability of tunneling will strongly depend on z_c and thus ionization will occur for a narrow range of distances as the atom approaches and leaves the surface.

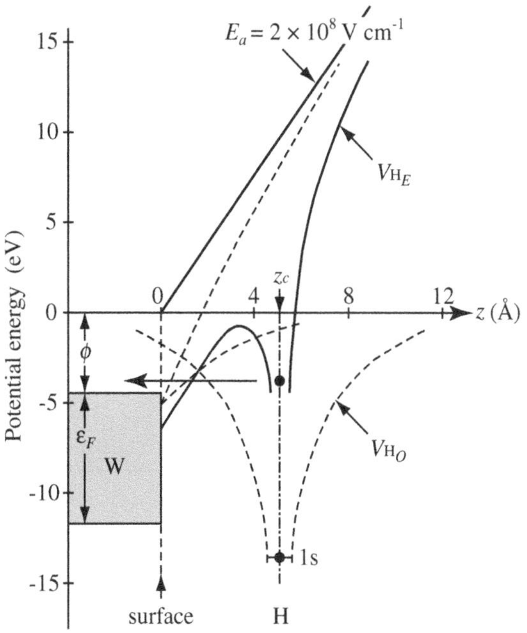

FIGURE 4.54 Potentials for a hydrogen atom on a W surface: (a) without external field (dashed line), (b) with the external field (solid line).

4.11.4 Adsorption

Atoms and molecules may be bound to surfaces in a process known as adsorption. Molecules may be physisorbed (weakly bound) or chemisorbed (strongly bound). In the former case there is a Van der Waals interaction between the molecule and the surface, whereas in the latter case the interaction is chemical, with an electron exchange, and therefore the value of the interaction potential is much larger. The schematic potential curves for both processes are shown in Figure 4.55. Physisorption allows formation of several layers of gas molecules and is nonspecific. In this case, the heat of adsorption is less than $2 \sim 3$ times the heat of evaporation. In the case of chemisorption the reaction is very specific. It may involve dissociation of molecules, and usually only a monolayer is formed. The heat of adsorption is more than $2 \sim 3$ greater than the heat of evaporation.

We describe the processes at the surface by using the surface coverage

$$\theta = \frac{\text{number of adsorbed molecules}}{\text{maximum uptake of molecules}}$$

and the sticking probability

$$s = \frac{\text{rate of adsorption}}{\text{rate of collisions with the surface}}.$$

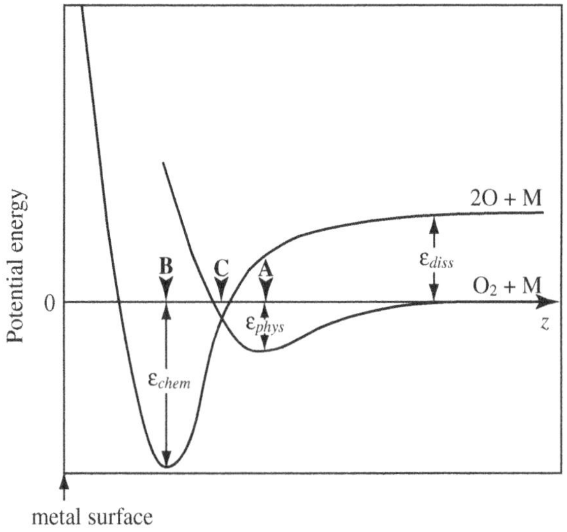

FIGURE 4.55 Potential energy diagrams for adsorption of an oxygen molecule on a metal surface.

The rate of adsorption should be proportional to the rate of collisions with the surface k_a, the pressure, and the fraction of vacant sites, $k_a p(1-\theta)$, and the rate of desorption should be proportional to the surface coverage and the specific rate of desorption k_d:

$$k_d\theta.$$

Under thermal equilibrium these two processes are balanced:

$$\frac{\theta}{1-\theta} = \frac{k_a}{k_d}p = bp, \tag{4.143}$$

where b is the equilibrium adsorption constant. As a result we obtain a correlation between the pressure of the external gas and the surface coverage as

$$\theta = \frac{bp}{1+bp}. \tag{4.144}$$

In the limit of low pressures, we obtain $\theta \sim p$, and at high pressures the correct limit $\theta \sim 1$ is obtained. At intermediate pressures one obtains $\theta \sim p^{1/n}$, which is known as Freundlich's isotherm. We typically have $n = 2$.

The lifetime of particles at the surface (i.e., the residence time) is another way to describe the kinetics of desorption:

$$\tau_s = \tau_0 \exp(\varepsilon_{phys}/kT), \tag{4.145}$$

where $\tau_0 \sim 10^{-12}$ s. Thus, one may estimate the residence times for small and large values of activation energy.

A number of processes proceed on the surface. Molecules may be dissociated upon adsorption, or atoms may recombine at the surface. In the case of hydrogen it was found that recombination of atoms into molecules leads to a release of excited molecules in high vibrational levels and that the properties of these molecules are dependent on the surface temperature. The recombination of two atoms of nitrogen to form a molecule at the surface leads to formation of a molecule in the metastable $A^3\Sigma$ state. However, at the surface, the metastable molecule is able to release an electron through the Auger process. Thus, surface recombination of atoms may provide a source of secondary electrons in gas discharge. In nitrogen discharges, it was found that the atoms may remain in afterglow for very long periods of time, on the order of hours. This is shown by the Lewis–Rayleigh afterglow, which is due to emission of excited states formed in the gas phase. The recombination of these atoms at the surface releases the electrons and makes it much easier to start the discharge even hours after the previous discharge.

There are two possible mechanisms for surface reactions. In the Langmuir–Hinshelwood mechanism, the reaction proceeds through adsorption of two atoms on the neighboring sites. In other words, two atoms that have been adsorbed come in proximity at the surface and react. In the Eley–Rideal mechanism, a single gas phase atom reacts with the atoms already adsorbed on the surface. As a result, the two processes have a different pressure dependence.

References

[1] French, A.P. and Taylor, E.F. 1978. *An Introduction to Quantum Physics.* Boston: MIT Press.

[2] Eisberg, R. and Resnick, R. 1974. *Quantum Physics of Atoms, Molecules, Solids, Nuclei and Particles.* New York: John Wiley & Sons.

[3] Massey, H.S.W., Burhop, E.H.S., and Gilbody, H.B. 1969. *Electronic and Ionic Impact Phenomena,* Vol 1–4. Oxford: Oxford Clarendon Press.

[4] Massey, H. 1979. *Atomic and Molecular Collisions.* London: Taylor and Francis.

[5] McDaniel, E.W. 1989. *Atomic Collisions: Electron and Photon Projectiles.* New York: John Wiley & Sons.

[6] O'Malley, T.F., Rosenberg, L., and Spruch, L. 1962. *Phys. Rev.* 125:1300.

[7] Opal, C.B., Peterson, W.K., and Beaty, E.C. 1971. *J.Chem. Phys.* 8:4100–4106.

[8] Sakai, Y., Nakamura, S., and Tagashira, H. 1985. *IEEE Trans. on Elect. Ins.* EI-20:134.

[9] Phelps, A.V. 1994. *J. Appl. Phys.* 76:747–753, *J. Phys. Chem. Ref. Data* 20:557.

[10] Phelps, A.V. 1990, 1991. *J. Phys. Chem. Ref. Data* 19:653; *J. Phys. Chem. Ref. Data* 20:1339.

[11] Golde, M.F., 1976. Gas kinetics and energy transfer. *Chem. Soc. London* 2:123–174.

[12] Thorne, A.P., 1988. *Spectrophysics.* London: Chapman and Hall.

[13] Thomas, E.W. 1984. In *Data Compendium for Plasma-Surface Interactions, Special Issue*, Eds. R.A. Langley, J. Bohdansky, E. Eckstein, P. Mioduszevski, J. Roth, E. Taglauer, E. W. Thomas, H. Verbeek, and K.L. Wilson, Nuclear Fusion, IAEA, Vienna.

CHAPTER 5

The Boltzmann Equation and Transport Equations of Charged Particles

5.1 INTRODUCTION

The Boltzmann equation is the basis for the kinetics of charged particles and neutral molecules in nonequilibrium and low-temperature plasmas that are maintained and controlled under a short-range two-body collision with feed gas molecules [1]. Of course, gas molecules have their own quantum states through their electronic configuration in a low-temperature plasma. Each of the atoms and molecules with a different quantum state exhibits unique characteristics in collision with an electron or ion. This provides a very interesting world of low-temperature plasmas with different functions reflected by the quantum structure even in a fixed external plasma condition. Therefore, it is essential to understand the terms of the collisions—including various kinds of two-body collisions—between electron and neutral molecules in the Boltzmann equation.

The Boltzmann equation is the transport equation in phase space, and usually we transfer the Boltzmann equation to the kinetic transport equation in real space in order to be readily able to calculate the plasma structure [2]. Accordingly, we derive the transport theory in detail. The purpose of this chapter is to provide a foundation, that is, the kinetic transport theory, and particularly electron transport theory, for the modeling and design of nonequilibrium and low-temperature plasmas.

5.2 THE BOLTZMANN EQUATION

The Boltzmann equation is associated with the famous Austrian physicist Ludwig Boltzmann (1844–1906), who published his kinetic treatment of gases in 1872. The form of the Boltzmann equation has not changed since its inception, but the equation for electrons requires a more detailed form of the collision integral between the electron and molecule.

5.2.1 Transport in Phase Space and Derivation of the Boltzmann Equation

The subject of this chapter is the theoretical representation of the transport of ensembles of charged particles in external electric ($\boldsymbol{E}$) and magnetic ($\boldsymbol{B}$) fields, which can be expressed as

$$\boldsymbol{F} = m\boldsymbol{\alpha}(\boldsymbol{r}, t) = e\boldsymbol{E}(\boldsymbol{r}, t) + e\boldsymbol{v} \times \boldsymbol{B}(\boldsymbol{r}, t), \tag{5.1}$$

where $\boldsymbol{\alpha}$ is the acceleration due to the force, and m and $e(> 0)$ are the mass and charge of the charged particle. Here, the number density of the swarm of the charged particle n is a function of position $\boldsymbol{r}$ and time t:

$$n = n(\boldsymbol{r}, t). \tag{5.2}$$

In addition, each of the particles has a different velocity $\boldsymbol{v}$ and the velocity changes over time. The transport of the charged particle is therefore designated in phase space, that is, both in the configuration (real) space and in the velocity space. The number of charged particles within a small volume defined by $d\boldsymbol{r}$ around a position $\boldsymbol{r}$ and within $d\boldsymbol{v}$ around a velocity $\boldsymbol{v}$ is designated dn, which can be defined as

$$dn = g(\boldsymbol{v}, \boldsymbol{r}, t)\, d\boldsymbol{v}\, d\boldsymbol{r}. \tag{5.3}$$

In Equation 5.3, in particular, when $\int dn = 1$ is satisfied, $g(\boldsymbol{v}, \boldsymbol{r}, t)$ is the probability density function and is called the velocity distribution function in physics (see Figure 5.1). The change of coordinates in the phase space after

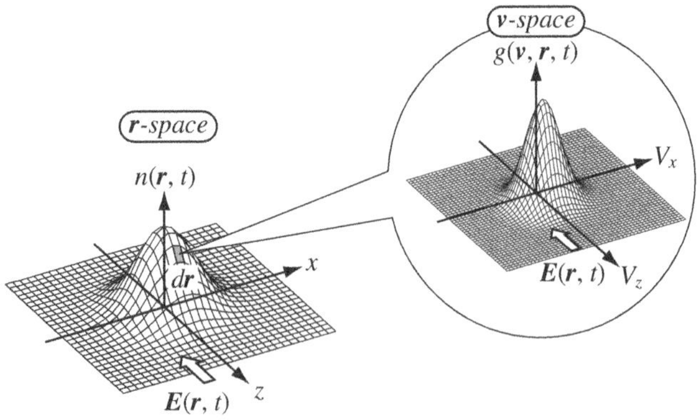

FIGURE 5.1 Example of the density distribution of charged particles $n(\boldsymbol{r}, t)$ in real space and the velocity distribution $g(\boldsymbol{r}, \boldsymbol{v}, t)$ in phase space.

a short interval dt as a result of the effects of both the self-motion and the external forces is $\boldsymbol{r}'(= \boldsymbol{r} + d\boldsymbol{r}) = \boldsymbol{r} + \boldsymbol{v}dt$, and $\boldsymbol{v}'(= \boldsymbol{v} + d\boldsymbol{v}) = \boldsymbol{v} + \boldsymbol{\alpha}dt$. That is, the number of charged particles changes in a small time increment dt to

$$dn' = g(\boldsymbol{v}', \boldsymbol{r}', t + dt)d\boldsymbol{v}'d\boldsymbol{r}'. \tag{5.4}$$

Here, the small element of phase space $d\boldsymbol{v}'\,d\boldsymbol{r}'$ after dt is connected to the initial element $d\boldsymbol{v}\,d\boldsymbol{r}$ by a Jacobian

$$d\boldsymbol{v}'d\boldsymbol{r}' = \frac{\partial(\boldsymbol{r}', \boldsymbol{v}')}{\partial(\boldsymbol{r}, \boldsymbol{v})}\,d\boldsymbol{v}\,d\boldsymbol{r}, \tag{5.5}$$

and from Equations 5.3 and 5.4 we obtain

$$\frac{\partial(\boldsymbol{r}', \boldsymbol{v}')}{\partial(\boldsymbol{r}, \boldsymbol{v})} = \begin{vmatrix} \frac{\partial x'}{\partial x} & \frac{\partial x'}{\partial y} & \cdot & \cdot & \frac{\partial x'}{\partial v_y} & \frac{\partial x'}{\partial v_z} \\ \frac{\partial y'}{\partial x} & \cdot & \cdot & \cdot & \cdot & \cdot \\ \cdot & \cdot & & & \cdot & \cdot \\ \cdot & \cdot & & & \cdot & \cdot \\ \frac{\partial v_y'}{\partial x} & \cdot & & & \cdot & \cdot \\ \frac{\partial v_z'}{\partial x} & \cdot & & & \cdot & \frac{\partial v_z'}{\partial v_z} \end{vmatrix} = 1. \tag{5.6}$$

The net change of the number of the particles during dt is given as

$$\begin{aligned} dn' - dn &= [g(\boldsymbol{v} + \boldsymbol{\alpha}dt, \boldsymbol{r} + \boldsymbol{v}dt, t + dt) - g(\boldsymbol{v}, \boldsymbol{r}, t)]\,d\boldsymbol{v}\,d\boldsymbol{r} \\ &= \left[\frac{\partial}{\partial t} + \boldsymbol{v}\cdot\frac{\partial}{\partial \boldsymbol{r}} + \boldsymbol{\alpha}\cdot\frac{\partial}{\partial \boldsymbol{v}}\right] g(\boldsymbol{v}, \boldsymbol{r}, t)\;d\boldsymbol{v}\,d\boldsymbol{r}\;dt, \end{aligned} \tag{5.7}$$

where we have used the Taylor expansion in $\boldsymbol{v}, \boldsymbol{r}$, and dt in order to obtain the second line of this equation. Equation 5.7 is equal to zero without the inflow or outflow from the element $d\boldsymbol{v}\,d\boldsymbol{r}$ due to a binary collision between the particle and the gas molecule. However, in general, we have to consider collisions that result in a change of the number of particles in any cell of the phase space, and in that case Equation 5.7 is not equal to zero. The effect of collisions is proportional to $d\boldsymbol{v}\,d\boldsymbol{r}\,dt$ and is described by using a function $J(g, F)$ as $dn' - dn = J(g, F)d\boldsymbol{v}\,d\boldsymbol{r}\,dt$. Here, $J(g, F)$ is named the collision operator (or collision integral), and $F = F(\boldsymbol{V}, \boldsymbol{r}, t)$ is the velocity distribution function of gas molecules. Usually gas molecules have a Maxwellian velocity distribution function at a gas temperature of T_g; in other words, gas molecules are in thermal equilibrium. In the realm of the so-called low-temperature or nonequilibrium plasmas, however, the charged particle is not in thermal equilibrium, and the velocity distribution satisfies

$$\frac{\partial}{\partial t}g(\boldsymbol{v}, \boldsymbol{r}, t) + \boldsymbol{v}\cdot\frac{\partial}{\partial \boldsymbol{r}}g(\boldsymbol{v}, \boldsymbol{r}, t) + \boldsymbol{\alpha}\cdot\frac{\partial}{\partial \boldsymbol{v}}g(\boldsymbol{v}, \boldsymbol{r}, t) = J(g, F). \tag{5.8}$$

Equation 5.8 demonstrates that the temporal change of the velocity distribution function of charged particles is, in principle, caused by three factors:

diffusion in configuration space (second term on the left-hand side), acceleration by external fields (third term on the left-hand side), and collisions (right-hand side).

Problem 5.2.1

Confirm Equation 5.6 (or $d\boldsymbol{v}\, d\boldsymbol{r} = d\boldsymbol{v}' d\boldsymbol{r}'$) by using the condition $\nabla_v \cdot \boldsymbol{F} = 0$ for external forces in Equation 5.1.

Equation 5.8 is known as the Boltzmann equation. The physical requirement of the independence of events among diffusion, external acceleration, and collision limits the interaction of a charged particle with another charged particle or gas molecule to a short-range interaction, that is, a two-body encounter.

On the other hand, there exists a system in which a long-range interaction of charged particles (i.e., a many-body collision) is dominant in plasmas. This occurs in a highly ionized plasma in which the Coulomb collision, represented by a large amount of small-angle scattering in collisions, is dominant. Then, Equation 5.8 is the Fokker–Planck equation. Under specific conditions the effect of collisions may be neglected, and then we have the so-called Vlasov or collisionless Boltzmann equation:

$$\frac{\partial}{\partial t} g(\boldsymbol{v}, \boldsymbol{r}, t) + \boldsymbol{v} \cdot \frac{\partial}{\partial \boldsymbol{r}} g(\boldsymbol{v}, \boldsymbol{r}, t) + \boldsymbol{\alpha} \cdot \frac{\partial}{\partial \boldsymbol{v}} g(\boldsymbol{v}, \boldsymbol{r}, t) = 0. \tag{5.9}$$

5.3 TRANSPORT COEFFICIENTS

When we integrate the velocity distribution function of charged particles $g(\boldsymbol{v}, \boldsymbol{r}, t)$ in velocity space we obtain macroscopic transport parameters in configuration space. By definition, the number density $n(\boldsymbol{r}, t)$ is determined from

$$n(\boldsymbol{r}, t) = \int g(\boldsymbol{v}, \boldsymbol{r}, t) d\boldsymbol{v}. \tag{5.10}$$

In general, the ensemble average of a quantity $\mathbf{A}(\boldsymbol{v}, \boldsymbol{r}, t)$ in velocity space is given by

$$\langle \mathbf{A}(\boldsymbol{r}, t) \rangle = \frac{\int \mathbf{A}(\boldsymbol{v}, \boldsymbol{r}, t) g(\boldsymbol{v}, \boldsymbol{r}, t) d\boldsymbol{v}}{\int g(\boldsymbol{v}, \boldsymbol{r}, t) d\boldsymbol{v}}. \tag{5.11}$$

Under the influence of an external field there appears a directional velocity $\boldsymbol{v}(\boldsymbol{r}, t)$ in charged particles, such that the mean velocity is not equal to zero ($\langle \boldsymbol{v} \rangle \neq 0$). It is convenient to divide the mean velocity into two components: the directional component $\boldsymbol{v}_d = \langle \boldsymbol{v}(\boldsymbol{r}, t) \rangle$ and the random component $\boldsymbol{v}_r$, as

$$\boldsymbol{v}(\boldsymbol{r}, t) = \boldsymbol{v}_d(\boldsymbol{r}, t) + \boldsymbol{v}_r(\boldsymbol{v}, \boldsymbol{r}, t), \tag{5.12}$$

where $\langle \boldsymbol{v}_r \rangle = 0$. The ensemble average of the velocity, often labeled as fluid velocity, is given by

$$\langle \boldsymbol{v}(\boldsymbol{r}, t) \rangle = \frac{1}{n} \int \boldsymbol{v} g(\boldsymbol{v}, t) d\boldsymbol{v}, \tag{5.13}$$

and the mean (kinetic) energy is

$$\langle \varepsilon(\boldsymbol{r},t)\rangle = \frac{1}{n}\int \frac{1}{2}mv^2 g(\boldsymbol{v},\boldsymbol{r},t)d\boldsymbol{v} = \frac{1}{2}m(v_d^2 + \langle v_r^2\rangle). \tag{5.14}$$

In the case where the number density has a spatial nonuniformity, $n(\boldsymbol{r},t)$, the velocity distribution function $g(\boldsymbol{v},\boldsymbol{r},t)$ may be represented in powers of the density gradient $\boldsymbol{\nabla}_r n$ as

$$\begin{aligned} g(\boldsymbol{v},\boldsymbol{r},t) &= g^0(\boldsymbol{v},t)n(\boldsymbol{r},t) + \boldsymbol{g}^1(\boldsymbol{v},t)\cdot\boldsymbol{\nabla}_r n(\boldsymbol{r},t) \\ &+ \mathbf{g}^2(\boldsymbol{v},t)\odot\boldsymbol{\nabla}_r^2 n(\boldsymbol{r},t) + \cdots . \end{aligned} \tag{5.15}$$

Here the coefficients in the density expansion, $\mathbf{g}^k$, are tensors of rank k and $\odot$ indicates a $k$fold scalar product. The functions independent of position are normalized as

$$\int \mathbf{g}^k(\boldsymbol{v},t)d\boldsymbol{v} = \begin{cases} 1; & k=0 \\ 0; & k\neq 0 \end{cases}. \tag{5.16}$$

The expansion 5.15 is known as hydrodynamic expansion, and its convergence is usually assumed in the Boltzmann equation of electrons (hydrodynamic approximation) [3]. In situations in which the convergence of 5.15 is not assured, the velocity distribution function is labeled as nonhydrodynamic. In hydrodynamic approximation the fluid velocity 5.13 is written as

$$\langle \boldsymbol{v}(\boldsymbol{r},t)\rangle = \frac{1}{n}\int \boldsymbol{v}g^0(\boldsymbol{v},t)d\boldsymbol{v} + \frac{1}{n}\int \boldsymbol{v}\boldsymbol{g}^1(\boldsymbol{v},t)d\boldsymbol{v}\cdot\boldsymbol{\nabla}_r n(\boldsymbol{r},t). \tag{5.17}$$

On the right-hand side the first term is the drift term due to the influence of external fields, and the second term is the diffusion term (tensor) under the density gradient, $\nabla_r n(\boldsymbol{r},t)$. Thus we have the following definition of the drift velocity and diffusion tensor, respectively:

$$\boldsymbol{v}_d(t) = \frac{1}{n}\int \boldsymbol{v}g^0(\boldsymbol{v},t)d\boldsymbol{v} \tag{5.18}$$

$$\mathbf{D}(t) = \frac{1}{n}\int \boldsymbol{v}\boldsymbol{g}^1(\boldsymbol{v},t)d\boldsymbol{v}. \tag{5.19}$$

When the external electric field is applied in the z-direction, for example, the diffusion tensor 5.19 has two components, the longitudinal (D_L) and transverse (D_T), and the tensor is represented by a diagonal matrix

$$\mathbf{D}(t) = \begin{vmatrix} D_T & 0 & 0 \\ 0 & D_T & 0 \\ 0 & 0 & D_L \end{vmatrix}. \tag{5.20}$$

Each of the components of the diffusion tensor is represented by the ensemble average of the first-order velocity distribution $\boldsymbol{g}^1(\boldsymbol{v},t)$ as

$$D_T(t) = \frac{1}{n}\int v_x g_x^1(\boldsymbol{v},t)d\boldsymbol{v} = \frac{1}{n}\int v_y g_y^1(\boldsymbol{v},t)d\boldsymbol{v}, \tag{5.21}$$

$$D_L(t) = \frac{1}{n}\int v_z g_z^1(\boldsymbol{v},t)d\boldsymbol{v}. \tag{5.22}$$

Problem 5.3.1
In the case where there is a magnetic field as well as an electric field, the drift tensor has six components. Derive each component of the drift velocity.

The total pressure tensor is defined as

$$\begin{aligned}\mathbf{P}_T(\boldsymbol{r},t) &= m\int \boldsymbol{v}\boldsymbol{v}g(\boldsymbol{v},\boldsymbol{r},t)d\boldsymbol{v},\\ &= m\int \boldsymbol{v}_r\boldsymbol{v}_r g(\boldsymbol{v},\boldsymbol{r},t)d\boldsymbol{v} + mn\boldsymbol{v}_d\boldsymbol{v}_d,\\ &= mn(\boldsymbol{r},t)\langle \boldsymbol{v}_r\boldsymbol{v}_r\rangle + mn(\boldsymbol{r},t)\boldsymbol{v}_d\boldsymbol{v}_d. \end{aligned} \quad (5.23)$$

The first term on the right-hand side (pressure tensor) shows the contribution of the thermal or random velocity, and the second term is due to drift (directional) velocity. As random velocity is usually much larger than the directional velocity, the total pressure tensor $\mathbf{P}_T$ is almost isotropic with equal components in the x-, y-, and z-directions, and each component (scalar pressure p) is written according to

$$\mathbf{P} = \begin{vmatrix} P_x & 0 & 0 \\ 0 & P_y & 0 \\ 0 & 0 & P_z \end{vmatrix}, \quad p = p_x = p_y = p_z = \frac{1}{3}mn\langle v_r^2\rangle. \quad (5.24)$$

The energy flux vector is defined as the energy carried by charged particles that cross unit area per unit time

$$\boldsymbol{Q}(\boldsymbol{r},t) = \int \frac{1}{2}mv^2\boldsymbol{v}g(\boldsymbol{v},\boldsymbol{r},t)d\boldsymbol{v}. \quad (5.25)$$

$\boldsymbol{Q}$ represents the third-order velocity moment and

$$\begin{aligned}\langle v^2\boldsymbol{v}\rangle &= \langle [(\boldsymbol{v}_d+\boldsymbol{v}_r)\cdot(\boldsymbol{v}_d+\boldsymbol{v}_r)]\,(\boldsymbol{v}_d+\boldsymbol{v}_r)\rangle\\ &= v_d^2\boldsymbol{v}_d + v_d^2\langle\boldsymbol{v}_r\rangle + 2\boldsymbol{v}_d\boldsymbol{v}_d\cdot\langle\boldsymbol{v}_r\rangle + 2\boldsymbol{v}_d\cdot\langle\boldsymbol{v}_r\boldsymbol{v}_r\rangle + \langle v_r^2\rangle\boldsymbol{v}_d + \langle v_r^2\boldsymbol{v}_r\rangle\\ &= v_d^2\boldsymbol{v}_d + 2\boldsymbol{v}_d\cdot\langle\boldsymbol{v}_r\boldsymbol{v}_r\rangle + \langle v_r^2\rangle\boldsymbol{v}_d + \langle v_r^2\boldsymbol{v}_r\rangle.\end{aligned}$$

Here we combine Equations 5.14 and 5.24 to obtain

$$\begin{aligned}\boldsymbol{Q}(\boldsymbol{r},t) &= \frac{1}{2}mn\langle v^2\boldsymbol{v}\rangle\\ &= \frac{1}{2}mn\left(v_d^2+\langle v_r^2\rangle\right)\boldsymbol{v}_d + mn\langle\boldsymbol{v}_r\boldsymbol{v}_r\rangle\cdot\boldsymbol{v}_d + \frac{1}{2}mn\langle v_r^2\boldsymbol{v}_r\rangle\\ &= n(\boldsymbol{r},t)\langle\varepsilon(\boldsymbol{r},t)\rangle\boldsymbol{v}_d + \mathbf{P}\cdot\boldsymbol{v}_d + \frac{1}{2}mn(\boldsymbol{r},t)\langle v_r^2\boldsymbol{v}_r\rangle.\end{aligned} \quad (5.26)$$

The first term in Equation 5.26 is the energy transported by drift, the second is the work performed, and the third is the energy transported by the random motion. This third term is called the thermal flux $\boldsymbol{q}$ and is defined as

$$\boldsymbol{q}(\boldsymbol{r},t) = \frac{1}{2}mn(\boldsymbol{r},t)\langle v_r^2\boldsymbol{v}_r\rangle. \quad (5.27)$$

Problem 5.3.2

The diffusion coefficients, D_L and D_T, are obtained by the trace of the trajectory of particles in the configuration space in accordance with

$$\begin{aligned} D_L(t) &= \frac{1}{2}\frac{d}{dt}\left\langle (z(t) - \langle z(t) \rangle)^2 \right\rangle \\ &= \langle z(t)v_z(t) \rangle - \langle z(t) \rangle \langle v_z(t) \rangle. \end{aligned} \tag{5.28}$$

$$\begin{aligned} D_T(t) &= \frac{1}{4}\frac{d}{dt}\langle x(t)^2 + y(t)^2 \rangle \\ &= \frac{1}{2}(\langle x(t)v_x(t) \rangle + \langle y(t)v_y(t) \rangle). \end{aligned} \tag{5.29}$$

In particular, in addition to $\boldsymbol{E} = E\boldsymbol{k}$ the magnetic field $\boldsymbol{B} = B\boldsymbol{i}$ is present, electrons have a complicated behavior, and each of the diagonal components of the diffusion tensor in $\boldsymbol{E} \times \boldsymbol{B}$ fields is expressed as

$$\begin{aligned} D_E &= \frac{1}{2}\frac{d}{dt}(< z^2 > - < z >^2), \\ D_{E\times B} &= \frac{1}{2}\frac{d}{dt}(< y^2 > - < y >^2), \\ D_B &= \frac{1}{2}\frac{d}{dt}(< x^2 > - < x >^2). \end{aligned}$$

In addition to the above components, the Hall diffusion coefficient, representing the sum of the off-diagonal components of the diffusion tensor, will appear and is given by

$$D_{Hall} = \frac{1}{2}\frac{d}{dt}(< y >< z > - < yz >).$$

Derive the above relationship from the statistical physics.

Exercise 5.3.1

An isolated swarm of charged particles with delta function in space is injected into gases under a uniform dc-field. After a finite collisional relaxation time, the swarm attains a quasi-equilibrium with time independent transport coefficients, $\boldsymbol{v}_d$ and $\mathbf{D}$. Discuss the transport and the coefficient in an initial stage before the relaxation.

Higher-order spatial moments about the center of mass of the swarm $< \boldsymbol{r} >$,

$$\mathbf{M_k}(t) = < (\boldsymbol{r}(t) - < \boldsymbol{r}(t) >)^{\mathbf{k}} >. \quad (\mathbf{k} = \mathbf{3}, \mathbf{4}..)$$

have a finite magnitude in the initial phase of the steep density distribution. In particular, $\mathbf{M_3}$ and $\mathbf{M_4}$ represent skewness and kurtosis of the density distribution, and the 3rd- and 4th-order transport coefficients, $\mathbf{D_3}$ and $\mathbf{D_4}$ are related to the skewness and kurtosis, respectively,

$$\begin{aligned} \mathbf{D_3} &= \frac{1}{3!}\frac{d}{dt}\mathbf{M_3}(t), \\ \mathbf{D_4} &= \frac{1}{4!}\frac{d}{dt}\left(\mathbf{M_4}(t) - \mathbf{3}(\mathbf{M_2}(t))^{\mathbf{2}}\right). \end{aligned}$$

Then, in an initial stage of the transport, the fluid velocity expressed by the drift velocity and diffusion tensor in Equation 5.17 will be extended to a form including the 3rd- and 4th-order transport coefficients.

5.4 THE TRANSPORT EQUATION

When we start from the general form of the Boltzmann Equation 5.8, multiply it by an arbitrary function $\mathbf{A}(\boldsymbol{v}, \boldsymbol{r}, t)$, and integrate it in velocity space, we obtain the transport equaton of particles in configuration space:

$$\int \mathbf{A}(\boldsymbol{v}, \boldsymbol{r}, t)\frac{\partial g(\boldsymbol{v}, \boldsymbol{r}, t)}{\partial t}d\boldsymbol{v} + \int \mathbf{A}(\boldsymbol{v}, \boldsymbol{r}, t)\boldsymbol{v} \cdot \frac{\partial g(\boldsymbol{v}, \boldsymbol{r}, t)}{\partial \boldsymbol{r}}d\boldsymbol{v}$$
$$+ \int \mathbf{A}(\boldsymbol{v}, \boldsymbol{r}, t)\boldsymbol{\alpha} \cdot \frac{\partial g(\boldsymbol{v}, \boldsymbol{r}, t)}{\partial \boldsymbol{v}}d\boldsymbol{v} = \int \mathbf{A}(\boldsymbol{v}, \boldsymbol{r}, t)J(g, F)d\boldsymbol{v}. \quad (5.30)$$

The first term may be expanded as

$$\begin{aligned} \int \mathbf{A}\frac{\partial g}{\partial t}d\boldsymbol{v} &= \frac{\partial}{\partial t}\int \mathbf{A}g d\boldsymbol{v} - \int \frac{\partial \mathbf{A}}{\partial t}g d\boldsymbol{v} \\ &= \frac{\partial}{\partial t}(n(\boldsymbol{r}, t)\langle \mathbf{A}\rangle) - n(\boldsymbol{r}, t)\left\langle \frac{\partial \mathbf{A}}{\partial t}\right\rangle, \end{aligned} \quad (5.31)$$

and the second term is equal to

$$\begin{aligned} \int \mathbf{A}\boldsymbol{v} \cdot \frac{\partial g}{\partial \boldsymbol{r}}d\boldsymbol{v} &= \int \frac{\partial}{\partial \boldsymbol{r}} \cdot (\boldsymbol{v}\mathbf{A}g)d\boldsymbol{v} - \int \left(\frac{\partial}{\partial \boldsymbol{r}} \cdot \boldsymbol{v}\mathbf{A}\right) g d\boldsymbol{v}, \\ &= \frac{\partial}{\partial \boldsymbol{r}} \cdot (n(\boldsymbol{r}, t)\langle \boldsymbol{v}\mathbf{A}\rangle) - n(\boldsymbol{r}, t)\left\langle \frac{\partial}{\partial \boldsymbol{r}} \cdot \boldsymbol{v}\mathbf{A}\right\rangle. \end{aligned} \quad (5.32)$$

The third term can be expanded as

$$\begin{aligned} \int \mathbf{A}\boldsymbol{\alpha} \cdot \frac{\partial g}{\partial \boldsymbol{v}}d\boldsymbol{v} &= \frac{\partial}{\partial \boldsymbol{v}} \cdot \int \boldsymbol{\alpha}\mathbf{A}g d\boldsymbol{v} - \int \left(\frac{\partial}{\partial \boldsymbol{v}} \cdot \boldsymbol{\alpha}\mathbf{A}\right) g d\boldsymbol{v} \\ &= \frac{\partial}{\partial \boldsymbol{v}} \cdot (n(\boldsymbol{r}, t)\langle \boldsymbol{\alpha}\mathbf{A}\rangle) - n(\boldsymbol{r}, t)\left\langle \frac{\partial}{\partial \boldsymbol{v}} \cdot \boldsymbol{\alpha}\mathbf{A}\right\rangle \\ &= -n(\boldsymbol{r}, t)\left\langle \boldsymbol{\alpha} \cdot \frac{\partial}{\partial \boldsymbol{v}}\mathbf{A}\right\rangle. \end{aligned} \quad (5.33)$$

Now we substitute Equations 5.31 to 5.33 into Equation 5.30 and obtain the general continuity equation as

$$\begin{aligned} &\frac{\partial}{\partial t}(n(\boldsymbol{r}, t)\langle \mathbf{A}(\boldsymbol{v}, \boldsymbol{r}, t)\rangle) - n(\boldsymbol{r}, t)\left\langle \frac{\partial \mathbf{A}(\boldsymbol{v}, \boldsymbol{r}, t)}{\partial t}\right\rangle + \frac{\partial}{\partial \boldsymbol{r}} \cdot (n(\boldsymbol{r}, t)\langle \boldsymbol{v}\mathbf{A}(\boldsymbol{v}, \boldsymbol{r}, t)\rangle) \\ &\qquad - n(\boldsymbol{r}, t)\left\langle \frac{\partial}{\partial \boldsymbol{r}} \cdot \boldsymbol{v}\mathbf{A}(\boldsymbol{v}, \boldsymbol{r}, t)\right\rangle - n(\boldsymbol{r}, t)\left\langle \boldsymbol{\alpha} \cdot \frac{\partial \mathbf{A}(\boldsymbol{v}, \boldsymbol{r}, t)}{\partial \boldsymbol{v}}\right\rangle \\ &\qquad = \int \mathbf{A}(\boldsymbol{v}, \boldsymbol{r}, t)J(g, F)d\boldsymbol{v}. \end{aligned} \quad (5.34)$$

5.4.1 Conservation of Number Density

If we use $\mathbf{A} = 1$ and replace it in the general form 5.34, the second, fourth, and fifth terms are equal to zero. The integral on the right-hand side is written as $n\,R_0$ by using the production rate R_0, and the general moment equation reduces to

$$\frac{\partial}{\partial t}n(\boldsymbol{r},t) + \frac{\partial}{\partial \boldsymbol{r}} \cdot (n(\boldsymbol{r},t)\langle \boldsymbol{v} \rangle) = n_e(\boldsymbol{r},t)R_0(\boldsymbol{r},t). \tag{5.35}$$

This equation represents the conservation of particle number density. The first term describes the temporal changes and the second describes the changes due to flow of the number density of charged particles. The term on the right-hand side gives the production and loss due to the collision between the particle and the neutral molecule.

Problem 5.4.1

Show that for the number density continuity of electrons, the term R_0 on the right-hand side of Equation 5.35 is expressed through the ionization rate R_i and electron attachment rate R_a as

$$R_0(\boldsymbol{r},t) = R_i(\boldsymbol{r},t) - R_a(\boldsymbol{r},t).$$

5.4.2 Conservation of Momentum

If we substitute $\mathbf{A} = m\boldsymbol{v}$ in Equation 5.34, the second and the fourth terms are equal to zero, and we obtain

$$\frac{\partial}{\partial t}(mn(\boldsymbol{r},t)\langle \boldsymbol{v} \rangle) + \frac{\partial}{\partial \boldsymbol{r}} \cdot (mn(\boldsymbol{r},t)\langle \boldsymbol{v}\boldsymbol{v} \rangle) - mn(\boldsymbol{r},t)\left\langle \boldsymbol{\alpha} \cdot \frac{\partial}{\partial \boldsymbol{v}}\boldsymbol{v} \right\rangle = m\int \boldsymbol{v}J d\boldsymbol{v}. \tag{5.36}$$

According to Equation 5.12 the velocity $\boldsymbol{v}$ is divided into two parts, the directional and thermal velocities, $\boldsymbol{v}_d$ and $\boldsymbol{v}_r$. The first term on the left-hand side of Equation 5.36 is replaced by using $\langle \boldsymbol{v} \rangle = \boldsymbol{v}_d$ and Equation 5.35 as

$$\begin{aligned} mn(\boldsymbol{r},t)\frac{\partial}{\partial t}\boldsymbol{v}_d \quad &+ \quad m\boldsymbol{v}_d\frac{\partial}{\partial t}n(\boldsymbol{r},t) = mn(\boldsymbol{r},t)\frac{\partial}{\partial t}\boldsymbol{v}_d - m\boldsymbol{v}_d\frac{\partial}{\partial \boldsymbol{r}} \cdot (n(\boldsymbol{r},t)\boldsymbol{v}_d) \\ &+ \quad mn_e(\boldsymbol{r},t)\boldsymbol{v}_d R_0, \end{aligned}$$

the second term of Equation 5.36 is expanded as

$$\begin{aligned} \frac{\partial}{\partial \boldsymbol{r}} \cdot (mn(\boldsymbol{r},t)\boldsymbol{v}_d\boldsymbol{v}_d) \quad &+ \quad \frac{\partial}{\partial \boldsymbol{r}} \cdot (mn(\boldsymbol{r},t)\langle \boldsymbol{v}_r\boldsymbol{v}_r \rangle) = \boldsymbol{v}_d\frac{\partial}{\partial \boldsymbol{r}} \cdot (mn(\boldsymbol{r},t)\boldsymbol{v}_d) \\ &+ \quad mn(\boldsymbol{r},t)\left(\boldsymbol{v}_d \cdot \frac{\partial}{\partial \boldsymbol{r}}\right)\boldsymbol{v}_d + \frac{\partial}{\partial \boldsymbol{r}} \cdot (mn(\boldsymbol{r},t)\langle \boldsymbol{v}_r\boldsymbol{v}_r \rangle), \end{aligned}$$

and finally, the third term is written as

$$-mn(\boldsymbol{r},t)\left\langle \boldsymbol{\alpha} \cdot \frac{\partial}{\partial \boldsymbol{v}}\boldsymbol{v} \right\rangle = -n(\boldsymbol{r},t)e(\boldsymbol{E} + \boldsymbol{v} \times \boldsymbol{B}).$$

As a result, the conservation of momentum is expressed as

$$mn(\boldsymbol{r},t)\frac{\partial}{\partial t}\boldsymbol{v}_d + mn(\boldsymbol{r},t)\boldsymbol{v}_d\left(\frac{\partial}{\partial \boldsymbol{r}}\cdot\boldsymbol{v}_d\right) + m\frac{\partial}{\partial \boldsymbol{r}}\cdot(n(\boldsymbol{r},t)\langle\boldsymbol{v}_r\boldsymbol{v}_r\rangle)$$
$$= n(\boldsymbol{r},t)e(\boldsymbol{E}+\boldsymbol{v}\times\boldsymbol{B}) - mn_e(\boldsymbol{r},t)\boldsymbol{v}_d R_0 + m\langle\boldsymbol{v}J\rangle. \quad (5.37)$$

Here the third term on the left-hand side is expressed by using the pressure tensor $\mathbf{P}$ (which is defined in Equation 5.24) as $\nabla_r\cdot\mathbf{P}$.

Problem 5.4.2
Show that the last term in the momentum conservation Equation 5.37 is approximated by using the rate of the momentum transfer collision R_m as

$$m\langle\boldsymbol{v}J\rangle = -mn(\boldsymbol{r},t)\boldsymbol{v}_d R_m. \quad (5.38)$$

Exercise 5.4.1
Consider the case that the directional velocity of charged particles is independent of space and time under a constant $\boldsymbol{E}$ without $\boldsymbol{B}$. Then, derive the current $en\boldsymbol{v}_d$.
From Equation 5.37 we obtain the relation under $R_m \gg R_0$

$$mn\boldsymbol{v}_d R_m = en\boldsymbol{E} - m\frac{\partial}{\partial \boldsymbol{r}}\cdot(n\langle\boldsymbol{v}_r\boldsymbol{v}_r\rangle).$$

The current is expressed by using the equation

$$\boldsymbol{J} = en\boldsymbol{v}_d = en\mu\boldsymbol{E} - e\frac{\partial}{\partial \boldsymbol{r}}(Dn),$$

where μ and D are equivalent as $\mu = e/mR_m$ and $D = \langle\boldsymbol{v}_r\boldsymbol{v}_r\rangle_z/R_m = \langle v_r^2\rangle/3R_m$, where the electric field is directed along the z-axis. Here μ is the mobility and D is the diffusion coefficient and these equations are consistent with simple expressions 2.4 and 2.5 in Chapter 2.

5.4.3 Conservation of Energy

If we replace $\mathbf{A}$ with $mv^2/2$ in Equation 5.34, the second and the fourth terms are equal to zero and the fifth term on the left-hand side is simplified as

$$-n(\boldsymbol{r},t)\left\langle\frac{1}{m}[e\boldsymbol{E}(\boldsymbol{r},t)+e\boldsymbol{v}\times\boldsymbol{B}(\boldsymbol{r},t)]\cdot\frac{\partial}{\partial \boldsymbol{v}}\frac{mv^2}{2}\right\rangle = -n(\boldsymbol{r},t)\langle e\boldsymbol{E}\cdot\boldsymbol{v}\rangle,$$

and as a result we obtain

$$\frac{\partial}{\partial t}\left(\frac{mn(\boldsymbol{r},t)}{2}\langle v^2\rangle\right) + \frac{\partial}{\partial \boldsymbol{r}}\cdot\left(\frac{mn(\boldsymbol{r},t)}{2}\langle v^2\boldsymbol{v}\rangle\right) - n(\boldsymbol{r},t)e\boldsymbol{E}\cdot\boldsymbol{v}_d$$
$$= \int\frac{mv^2}{2}J d\boldsymbol{v}. \quad (5.39)$$

The first term is the change in the mean energy $\varepsilon(\boldsymbol{r},t)$, and the second is the change in the energy flow. When we substitute Equation 5.26 in the second term we obtain

$$\frac{\partial}{\partial t}(n(\boldsymbol{r},t)\langle\varepsilon(\boldsymbol{r},t)\rangle) + \frac{\partial}{\partial \boldsymbol{r}}\cdot(n(\boldsymbol{r},t)\langle\varepsilon(\boldsymbol{r},t)\rangle\boldsymbol{v}_d) + \frac{\partial}{\partial \boldsymbol{r}}\cdot(\mathbf{P}\cdot\boldsymbol{v}_d)$$
$$+ \frac{\partial}{\partial \boldsymbol{r}}\cdot\boldsymbol{q} - n(\boldsymbol{r},t)e\boldsymbol{E}\cdot\boldsymbol{v}_d = \int\frac{mv^2}{2}Jd\boldsymbol{v}. \quad (5.40)$$

Problem 5.4.3

Derive the collisional integral term in the energy conservation Equation 5.40 for electrons that is given by

$$= \frac{2m}{M}\left\langle\varepsilon(\boldsymbol{r},t)NQ_m(\varepsilon)\left(\frac{2\varepsilon}{m}\right)^{1/2}\right\rangle n_e(\boldsymbol{r},t)$$
$$+\left(\sum\varepsilon_jR_j+\varepsilon_iR_i+\left\langle\varepsilon NQ_a(\varepsilon)\left(\frac{2\varepsilon}{m}\right)^{1/2}\right\rangle\right)n_e(\boldsymbol{r},t), \quad (5.41)$$

where R_m, R_j, R_i, and R_a are the rates of elastic momentum transfer, electronic excitation, ionization, and electron attachment collisions, respectively. Also ε_j, ε_i, and ε_a are the corresponding threshold energies.

5.5 COLLISION TERM IN THE BOLTZMANN EQUATION

5.5.1 Collision Integral

So far we have not specified the exact structure of the collision term on the right-hand side of the Boltzmann Equation 5.8. In Chapter 4 we have discussed specific types of collisions and how to represent their quantitative properties by the differential cross section $\sigma(\theta,\phi;\varepsilon)$ and the integral cross sections $Q(\varepsilon)$. The velocity distribution of neutral gas molecules is represented by $F(\boldsymbol{V}',\boldsymbol{r},t)$, where gas molecules have a mass M and velocity $\boldsymbol{V}'$ before collision. The charged particles of mass m and velocity $\boldsymbol{v}'$ before collision are described by the corresponding velocity distribution function $g(\boldsymbol{v}',\boldsymbol{r},t)$. When we consider a small element of the phase space $d\boldsymbol{v}\,d\boldsymbol{r}$, then the number of the charged particles that enter this element during a short time dt is equal to

$$J_{in} = \int_\Omega\int_{v'} F(\boldsymbol{V}',\boldsymbol{r},t)d\boldsymbol{V}'g(\boldsymbol{v}',\boldsymbol{r},t)d\boldsymbol{v}'v'_\gamma\sigma(\theta',\phi';v'_\gamma)d\Omega'd\boldsymbol{r}\,dt. \quad (5.42)$$

During the same time dt, the number of charged particles leaving the same element $d\boldsymbol{v}\,d\boldsymbol{r}$ is

$$J_{out} = \int_\Omega\int_v F(\boldsymbol{V},\boldsymbol{r},t)d\boldsymbol{V}g(\boldsymbol{v},\boldsymbol{r},t)d\boldsymbol{v}v_\gamma\sigma(\theta,\phi;v_\gamma)d\Omega d\boldsymbol{r}\,dt, \quad (5.43)$$

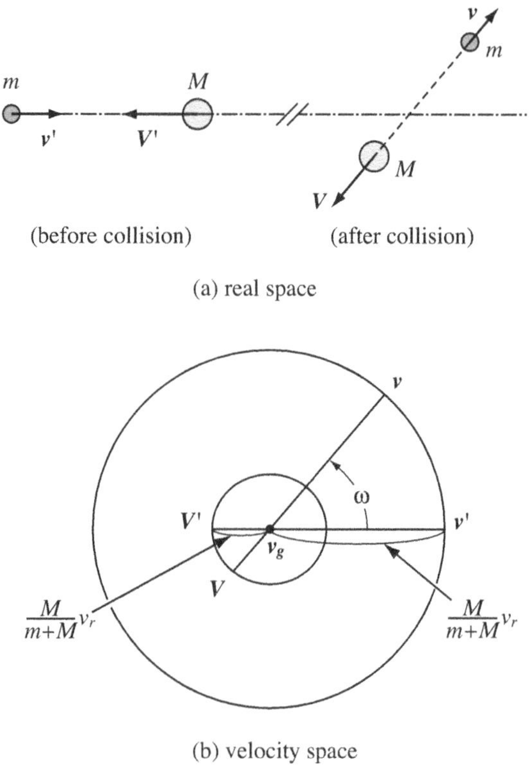

FIGURE 5.2 The definition of a collision in (a) a laboratory frame of reference (in real space) and (b) a center-of-mass frame of reference (in velocity space).

where $v_{\gamma'}$ and v_γ are the relative speed between the gas molecule and the charged particle just before and after the collision. The resulting change in the number of charged particles in the element $d\boldsymbol{v}\, d\boldsymbol{r}$ in time dt is

$$J d\boldsymbol{v}\, d\boldsymbol{r}\; dt = \{J_{in} - J_{out}\}. \tag{5.44}$$

5.5.2 Collision Integral between an Electron and a Gas Molecule

In Figure 5.2 we show a schematic representation of a collision between a charged particle (lowercase symbols) and a gas molecule (uppercase symbols). The properties before collisions are denoted by ('). In the laboratory frame of reference, a complete set of momentum and energy balance equations should be solved to describe the collisions. However, the calculation is simplified in the center-of-mass frame where we subtract the motion of the center of mass.

We have learned in Chapter 2 that even under a substantial electric field in gases, the directional velocity of charged particles $\boldsymbol{v}_d$ is much smaller than the random component $\boldsymbol{v}_r$. This is particularly valid for electrons in a low-temperature plasma, and we mainly concentrate on the electron transport under these physical circumstances. In practice, this fact means that the electron velocity distribution is nearly spherical with a small perturbation in the field direction. Thus it is convenient to expand the velocity distribution in spherical harmonics in velocity space

$$g(\boldsymbol{v}) = \sum_{mn} g_{mn}(v) Y^e_{mn}(\theta, \varphi), \tag{5.45}$$

where $Y^e_{mn}(\theta, \varphi) = P^m_n(\cos\theta)\cos m\varphi$ are the spherical harmonics by Morse–Feshbach notation [4]; this expansion is discussed further in Section 5.6.1. The spherical harmonic expansion helps us take care of the dependence on three components of the vector $\boldsymbol{v}$ by taking advantage of the cylindrical symmetry of the problem and turning the solution of the velocity dependence to a series of equations and corresponding distribution functions that depend on a single variable, the magnitude of the velocity $\boldsymbol{v}$. Furthermore, it is possible to use the energy of particles as a variable. Solving the velocity dependence is the critical step in solving the Boltzmann equation, and quite often the equations for the velocity (energy) dependence are referred to as the Boltzmann equation.

5.5.2.1 *Elastic Collision Term* J_{elas}

The elastic collision term is established by considering the momentum and energy conservation equations. In this process we are able to take advantage of the following simplifications:

i. The relative velocity between the electron and the gas molecule is not changed after the collision;

ii. The velocity of gas molecules is much less than the velocity of electrons, $|\boldsymbol{V}| \ll |\boldsymbol{v}|$, and we represent the velocity distribution of gas molecules with the number density N by $F(\boldsymbol{V}) = N\delta(\boldsymbol{V})$, where δ is Dirac's delta function; and

iii. From the momentum and energy conservation equations before and after collisions, $\boldsymbol{v}'$ and $\boldsymbol{v}$, we have the relation

$$\frac{v'^2 - v^2}{v^2} = \frac{2m}{M+m}(1 - \cos\omega), \tag{5.46}$$

where ω is the scattering angle as defined in Figure 5.2b.

From Equation 5.46 we obtain $d\boldsymbol{v}'/d\boldsymbol{v} = (v'/v)^3$, and thus we replace the small velocity element $d\boldsymbol{v}'$ by $d\boldsymbol{v}$. As a result, we obtain the elastic term of electrons as follows:

$$
\begin{aligned}
J_{elas}d\boldsymbol{v} &= \int_{\Omega'}\int_{V'} N\delta(\boldsymbol{V}')g(\boldsymbol{v}')v_\gamma\sigma(v_\gamma,\omega)d\Omega' d\boldsymbol{v}' d\boldsymbol{V}' \\
&\quad - \int_{\Omega}\int_{V} N\delta(\boldsymbol{V})g(\boldsymbol{v})v_\gamma\sigma(v_\gamma,\omega)d\Omega d\boldsymbol{v}\, d\boldsymbol{V} \\
&= N\left(\int_{\Omega'} \frac{v'^3}{v^3} g(\boldsymbol{v}')v'\sigma(v',\omega)d\Omega' - \int_{\Omega} g(\boldsymbol{v})v\sigma(v,\omega)d\Omega\right) d\boldsymbol{v}. \quad (5.47)
\end{aligned}
$$

The velocity distribution $g(\boldsymbol{v})$ is expanded in spherical harmonics

$$
g(\boldsymbol{v}) = \sum_{mn} g_{mn}(v)Y^e_{mn}(\theta,\varphi), \tag{5.48}
$$

and in order to divide the differential cross section into spherical (σ_0) and forward-directed (σ_1) parts, and the like, we expand the differential cross section $\sigma(\theta,\varphi;\varepsilon)$ in terms of the Legendre function $P_n(\cos\theta)$ according to

$$
\sigma(v,\omega) = \sum_{n'} \sigma_{n'}(v)P_{n'}(\cos\omega). \tag{5.49}
$$

By using the addition theorem for the spherical harmonics (see Equation 5.73) we exchange θ' to θ in Equation 5.47.

$$
\begin{aligned}
& N\frac{v'^4}{v^3}\sum_{mn}\sum_{n'} g_{mn}(v')\sigma_{n'}(v')\left[\int Y^e_{mn}(\theta',\varphi')P_{n'}(\cos\theta')P_n(\cos\theta)d\Omega'\right. \\
& + 2\sum_{m=1}^{n'}\frac{(n-m)!}{(n+m)!}\int_{\Omega} Y^e_{mn}(\theta',\varphi')\left\{Y^e_{mn}(\theta',\varphi')Y^e_{mn}(\theta,\varphi)\right. \\
& \left.\left. + Y^0_{mn}(\theta',\varphi')Y^0_{mn}(\theta,\varphi)\right\} d\Omega'\right] = N\frac{v'^4}{v^3}\left[\sum_n g_{0n}(v')\sigma_n(v')\frac{4\pi}{2n+1}P_n(\cos\theta)\right. \\
& \left. + \sum_{m=1}^{n}\sum_n g_{mn}(v')\sigma(v')\frac{4\pi}{2n+1}Y^e_{mn}(\theta,\varphi)\right].
\end{aligned}
$$

Here we consider the orthogonality property of the Legendre function, $4\pi/(2n+1) = \int P_n(\cos\omega)P_{n'}(\cos\omega)d\Omega$ (see Equation 5.74). Then the elastic collision term is given as

$$
\begin{aligned}
& J_{elas} \\
& = N\sum_{mn}\int_{\Omega} Y^e_{mn}(\theta,\varphi)\frac{1}{v^3}\left[v'^4 g_{mn}(v')\sigma(v',\omega)P_n(\cos\omega) - v^4 g_{mn}(v)\sigma(v,\omega)\right] d\Omega \\
& = N\sum_{mn}\int_{\Omega} Y^e_{mn}(\theta,\varphi)\frac{1}{v^3}\left[v'^4 g_{mn}(v')\sigma(v',\omega) - v^4 g_{mn}(v)\sigma(v,\omega)\right] P_n(\cos\omega)d\Omega \\
& -N\sum_{mn}\int_{\Omega} Y^e_{mn}(\theta,\varphi)g_{mn}(v)v\sigma(v,\omega)\left\{1 - P_n(\cos\omega)\right\} d\Omega. \quad (5.50)
\end{aligned}
$$

When we compare the magnitudes of velocities before and after the elastic collision by using Equation 5.46, we conclude that both of the velocities have nearly the same magnitude, and $\Delta v^2 = v'^2 - v^2$. Therefore we are able to expand the first term in the right-hand side of Equation 5.50 by the Taylor series with the variable $\Delta v^2 = v'^2 - v^2$ as

$$J_{elas} \cong N\sum_{mn}\int_{\Omega} Y^e_{mn}(\theta,\varphi)\frac{1}{v}\frac{2m}{M+m}(1-\cos\omega)\frac{\partial[v^4 g_{mn}(v)\sigma(v,\omega)]}{\partial(v^2)}P_n(\cos\omega)d\Omega$$
$$-N\sum_{mn} Y^e_{mn}(\theta,\varphi)g_{mn}(v)v\sigma(v,\omega)\left\{1-P_n(\cos\omega)\right\}d\Omega. \tag{5.51}$$

In the particular case of the isotropic part of the distribution ($m = 0$, $n = 0$), by considering $P_0(\cos\omega) = 1$, Equation 5.51 reduces to

$$J^{00}_{elas} = N\frac{2m}{M+m}\frac{1}{2v^2}\frac{\partial}{\partial v}\left\{v^4 g_{00}(v)Q_m(v)\right\}. \tag{5.52}$$

Here, $Q_m(v)$ is the momentum transfer cross section defined as

$$Q_m(v) = \int_{\Omega}(1-\cos\omega)\sigma(v,\omega)d\Omega.$$

For thermal equilibrium, electrons have the Maxwellian distribution. For zero field, the temperature of electrons T_e are equal to T_g when the gas temperature is not zero; that is, $T_g \neq 0$. However, under all circumstances collisions with thermal gas molecules lead to an additional velocity of electrons that is on the order of $O(V)$. Thus the total velocity has two components, one for zero temperature and an additional component due to thermal motion ($T_g \neq 0$):

$$v = v_0 + O(V).$$

For thermal equilibrium the distribution function has to be a Maxwellian, and therefore $J^{00}_{elas} = 0$. As a result, we obtain the order of V'^2 according to the principle of equipartitioning of energy:

$$O(V'^2) = -g_{00}\left(v_0^2\right) \Bigg/ \frac{\partial g_{00}\left(v_0^2\right)}{\partial(v^2)} = \frac{2kT_g}{m}. \tag{5.53}$$

Finally, we obtain the elastic collision integral for a general case of nonzero gas temperature ($T_g \neq 0$):

$$J^{00}_{elas} = N\frac{m}{M+m}\frac{1}{v^2}\frac{\partial}{\partial v}\left[v^4 Q_m(v)\left(g_{00}(v) + \frac{kT_g}{mv}\frac{\partial}{\partial v}g_{00}(v)\right)\right]. \tag{5.54}$$

5.5.2.2 Excitation Collision Term J_{ex}

Inelastic collisions may be represented in the Boltzmann theory under the following assumptions:

i. The change of kinetic energy in jth inelastic collisions ε_j usually satisfies the relation $\varepsilon_j \gg kT_g$ with gas molecule $F(\boldsymbol{V}) = N\delta(\boldsymbol{V})$; and

ii. From the energy conservation we have

$$\frac{1}{2}mv'^2 = \frac{1}{2}mv^2 + \varepsilon_j. \tag{5.55}$$

From Equation 5.55, that is, $v'dv' = vdv$, it is possible to transform the small velocity element $d\boldsymbol{v}'$ to $d\boldsymbol{v}$ by $d\boldsymbol{v}' = (v'/v)d\boldsymbol{v}$, and finally, we obtain the collision operator for excitation

$$\begin{aligned} J_{exj}d\boldsymbol{v} &= N\int_{\Omega'}\int_{V} g(\boldsymbol{v}')v'\sigma_j(v',\omega)\delta(\boldsymbol{V}')d\Omega' d\boldsymbol{v}' d\boldsymbol{V}' \\ &\quad -N\int_{\Omega}\delta(\boldsymbol{V})g(\boldsymbol{v})v\sigma_j(v,\omega)d\Omega d\boldsymbol{v}\, d\boldsymbol{V} \\ &= N\frac{1}{v}\left[\int g(\boldsymbol{v}')v'^2\sigma_j(v',\omega)d\Omega' - g(v)v^2\int \sigma_j(v,\omega)d\Omega\right] d\boldsymbol{v}. \end{aligned} \tag{5.56}$$

which is further simplified by using the spherical harmonic expansion for $g(\boldsymbol{v})$,

$$J_{exj} = N\frac{1}{v}\sum Y_{mn}(\theta,\varphi)\left[g_{mn}(v')v'^2\int_{\Omega}\sigma_j(v',\omega)P_n(\cos\omega)d\Omega - g_{mn}(v)v^2Q_j(v)\right], \tag{5.57}$$

where the integrated cross section is defined on the basis of the corresponding differential cross section as

$$Q_j(v) = \int_{\Omega}\sigma_j(v,\omega)d\Omega.$$

5.5.2.3 Ionization Collision Term J_{ion}

Ionization is a specific process by which a new electron is created in the inelastic transition to the ionization continuum. Ionization by electrons is a nonconservative process as the number density of electrons is not conserved. With regard to the ionization collision operator, the following simplification is made:

i. The change of kinetic energy in ionization ε_i usually satisfies the relation $\varepsilon_i \gg kT_g$ with gas molecule $F(\boldsymbol{V}) = N\delta(\boldsymbol{V})$.

ii. In the principle of indistinguishability, the incoming and newly produced electrons at ionization cannot be distinguished after collision, but experimentally we know that there exist a pair of electrons with high and low energy. We assume that the rest of the kinetic energy is shared by two electrons according to the ratio $(1-\Delta)$:Δ. Then, the velocity element can be expressed as

$$d\boldsymbol{v}' = \frac{1}{\Delta}\frac{v'}{v}d\boldsymbol{v}_2 \quad \text{or} \quad d\boldsymbol{v}' = \frac{1}{(1-\Delta)}\frac{v'}{v}d\boldsymbol{v}_1. \tag{5.58}$$

iii. From the energy conservation, we have

$$\frac{1}{2}mv'^2 = \frac{1}{2}mv_1^2 + \frac{1}{2}mv_2^2 + \varepsilon_i. \tag{5.59}$$

On the basis of these assumptions and the spherical harmonic expansion, we derive the collision integral for ionization as

$$\begin{aligned} J_{ion} &= N\frac{1}{(1-\Delta)v}\int_\Omega g(v')v'^2\sigma_i(v',\omega)d\Omega + N\frac{1}{\Delta v}\int_\Omega g(v'')v''^2\sigma_i(v'',\omega)d\Omega \\ &\quad -N\frac{1}{v}g(v)v^2\int_\Omega \sigma_i(v,\omega)d\Omega \\ &= N\sum_{mn} Y_{mn}(\theta,\varphi)\frac{1}{v}\left\{\frac{1}{(1-\Delta)}\int_\Omega g_{mn}(v')v'^2\sigma_i(v',\omega)P_n(\cos\omega)d\Omega\right. \\ &\quad \left. +\frac{1}{\Delta}\int_\Omega g_{mn}(v'')v''^2\sigma_i(v'',\omega)P_n(\cos\omega)d\Omega - Ng_{mn}v^2Q_i(v)\right\}, \end{aligned} \tag{5.60}$$

where the integrated ionization cross section is defined as

$$Q_i(v) = \int \sigma_i(v,\omega)d\Omega.$$

5.5.2.4 *Electron Attachment Collision Term J_{att}*

Electron attachment is a very specific and nonconservative process wherein an electron with energy ε is lost in collision with threshold energy ε_a. The collision operator is simplified under $J_{in} = 0$ as

$$\begin{aligned} J_{atta} &= -Ng(\boldsymbol{v})v\int_\Omega \sigma_a(v,\omega)d\Omega \\ &= -N\sum_{mn} Y_{mn}^e(\theta,\varphi)g_{mn}(v)vQ_a(v), \end{aligned} \tag{5.61}$$

where the integrated attachment cross section is given by

$$Q_a(v) = \int_\Omega \sigma_a(v,\omega)d\Omega.$$

5.6 BOLTZMANN EQUATION FOR ELECTRONS

Charged particles in gases under the influence of an electric field are not in thermal equilibrium but may be in quasi-equilibrium in the case where energy gained from the field is dissipated in inelastic (and even elastic) collisions with gas molecules. Under these conditions, transport coefficients of charged particles are well defined and the density gradient is relatively small. It is therefore possible to separate the spatial and velocity parts of the distribution function. This approximation is termed a "hydrodynamic description" or "density gradient expansion" [3]. We have already mentioned that the directed velocity of charged particles, and of electrons in particular, is much smaller than the random velocity. This is the rationale for applying the spherical harmonic expansion to simplify the theory in the velocity space. Both components of the distribution function may be dependent on time and may have a complex time dependence or relaxation to quasi-stationary conditions. In this section we describe the detailed equations arising from the Boltzmann equation. The first step is to apply both density gradient and spherical harmonics expansions to the velocity distribution function $g(\boldsymbol{v},\boldsymbol{r},t)$:

$$
\begin{aligned}
g(\boldsymbol{v},\boldsymbol{r},t) &= \sum_k \boldsymbol{g}^k(\boldsymbol{v},t) \otimes (\boldsymbol{\nabla}_r)^k n(\boldsymbol{r},t) && (5.62)\\
&= \sum_k \sum_{mn} g^k_{mn}(v,t) Y^e_{mn}(\theta,\varphi) \otimes (\boldsymbol{\nabla}_r)^k n(\boldsymbol{r},t), && (5.63)
\end{aligned}
$$

where tensor $\mathbf{g}^k(\boldsymbol{v},t)$ has components that are velocity distributions of different order in density gradient expansion, and $Y^e_{mn}(\theta,\varphi)$ are spherical harmonics in Morse–Feshbach representation [4].

5.6.1 Spherical Harmonics and Their Properties

By definition, spherical harmonics have the following form:

$$
Y^m_n(\theta,\varphi) = (-1)^m \left(\frac{2n+1}{4\pi}\frac{(n-m)!}{(n+m)!}\right)^{1/2} P^m_n(\cos\theta) e^{im\varphi}. \qquad (5.64)
$$

Their orthogonality is given by

$$
\int_\varphi \int_\theta Y^{m_1}_{n_1}(\theta,\varphi) Y^{m_2}_{n_2}(\theta,\varphi) d\Omega = \delta_{n_1,n_2}\delta_{m_1,m_2}. \qquad (5.65)
$$

The dependence on two angles (θ,ϕ) may be separated by using $Y^m_n(\theta,\varphi) = \Theta(\theta)\Psi_m(\varphi)$, where

$$
\Psi_m(\varphi) = \frac{1}{\sqrt{2\pi}} e^{im\varphi} \qquad (5.66)
$$

and

$$
\Theta(\theta) = (-1)^m \left(\frac{2n+1}{2}\frac{(n-m)!}{(n+m)!}\right)^{1/2} P^m_n(\cos\theta), \qquad -n \le m \le n. \quad (5.67)
$$

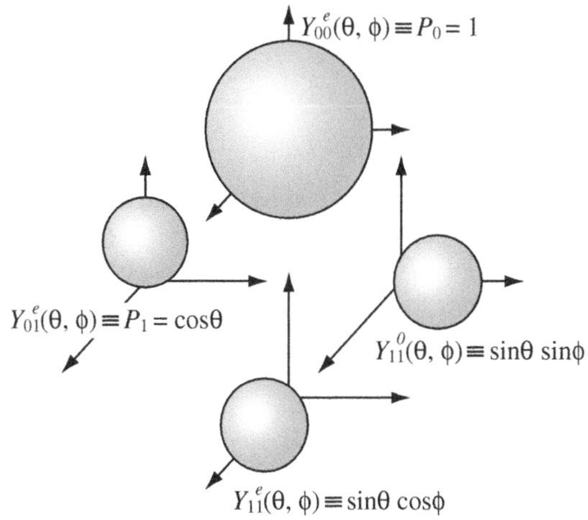

FIGURE 5.3 Four spherical harmonics of the lowest order in Morse–Feshbach notation.

Here, $P_n^m(\cos\theta)$ is the associated Legendre polynomial. The spherical harmonics is written in Morse–Feshbach representation as

$$Y^e_{mn}(\theta,\varphi) = P_n^m(\cos\theta)\cos m\varphi \tag{5.68}$$

and

$$Y^0_{mn}(\theta,\varphi) = P_n^m(\cos\theta)\sin m\varphi. \tag{5.69}$$

A few terms of Y^e_{mn} and Y^0_{mn} are shown in Figure 5.3 and thus the velocity distribution function is expanded as $g(\boldsymbol{v},t) = \sum g_{mn}(v,t)Y^e_{mn}(\theta,\varphi)$ (see Equation 5.63). Normalization of the distribution function requires the following integrals of spherical harmonics:

$$\int_0^{2\pi}\int_0^{\pi}\left[Y^e_{mn}(\theta,\varphi) \quad \text{or} \quad Y^0_{mn}(\theta,\varphi)\right]^2 d\Omega$$
$$= \begin{cases} \dfrac{4\pi}{2(2n+1)}\dfrac{(n+m)!}{(n-m)!} & , \quad n = 1,2,3,\ldots \\ 4\pi & , \quad n = 0. \end{cases} \tag{5.70}$$

Recurrent relations for these functions are

$$(2n+1)\cos\theta Y^e_{mn}(\theta,\varphi) = (n+m)Y^e_{m(n-1)}(\theta,\varphi)$$
$$+(n-m+1)Y^e_{m(n+1)}(\theta,\varphi) \tag{5.71}$$

and

$$
\begin{aligned}
&(2n+1)\cos^2\theta\frac{\partial}{\partial\cos\theta}Y^e_{mn}(\theta,\varphi)\\
&= (n+1)(n+m)Y^e_{m(n-1)}(\theta,\varphi) \quad - \quad (n-m+1)Y^e_{m(n+1)}(\theta,\varphi), \quad (5.72)
\end{aligned}
$$

and these are also useful. Because the spherical harmonics in Morse–Feshbach notation is connected to the associated Legendre polynomials by $Y^e_{mn} = P^m_n \cos m\varphi$ and $Y^e_{0n} = P_n$, P_n and P^m_n also satisfy the relations 5.71 and 5.72. The addition theorem for spherical harmonics is given by

$$
\begin{aligned}
P_n(\cos\omega) &= P_n(\cos\theta_1)P_n(\cos\theta_2)\\
&+2\sum_{m=1}^{n}\frac{(n-m)!}{(n+m)!}P^m_n(\cos\theta_1)P^m_n(\cos\theta_2)\cos m(\varphi_1-\varphi_2)\\
&= P_n(\cos\theta_1)P_n(\cos\theta_2)\\
&+2\sum_{m=1}^{n}\frac{(n-m)!}{(n+m)!}\left[Y^e_{mn}(\theta_1,\varphi_1)Y^e_{mn}(\theta_2,\varphi_2)+Y^o_{mn}(\theta_1,\varphi_1)Y^o_{mn}(\theta_2,\varphi_2)\right].
\end{aligned}
\quad (5.73)
$$

Some of the low-order Legendre and associated Legendre polynomials are given in Tables 5.1 and 5.2, respectively. The orthogonality property of the Legendre

TABLE 5.1 Legendre Polynomials

P_n	
$P_0(\cos\theta)$	1
$P_1(\cos\theta)$	$\cos\theta$
$P_2(\cos\theta)$	$(3\cos^2\theta-1)/2$
$P_3(\cos\theta)$	$(5\cos^3\theta-3\cos\theta)/2$
$P_4(\cos\theta)$	$(35\cos^4\theta-30\cos^2\theta+3)/8$

TABLE 5.2 Associated Legendre Polynomials

P^m_n	
$P^1_1(\cos\theta)$	$\sin\theta$
$P^1_2(\cos\theta)$	$3\cos\theta\sin\theta$
$P^2_2(\cos\theta)$	$3\sin^2\theta$
$P^1_3(\cos\theta)$	$3(5\cos^2\theta-1)\sin\theta/2$
$P^2_3(\cos\theta)$	$15\cos\theta\sin^2\theta$
$P^3_3(\cos\theta)$	$15\sin^3\theta$

function is given by

$$\int_0^{\pi} P_n(\cos\theta)P_{n'}(\cos\theta)\sin\theta d\theta = \begin{cases} \dfrac{2}{2n+1} \quad , & n = n' \\ 0 \quad , & n \neq n' \end{cases} . \tag{5.74}$$

If we expand a function $f(\cos\theta)$ in Legendre polynomials as

$$f(\cos\theta) = \sum_{n=0}^{\infty} a_n P_n(\cos\theta), \tag{5.75}$$

the expansion coefficients $a_n (n = 0, 1, 2, \ldots)$ are given by

$$a_n = \frac{2n+1}{2}\int_0^{\pi} f(\cos\theta)P_n(\cos\theta)\sin\theta d\theta, \tag{5.76}$$

and therefore we have

$$f(\cos\theta) = \sum_{n=0}^{\infty} \frac{2n+1}{2}\left(\int_0^{\pi} f(\cos\theta')P_n(\cos\theta')\sin\theta' d\theta'\right) P_n(\cos\theta). \tag{5.77}$$

If we have a function of both azimuthal and polar angles $f(\theta, \varphi)$, then the expansion is in terms of $Y_{mn}^e(\theta, \varphi)$:

$$f(\theta, \varphi) = \sum a_{mn} Y_{mn}^e(\theta, \varphi), \tag{5.78}$$

where

$$a_{mn} = \frac{2(2n+1)}{4\pi}\frac{(n-m)!}{(n+m)!}\int_0^{2\pi}\int_0^{\pi} f(\theta, \varphi)Y_{mn}^e(\theta, \varphi)d\Omega. \tag{5.79}$$

5.6.2 Velocity Distribution of Electrons

When we expand the velocity distribution function of electrons $g(\boldsymbol{v}, \boldsymbol{r}, t)$ in the Boltzmann Equation 5.8 by using the spatial density-gradient expansion 5.63, we obtain the chain of difference differential equations. The three lowest terms, $g^0(\boldsymbol{v}, t)$, $\boldsymbol{g}^1(\boldsymbol{v}, t)$, and $\mathbf{g}^2(\boldsymbol{v}, t)$, are defined by the equations

$$\frac{\partial}{\partial t} g^0(\boldsymbol{v}, t) + \boldsymbol{\alpha}(t) \cdot \frac{\partial}{\partial \boldsymbol{v}} g^0(\boldsymbol{v}, t) + R_0(t) g^0(\boldsymbol{v}, t) = J(g^0, F), \tag{5.80}$$

$$\begin{aligned} \frac{\partial}{\partial t}\boldsymbol{g}^1(\boldsymbol{v}, t) \quad &+ \quad \boldsymbol{\alpha}(t) \cdot \frac{\partial}{\partial \boldsymbol{v}}\boldsymbol{g}^1(\boldsymbol{v}, t) + R_0(t)\boldsymbol{g}^1(\boldsymbol{v}, t) \\ &= \quad J(g^1, F) + \boldsymbol{v} g^0(\boldsymbol{v}, t) - \boldsymbol{v}_d(t) g^0(\boldsymbol{v}, t), \end{aligned} \tag{5.81}$$

and

$$\begin{aligned}\frac{\partial}{\partial t}\mathbf{g}^2(\boldsymbol{v},t) &+ \boldsymbol{\alpha}(t)\cdot\frac{\partial}{\partial \boldsymbol{v}}\mathbf{g}^2(\boldsymbol{v},t) + R_0(t)\mathbf{g}^2(\boldsymbol{v},t) \\ &= J(g^2,F) + \boldsymbol{v}\boldsymbol{g}^1(\boldsymbol{v},t) - \boldsymbol{v}_d(t)\boldsymbol{g}^1(\boldsymbol{v},t) + \mathbf{D}(t)g^0(\boldsymbol{v},t), \quad (5.82)\end{aligned}$$

where R_0 is the effective production rate of electrons usually equal to $R_i - R_a$. $\boldsymbol{v}_d$ and $\mathbf{D}$ are the drift velocity and the diffusion tensor, respectively. Equation 5.80 describing $g^0(\boldsymbol{v},t)$ is independent of and uncoupled with other distributions, $\boldsymbol{g}^1$ and $\mathbf{g}^2$, and so on, and thus it enables us to solve the set of equations successfully as $g^0(\boldsymbol{v},t) \to \boldsymbol{g}^1(\boldsymbol{v},t) \to \mathbf{g}^2(\boldsymbol{v},t)$.

5.6.2.1 *Velocity Distribution under Uniform Number Density:* g^0

The lowest-order term in the density-gradient expansion of the velocity distribution function $g^0(\boldsymbol{v},t)$ is calculated from Equation 5.80. g^0 is axisymmetric with respect to the external field $\boldsymbol{E}$ or $\boldsymbol{\alpha} = e\boldsymbol{E}/m$, and usually it is further expanded in spherical harmonics for $m = 0$, that is, for Legendre polynomials $P_n(\theta)$, according to the equation

$$g^0(\boldsymbol{v},t) = \sum_{n=0} g^0_{0n}(v,t)Y^e_{0n}(\theta,\varphi) = \sum_n g^0_n(v,t)P_n(\theta). \quad (5.83)$$

Then, the second term in the right-hand side of Equation 5.80 is rearranged by the aid of the addition theorem of the Legendre function as

$$\begin{aligned}\boldsymbol{\alpha}(t)\cdot&\frac{\partial}{\partial \boldsymbol{v}}g^0(\boldsymbol{v},t) \\ &= \alpha_z(t)\left(\cos\theta\frac{\partial g^0}{\partial v} + \frac{\sin^2\theta}{v}\frac{\partial g^0}{\partial\cos\theta}\right) \\ &= \sum_n \alpha_z(t)\cos\theta P_n(\theta)\frac{\partial g^0_n(v)}{\partial v} + \sum_n \alpha_z(t)\frac{\sin^2\theta}{v}g^0_n(v)\frac{\partial P_n(\theta)}{\partial\cos\theta}.\end{aligned}$$

In the above equation, $\cos\theta P_n$ and $\sin^2\theta\partial P_n/\partial\cos\theta$ are replaced by using Equations 5.71 and 5.72, and the equation is simplified to

$$\begin{aligned}\boldsymbol{\alpha}(t)\cdot&\frac{\partial}{\partial \boldsymbol{v}}g^0(\boldsymbol{v},t) \\ &= \alpha_z(t)\sum_n\left(\frac{n+1}{2n+1}P_{n+1}(\theta) + \frac{n}{2n+1}P_{n-1}(\theta)\right)\frac{\partial g^0_n(v)}{\partial v} \\ &+ \alpha_z(t)\sum_n\frac{g^0_n(v)}{v}\left(-\frac{n(n+1)}{2n+1}P_{n+1}(\theta) + \frac{n(n+1)}{2n+1}P_{n-1}(\theta)\right).\end{aligned}$$

The above equation is further arranged after binding each term associated with $P_n(\theta)$ as

$$\begin{aligned}
\boldsymbol{\alpha}(t)\cdot\frac{\partial}{\partial \boldsymbol{v}}g^0(\boldsymbol{v},t)
&= \alpha_z(t)\sum_n\left(\frac{n}{2n-1}\frac{\partial g^0_{n-1}(v)}{\partial v}+\frac{n+1}{2n+3}\frac{\partial g^0_{n+1}(v)}{\partial v}\right.\\
&\quad\left.-\frac{(n-1)n}{2n-1}\frac{g^0_{n-1}(v)}{v}+\frac{(n+1)(n+2)}{2n+3}\frac{g^0_{n+1}(v)}{v}\right)P_n(\theta)\\
&= \sum_n\left\{\alpha_z(t)\frac{n}{2n-1}\left(\frac{\partial}{\partial v}-\frac{n-1}{v}\right)g^0_{n-1}(v,t)\right.\\
&\quad\left.+\,\alpha_z(t)\frac{n+1}{2n+3}\left(\frac{\partial}{\partial v}+\frac{n+2}{v}\right)g^0_{n+1}(v)\right\}P_n(\theta). \qquad (5.84)
\end{aligned}$$

Finally, we derive a hierarchy of equations in terms of spherical harmonics $g^0_n(\boldsymbol{v},t)(n=0,1,2,3,\ldots)$ for the zeroth order in the density-gradient expansion of the velocity distribution function of electrons $g^0(\boldsymbol{v},t)$:

$$\begin{aligned}
\frac{\partial}{\partial t}g^0_n(v,t) &+ \alpha_z(t)\frac{n}{2n-1}\left(\frac{\partial}{\partial v}-\frac{n-1}{v}\right)g^0_{n-1}(v,t)\\
&+ \alpha_z(t)\frac{n+1}{2n+3}\left(\frac{\partial}{\partial v}+\frac{n+2}{v}\right)g^0_{n+1}(v,t)\\
&+ R_0(t)g^0_n(v,t)-J(g^0,F)=0. \qquad (5.85)
\end{aligned}$$

In Figure 5.4 we show one example of the solution for $g^0(v)$ in Ar at 100 Td. Cylindrical symmetry is assumed and the exact solution is plotted in velocity space (v_z, v_r) where $v_r=\sqrt{v_x^2+v_y^2}$. We also show the relative magnitude of various components in the Legendre polynomial expansion, $g^0_n(v)$ $(n=0,1,2,\ldots)$.

5.6.2.2 *Velocity Distribution Proportional to* $\nabla_r n(r,t)$: g^1

First, we divide the vector function $\boldsymbol{g}^1(\boldsymbol{v},t)$ into spatial components, g^1_x, g^1_y, and g^1_z:

$$\boldsymbol{g}^1(\boldsymbol{v},t)=g^1_x(\boldsymbol{v},t)\boldsymbol{i}+g^1_y(\boldsymbol{v},t)\boldsymbol{j}+g^1_z(\boldsymbol{v},t)\boldsymbol{k}. \qquad (5.86)$$

When the external force is along the z-axis, we have $\boldsymbol{\alpha}(t)=-\alpha_z(t)\boldsymbol{k}$. From Equation 5.81 we obtain

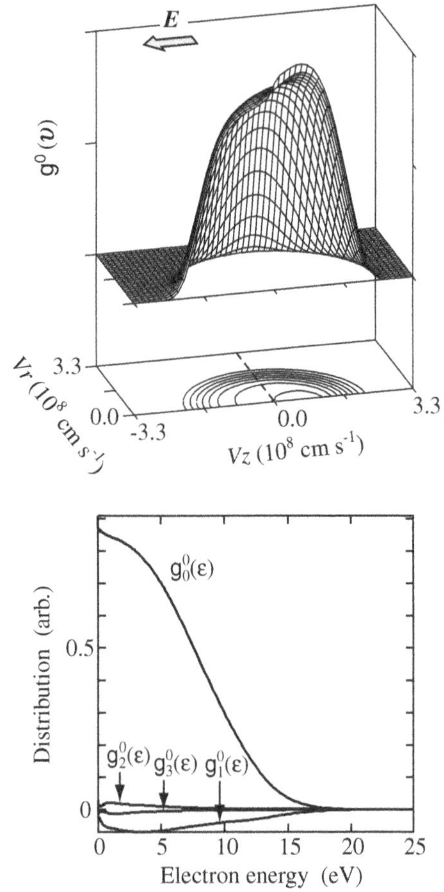

FIGURE 5.4 Lowest-order solution for the velocity distribution function for electrons at 100 Td in Ar: (a) $g^0(\boldsymbol{v})$ and (b) $g_n^0(v)$ $(n = 0, 1, 2, \ldots)$.

$$\begin{vmatrix} \dfrac{\partial}{\partial t} g_x^1(\boldsymbol{v},t) \\ \\ \dfrac{\partial}{\partial t} g_y^1(\boldsymbol{v},t) \\ \\ \dfrac{\partial}{\partial t} g_z^1(\boldsymbol{v},t) \end{vmatrix} + \begin{vmatrix} \alpha_z(t)\dfrac{\partial}{\partial v_z} g_x^1(\boldsymbol{v},t) \\ \\ \alpha_z(t)\dfrac{\partial}{\partial v_z} g_y^1(\boldsymbol{v},t) \\ \\ \alpha_z(t)\dfrac{\partial}{\partial v_z} g_z^1(\boldsymbol{v},t) \end{vmatrix} + R_0 \begin{vmatrix} g_x^1 \\ \\ g_y^1 \\ \\ g_z^1 \end{vmatrix} + \begin{vmatrix} v_x g^0 \\ \\ v_y g^0 \\ \\ (v_z - v_d) g^0 \end{vmatrix} = \begin{vmatrix} J(g_x^1, F) \\ \\ J(g_y^1, F) \\ \\ J(g_z^1, F) \end{vmatrix}. \tag{5.87}$$

Here, the component along the external field (i.e., the z-axis), $g_z^1(\boldsymbol{v},t)$, has a directional component of velocity v_d, and the other two components v_x and v_y

are symmetric to each other. The longitudinal component $g_z^1(\boldsymbol{v},t)$ is symmetric with respect to the external field direction (i.e., the z-axis) and is expanded by spherical harmonics for $m=0$:

$$g_z^1(\boldsymbol{v},t) = \sum_n g_{z0n}^1(v,t)Y_{0n}^e(\theta,\varphi) = \sum_n g_{z_n}^1(v,t)P_n(\theta). \tag{5.88}$$

The term $(v_z - v_d)g^0(\boldsymbol{v},t)$ in Equation 5.87 is expanded by using Equation 5.71:

$$\begin{aligned}
&(v_z - v_d)g^0(v,t) \\
&= \sum_n (v\cos\theta - v_d)g_n^0 P_n(\theta) \\
&= \sum_n \left(v\frac{n}{2n+1}P_{n-1}(\theta)g_n^0 + v\frac{n+1}{2n+1}P_{n+1}(\theta)g_n^0 - v_d g_n^0 P_n(\theta)\right) \\
&= \sum_n \left(\frac{n+1}{2n+3}vg_{n+1}^0(v,t) + \frac{n}{2n-1}vg_{n-1}^0(v,t) - v_d g_n^0(v,t)\right)P_n(\theta),
\end{aligned} \tag{5.89}$$

where we use recurrent relations for $P_n(\theta)$.

The governing equations for $g_{z_n}^1(\boldsymbol{v},t)$ are obtained by considering Equation 5.89 and the master Equation 5.85 for g^0:

$$\begin{aligned}
&\frac{\partial}{\partial t}g_{z_n}^1(v,t) + \alpha_z(t)\frac{n}{2n-1}\left(\frac{\partial}{\partial v} - \frac{n-1}{v}\right)g_{z_{n-1}}^1(v,t) \\
&\quad + \alpha_z(t)\frac{n+1}{2n+3}\left(\frac{\partial}{\partial v} + \frac{n+2}{v}\right)g_{z_{n+1}}^1(v,t) + R_0(t)g_{z_n}^1(v,t) - J(g_z^1,F) \\
&\quad = -\frac{n+1}{2n+3}vg_{n+1}^0(v,t) - \frac{n}{2n-1}vg_{n-1}^0(v,t) + v_d g_n^0(v,t).
\end{aligned} \tag{5.90}$$

The hierarchy of the first-order functions $g_{z_n}^1(v,t)(n=0,1,2,\ldots)$ is derived numerically by using the known functions $g_n^0(v,t)(n=0,1,2,\ldots)$ under uniform density from Equation 5.85. One example of g_z^1 is shown in Figure 5.5.

The perpendicular components $g_x^1(\boldsymbol{v},t)$ and $g_y^1(\boldsymbol{v},t)$ are determined by equations in the first and second row in Equation 5.87. The appropriate expansion for $g_x^1(\boldsymbol{v},t)$ is the spherical harmonics for $m=1$ in Equation 5.68,

$$g_x^1(\boldsymbol{v},t) = \sum_n g_{x1n}^1(v,t)Y_{1n}^e(\theta,\varphi) = \sum_n g_{x_n}^1(v,t)P_n^1(\theta)\cos\varphi. \tag{5.91}$$

The term $\alpha_z(t)\partial g_x^1(\boldsymbol{v},t)/\partial v_z$ in Equation 5.87 is expanded as

$$\begin{aligned}
\alpha_z(t)\frac{\partial}{\partial v_z}g_x^1(\boldsymbol{v},t) &= \alpha_z(t)\left(\cos\theta\frac{\partial g_x^1}{\partial v} + \frac{\sin^2\theta}{v}\frac{\partial g_x^1}{\partial\cos\theta} + \frac{\partial\varphi}{\partial v_z}\frac{\partial g_x^1}{\partial\varphi}\right) \\
&= \sum_n \alpha_z(t)\cos\theta P_n^1(\theta)\cos\varphi\frac{\partial g_{x_n}^1}{\partial v} + \sum_n \alpha_z(t)\frac{g_{x_n}^1}{v}\cos\varphi\sin^2\theta\frac{\partial P_n^1(\theta)}{\partial\cos\theta}.
\end{aligned}$$

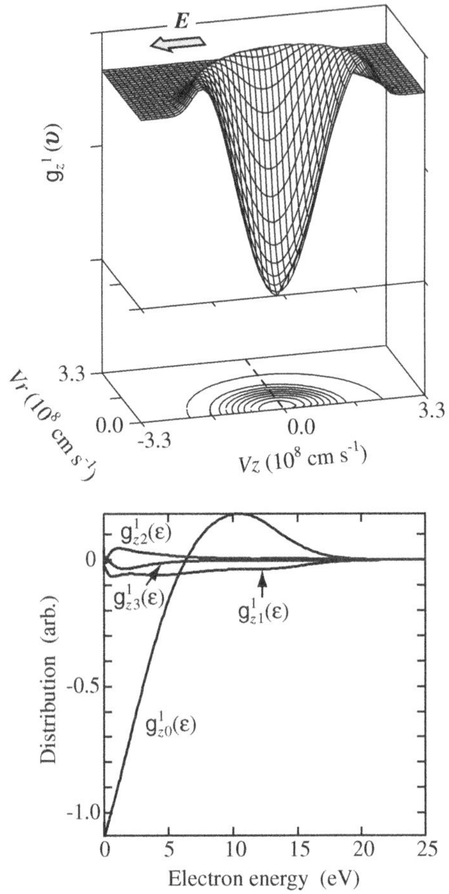

FIGURE 5.5 The longitudinal component (z) of the first-order solution for the velocity distribution function for electrons at 100 Td in Ar: (a) $g_z^1(\boldsymbol{v})$ and (b) $g_{zn}^1(\varepsilon)$ ($n = 0, 1, 2, \ldots$).

When we use Equations 5.71 and 5.72 to replace $\cos\theta P_n^1(\theta)$ and $\sin^2\theta \partial P_n^1(\theta)/\partial\cos\theta$, we obtain

$$
\begin{aligned}
&= \alpha_z(t)\cos\varphi \sum_{n=1} \left(\frac{n}{2n+1} P_{n+1}^1(\theta) + \frac{n+1}{2n+1} P_{n-1}^1(\theta) \right) \frac{\partial g_{x_n}^1(v,t)}{\partial v} \\
&\quad + \alpha_z(t)\cos\varphi \sum_{n=1} \left(-\frac{n^2}{2n+1} P_{n+1}^1(\theta) + \frac{(n+1)^2}{2n+1} P_{n-1}^1(\theta) \right) \frac{g_{x_n}^1(v,t)}{v},
\end{aligned}
$$

and after we order the right-hand side of the above equation in terms of $P_n^1(\theta)$, we have

$$
\begin{aligned}
&= \alpha_z(t)\sum_{n=1}\left(\frac{n-1}{2n-1}\frac{\partial g^1_{x_{n-1}}(v,t)}{\partial v}+\frac{n+2}{2n+3}\frac{\partial g^1_{x_{n+1}}(v,t)}{\partial v}\right.\\
&\quad\left.-\frac{(n-1)^2}{2n-1}\frac{g^1_{x_{n-1}}(v,t)}{v}+\frac{(n+2)^2}{2n+3}\frac{g^1_{x_{n+1}}(v,t)}{v}\right)P^1_n(\theta)\cos\varphi\\
&= \sum_{n=1}\left\{\alpha_z(t)\frac{n-1}{2n-1}\left(\frac{\partial}{\partial v}-\frac{n-1}{v}\right)g^1_{x_{n-1}}(v,t)\right.\\
&\quad\left.+\alpha_z(t)\frac{n+2}{2n+3}\left(\frac{\partial}{\partial v}+\frac{n+2}{v}\right)g^1_{x_{n+1}}(v,t)\right\}P^1_n(\theta)\cos\varphi. \qquad (5.92)
\end{aligned}
$$

The term $v_x g^0(\boldsymbol{v},t)$ is expanded by the aid of the recurrence formula in Legendre polynomials:

$$
\begin{aligned}
v_x g^0(v,t) &= \sum_{n=0} v\sin\theta\cos\varphi P_n(\theta)g^0_n(v,t)\\
&= \sum_{n=0} v g^0_n(v,t)\cos\varphi\left(\frac{1}{2n+1}P^1_{n+1}(\theta)-\frac{1}{2n+1}P^1_{n-1}(\theta)\right)\\
&= \sum\left(v\frac{1}{2n-1}g^0_{n-1}(v,t)-v\frac{1}{2n+3}g^0_{n+1}(v,t)\right)P(\theta)\cos\varphi. \qquad (5.93)
\end{aligned}
$$

Finally, when we substitute all the terms in Equations 5.92 and 5.93 into the first row of Equation 5.87, we have the governing equation of $g^1_{x_n}(v,t)(n = 1,2,\ldots)$:

$$
\begin{aligned}
&\frac{\partial}{\partial t}g^1_{x_n}(v,t)+\alpha_z(t)\frac{n-1}{2n-1}\left(\frac{\partial}{\partial v}-\frac{n-1}{v}\right)g^1_{x_{n-1}}(v,t)\\
&+\alpha_z(t)\frac{n+2}{2n+3}\left(\frac{\partial}{\partial v}+\frac{n+2}{v}\right)g^1_{x_{n+1}}(v,t)+R_0(t)g^1_{x_n}(v,t)-J(g^1_x,F)\\
&= -\frac{1}{2n-1}vg^0_{n-1}(v,t)+\frac{1}{2n+3}vg^0_{n+1}(v,t). \qquad (5.94)
\end{aligned}
$$

We numerically obtain $g^1_{x_n}(v,t)(n = 1,2,\ldots)$ by using the known function $g^0_n(v,t)(n = 0,1,2,\ldots)$ derived from Equation 5.85. One example of g^1_x is shown in Figure 5.6.

Because of the symmetry of the problem when no magnetic field is present, we need solve for only one transverse term, g^1_x. Thus, by using Equation 5.94 we have three series of equations that must be solved consecutively. First we obtain the solution for g^0 from the governing Equation 5.85, then the longitudinal component g^1_z from Equation 5.90, and finally the transverse component g^1_x from Equation 5.94. Each of these governing equations actually represents a hierarchy of equations in the expansion of spherical harmonics.

In Table 5.3 we show expanded collision integrals for different terms in the gradient and spherical harmonic expansions. The expansion in terms of

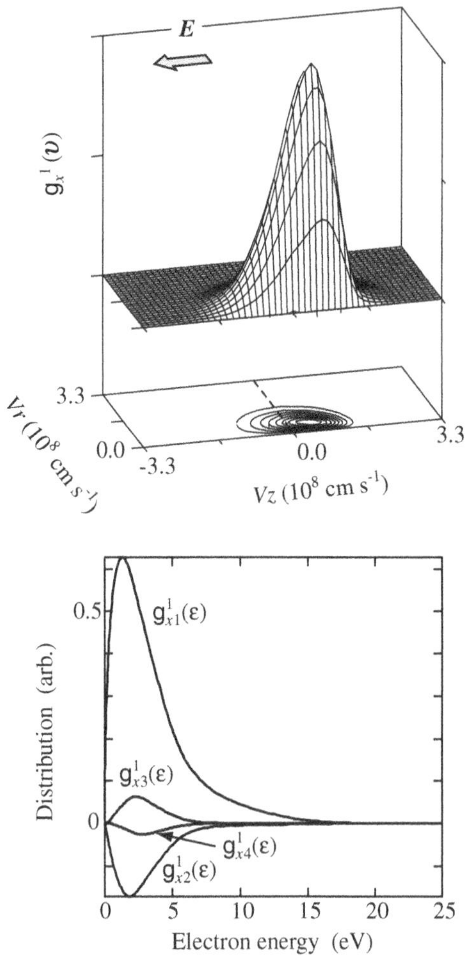

FIGURE 5.6 The transverse component (x) of the first-order solution for the velocity distribution function for electrons at 100 Td in Ar: (a) $g_x^1(\boldsymbol{v})$ and (b) $g_{xn}^1(v)$ ($n = 1, 2, \ldots$).

number density gradients may be extended to higher terms. In addition, the expansion in terms of spherical harmonics may be extended to an order sufficient to capture the angular distribution in velocity space accurately. The lowest-order solution would be to use only the lowest-order term in gradient expansion and the first two terms ($n = 0, 1$) in spherical harmonic expansion. This is known as a two-term expansion and is discussed later. The solution of equations of different order may proceed by expansion in terms of a particular velocity distribution, which results in additional requirements in the expansion and convergence of the solution, or by applying finite difference or

TABLE 5.3 Each of the Collision Terms Appearing in $J(g, F)$

Collision Type	Collision Integral	Expanded Collision Term
Elastic	$J_m(g_0^0)$	$\frac{m}{M}\frac{1}{v^2}\frac{\partial}{\partial v}\left\{NQ_m(v)v^4\left(g_0^0+\frac{kT_g}{mv}\frac{\partial g_0^0}{\partial v}\right)\right\}$
		$=\frac{m}{4\pi}\frac{2m}{M}\left[\left(\varepsilon-\frac{1}{2}kT_g\right)\frac{\partial}{\partial\varepsilon}NQ_m(\varepsilon)+\left(\frac{3}{2}-\frac{kT_g}{4\varepsilon}\right)NQ_m\right]f_0^0$
		$+\frac{m}{4\pi}\frac{2m}{M}\left[(\varepsilon+kT_g)NQ_m(\varepsilon)+kT_g\varepsilon\frac{\partial}{\partial\varepsilon}NQ_m(\varepsilon)\right]\frac{\partial}{\partial\varepsilon}f_0^0$
		$+\frac{m}{4\pi}\frac{2m}{M}kT_g\varepsilon NQ_m(\varepsilon)\frac{\partial}{\partial\varepsilon}f_0^0$
	$J_m(g_1^0)$	$-NQ_m(v)vg_1^0+\frac{m}{M}\frac{1}{v^2}\frac{\partial}{\partial v}\left\{(NQ_v(v)-NQ_m(v))\,v^4g_1^0\right\}$
		$=\frac{m}{4\pi}NQ_m(\varepsilon)f_1^0$
		$+\frac{m}{4\pi}\frac{2m}{M}\left[\frac{3}{2}\left(NQ_v(\varepsilon)-NQ_m(\varepsilon)\right)\right.$
		$\left.+\varepsilon\frac{\partial}{\partial\varepsilon}\left[NQ_v(\varepsilon)-NQ_m(\varepsilon)\right]\right]f_1^0$
		$+\frac{2m}{M}\varepsilon\left(NQ_v(\varepsilon)-NQ_m(\varepsilon)\right)\frac{\partial}{\partial\varepsilon}f_1^0$
	$J_m(g_2^0)$	$-\frac{3}{2}NQ_v(v)vg_2^0=\frac{m}{4\pi}\left(-\frac{3}{2}\right)NQ_v(\varepsilon)$
	$J_m(g_{x11}^0)$	$-NQ_m(v)vg_{x11}^1=-\frac{m}{4\pi}NQ_m(\varepsilon)f_{x11}^1$
	$J_m(g_{x12}^0)$	$-\frac{3}{2}NQ_v(v)vg_{x12}^1=\frac{m}{4\pi}\left(-\frac{3}{2}\right)NQ_v(\varepsilon)f_{x12}^1$
Excitation	$J_j(g_0^0)$	$\frac{1}{v}\left\{v'^2NQ_j(v')g_0^0(v')-v^2NQ_j(v)g_0^0(v)\right\}$
		$=\frac{m}{4\pi}\frac{1}{\sqrt{\varepsilon}}\frac{\partial}{\partial\varepsilon}\int_{\varepsilon}^{\varepsilon+\varepsilon_j}\sqrt{\varepsilon}NQ_j(\varepsilon)f_0^0(\varepsilon)d\varepsilon$
	$J_j(g_1^0 or g_2^0)$	$-NQ_j(v)vg_1^0\left(or g_2^0\right)=-\frac{m}{4\pi}NQ_j(\varepsilon)f_1^0\left(or f_2^0\right)$
	$J_j(g_{x11}^0 or g_{x12}^0)$	$-NQ_j(v)vg_{x11}^1\left(or g_{x12}^1\right)=-\frac{m}{4\pi}NQ_j(\varepsilon)f_{x11}^1\left(or f_{x12}^1\right)$
Ionization	$J_i(g_0^0)$	$\frac{1}{v}\left\{\frac{1+k}{k}v'^2NQ_i(v')g_0^0+(1+k)v'^2NQ_i(v')g_0^0-v^2NQ_i(v)g_0^0\right\}$
		$=\frac{m}{4\pi}\frac{1}{\sqrt{\varepsilon}}\left(\frac{\partial}{\partial\varepsilon}\int_{\varepsilon}^{(1+k)\varepsilon+\varepsilon_i}\sqrt{\varepsilon}NQ_i(\varepsilon)f_0^0(\varepsilon)d\varepsilon\right.$
		$\left.+\frac{\partial}{\partial\varepsilon}\int_0^{\frac{1+k}{k}\varepsilon+\varepsilon_i}\sqrt{\varepsilon}NQ_i(\varepsilon)f_0^0(\varepsilon)d\varepsilon\right)$
	$J_i(g_1^0 or g_2^0)$	$-NQ_i(v)vg_1^0\left(or g_2^0\right)=-\frac{m}{4\pi}NQ_i(\varepsilon)f_1^0\left(or f_2^0\right)$
	$J_i(g_{x11}^0 or g_{x12}^0)$	$-NQ_i(v)vg_{x11}^1\left(or g_{x12}^1\right)=-\frac{m}{4\pi}NQ_i(\varepsilon)f_{x11}^1\left(or f_{x12}^1\right)$
Attachment	$J_a(g_0^0)$	$-NQ_a(v)vg_0^0=-\frac{m}{4\pi}NQ_a(\varepsilon)f_0^0$
	$J_a(g_1^0 or g_2^0)$	$-NQ_a(v)vg_1^0\left(or g_2^0\right)=-\frac{m}{4\pi}NQ_a(\varepsilon)f_1^0\left(or f_2^0\right)$
	$J_a(g_{x11}^0 or g_{x12}^0)$	$-NQ_a(v)vg_{x11}^1\left(or g_{x12}^1\right)=-\frac{m}{4\pi}NQ_a(\varepsilon)f_{x11}^1\left(or f_{x12}^1\right)$

similar numerical techniques. Such techniques may be applied at any level of expansion even before the density gradient or spherical harmonic expansions. However, if no expansions are made the numerical solution will need to be made in seven dimensions in $(\boldsymbol{v}, \boldsymbol{r}, t)$, which under cylindrical symmetry may reduce to five. Even at a very poor resolution of 100 points per dimension, the number of mesh points is formidable, though it may be within the range of modern computers. Nevertheless, expansions make the problem more complex but numerically much easier to handle.

5.6.3 Electron Transport Parameters

In Section 5.3, we defined the general form of transport parameters and their definitions. In this section, we evaluate the expressions for transport coefficients in terms of the velocity distribution function $g(\boldsymbol{v}, \boldsymbol{r}, t)$. First we write down the expanded form of the distribution function in configuration and velocity space:

$$
\begin{aligned}
g(\boldsymbol{v}, \boldsymbol{r}, t) &= \sum_k \mathbf{g}^k(\boldsymbol{v}, t) \otimes (\boldsymbol{\nabla}_r)^k\, n(\boldsymbol{r}, t) \\
&= \sum_k \sum_{mn} g^k_{mn}(v, t) Y^e_{mn}(\theta, \varphi) \otimes (\boldsymbol{\nabla}_r)^k\, n(\boldsymbol{r}, t) \\
&= \left(\sum_{n=0} g^0_n(v, t) P_n(\theta) \right) n(\boldsymbol{r}, t) \\
&\quad + \left(\sum_{n=1} g^1_{x_{1n}}(v, t) Y^e_{1n}(\theta, \varphi)\boldsymbol{i} + \sum_{n=1} g^1_{y_{1n}}(v, t) Y^e_{1n}(\theta, \varphi)\boldsymbol{j} \right. \\
&\quad \left. + \sum_{n=1} g^1_{z_n}(v, t) P_n(\theta)\boldsymbol{k} \right) \cdot \frac{\partial}{\partial \boldsymbol{r}} n(\boldsymbol{r}, t) + O\left(\nabla_r^2 n\right) \\
&= \left(g^0_0(v, t) + g^0_1(v, t)\cos\theta + g^0_2(v, t)\frac{3\cos^2\theta - 1}{2} + \cdots \right) n(\boldsymbol{r}, t) \\
&\quad + \left(g^1_{x_1}(v, t)\sin\theta\cos\varphi + g^1_{x_2}(v, t) 3\cos\theta\sin\theta\cos\varphi + \cdots \right) \boldsymbol{i} \cdot \frac{\partial}{\partial \boldsymbol{r}} n(\boldsymbol{r}, t) \\
&\quad + \left(g^1_{y_1}(v, t)\sin\theta\cos\varphi + g^1_{y_2}(v, t) 3\cos\theta\sin\theta\cos\varphi + \cdots \right) \boldsymbol{j} \cdot \frac{\partial}{\partial \boldsymbol{r}} n(\boldsymbol{r}, t) \\
&\quad + \left(g^1_{z_0}(v, t) + g^1_{z_1}(v, t)\cos\theta + g^1_{z_2}(v, t)\frac{3\cos^2\theta - 1}{2} + \cdots \right) \boldsymbol{k} \cdot \frac{\partial}{\partial \boldsymbol{r}} n(\boldsymbol{r}, t) \\
&\quad + O\left(\nabla_r^2 n\right). \qquad (5.95)
\end{aligned}
$$

Exercise 5.6.1

Discuss and develop equations for two-term approximation in the case of electron transport in gases. Discuss the order of errors and applicability.

Two-term approximation consists of taking the first two terms in the spherical harmonic expansion in the lowest order of gradient expansion ($k = 0$) in a steady state. From Equation 5.95 we directly obtain

$$g(\boldsymbol{v}) = \left[g^0_0(v) + g^0_1(v)\cos\theta + O\left(g^0_2\right) \right] n_e,$$

where the order of error in spherical expansion is given by $O(g_2^0)$, and higher-order terms in the gradient expansion are neglected. In order for this solution to be accurate (i.e., in order that the spherical harmonic expansion converges), we need to satisfy the condition $g_0^0(v,t) \gg g_1^0(v,t)$. The corresponding equations of the two-term expansion are given by

$$-\frac{m}{M}\frac{1}{v^2}\frac{\partial}{\partial v}\left(NQ_m(v)v^4 g_0^0\right) + R_0 g_0^0(v) + \frac{1}{3}\frac{eE}{m}\left(\frac{\partial}{\partial v} + \frac{2}{v}\right) g_1^0(v) = 0, \quad (5.96)$$

where $g_1^0(v)$ is given as

$$g_1^0(v) = -\frac{1}{\sum NQ_k(v)v}\frac{eE}{m}\frac{\partial g_0^0(v)}{\partial v}. \quad (5.97)$$

Finally, it is noticed that the velocity distribution of electrons expressed by the two-term theory, Equations 5.96 and 5.97, is nearly isotropic. On the other hand, an anisotropic velocity distribution of electrons is realized under a higher E/N or high degree of inelastic collsions that strengthen the directional component of electrons to the external field $\boldsymbol{E}$. Thus we conclude that the two-term theory is of acceptable accuracy for situations in which the cross section for inelastic processes is much smaller than that for elastic processes. It has often been assumed that the theory is exact for atomic gases at low E/N when only elastic collisions occur.

First we must normalize the distribution function $\int g(\boldsymbol{v},\boldsymbol{r},t)d\boldsymbol{v} = 1$. Bearing this normalized function in mind, and on the basis of the definition of the directed or drift velocity in Equation 5.18, we derive the equation for the drift velocity $\boldsymbol{v}_d(t)$ as

$$\begin{aligned} \boldsymbol{v}_d(t) &= \int \boldsymbol{v} g^0(\boldsymbol{v},t)d\boldsymbol{v} \\ &= \sum_n \int_0^\pi v\cos\theta g_n^0(v,t)P_n(\theta)v^2 dv d\Omega \\ &= \frac{4\pi}{3}\int v^3 g_1^0(v,t)dv. \end{aligned} \quad (5.98)$$

We can derive the expression for the transverse diffusion coefficient $D_T(t)$ and for the longitudinal diffusion coefficient $D_L(t)$ from Equations 5.21 and 5.22 as

$$\begin{aligned} D_T(t) &= \int v_x g_x^1(\boldsymbol{v},t)d\boldsymbol{v} = \int v_y g_y^1(\boldsymbol{v},t)d\boldsymbol{v} \\ &= \sum_n \int v\sin\theta\cos\varphi g_{x_{1n}}^1(v,t)Y_{1n}^e(\theta,\varphi)v^2 dv d\Omega \\ &= \frac{4\pi}{3}\int v^3 g_{x_1}^1(v,t)dv, \end{aligned} \quad (5.99)$$

and

$$\begin{aligned} D_L(t) &= \int v_z g_z^1(\boldsymbol{v}, t) d\boldsymbol{v} \\ &= \sum_n \int v \cos\theta g_{z_n}^1(v, t) P_n(\theta) v^2 dv d\Omega \\ &= \frac{4\pi}{3} \int v^3 g_{z_1}^1(v, t) dv. \end{aligned} \tag{5.100}$$

The effective production rate of electrons is described by combining the ionization and electron attachment rates, R_i and R_a:

$$\begin{aligned} R_0(t) &= R_i(t) - R_a(t) \\ &= N \int [Q_i(v) - Q_a(v)] v g(\boldsymbol{v}, t) d\boldsymbol{v} \\ &= 4\pi N \int [Q_i(v) - Q_a(v)] v^3 g_0^0(v, t) dv. \end{aligned} \tag{5.101}$$

The rates of excitation of electronically excited levels are also defined in a similar way as the ionization or attachment rates: $R_j(t) = 4\pi N \int Q_j(v) v^3 g_0^0(v, t) dv$. The ensemble average of the electron energy $\langle \varepsilon(t) \rangle$ as defined by Equation 5.14 can be described as

$$\begin{aligned} \langle \varepsilon(\boldsymbol{r}, t) \rangle & \int \frac{1}{2} m v^2 g(\boldsymbol{v}, \boldsymbol{r}, t) d\boldsymbol{v} \\ &= \sum_n \int \frac{1}{2} m v^2 g_n^0(v, t) P_n(\theta) v^2 dv d\Omega \\ & \quad - \nabla_z n \boldsymbol{k} \sum_n \int \frac{1}{2} m v^2 g_{z_n}^1(v, t) P_n(\theta) v^2 dv d\Omega \\ & \quad - \nabla_x n \boldsymbol{i} \sum_n \int \frac{1}{2} m v^2 g_{x_{1n}}^1(v, t) Y_{1n}^e(\theta, \varphi) v^2 dv d\Omega \\ &= 2\pi m \int v^4 g_0^0(v, t) dv - \left(2\pi m \int v^4 g_{z_0}^1(v, t) dv \right) \nabla_z n \boldsymbol{k}. \end{aligned} \tag{5.102}$$

Problem 5.6.1

Derive the following expression of the diffusion tensor of electrons in gases without inelastic collisions in a direct-current field:

$$\mathbf{D} = \int \frac{\boldsymbol{v}(\boldsymbol{v} - \boldsymbol{v}_d) g^0(\boldsymbol{v})}{N Q_m(v) v (1 + \Delta(v))} d\boldsymbol{v} + \int \frac{\boldsymbol{v} \frac{e\boldsymbol{E}}{m} \cdot \frac{d}{d\boldsymbol{v}} \boldsymbol{g}^1(\boldsymbol{v})}{N Q_m(v) v (1 + \Delta(v))} d\boldsymbol{v}, \tag{5.103}$$

where $\Delta(v)$ is a function of M and T_g of gas molecules,

$$\Delta(v) = \frac{kT_g}{Mv} \frac{d}{dv} \ln[N Q_m(v) v].$$

References

[1] Chapman, S. and Cowling, T.G. 1970. *The Mathematical Theory of Non-Uniform Gases*, 3rd ed., Cambridge: Cambridge University Press.

[2] Shkarofsky, I.P., Johnston, T.W., and Bachynski, M.P. 1966. *The Particle Kinetics of Plasmas.* Reading, MA: Addison-Wesley.

[3] Kumar, K., Skullerud, H.R., and Robson, R.E. 1980. Kinetic theory of charged particles swarms in neutral gases. *Aust. J. Phys.* 33:343–448.

[4] Morse, P.M. and Feshbach, H. 1953. *Methods of Theoretical Physics.* New York: McGraw-Hill.

CHAPTER 6

General Properties of Charged Particle Transport in Gases

6.1 INTRODUCTION

Swarm parameters are generally applied to plasma modeling and simulations. At the same time, the nonequilibrium regime in discharges is well represented under a broad range of conditions by using the Boltzmann equation, as discussed in the previous chapter. In this chapter we discuss the basic features of transport coefficients (swarm parameters) of electrons and also, to a lesser degree, of ions. Together with transport data we compile cross sections of the electron for interesting gases. First we discuss the basic characteristics of the E/N dependence of the coefficient for swarms in the hydrodynamic regime. We consider swarms in direct current (DC) fields and also in time-varying radio frequency (rf) fields. Collective phenomena that cannot be easily described on the basis of single electron dynamics are labeled kinetic phenomena, and some special examples are described. Differences between the transport in conservative and nonconservative situations (i.e., when there are number-changing collisions such as ionization and attachment) are briefly considered.

6.2 ELECTRON TRANSPORT IN DC-ELECTRIC FIELDS

Swarm experiments were initially developed to study the motion of charged particles in gases. These experiments were made at relatively high pressures to ensure that nonhydrodynamic effects were negligible. The discussion of experimental techniques developed to measure transport coefficients is beyond the scope of this book [1]. At present we have an opportunity of highly accurate transport data available for low-temperature plasma modeling. Data on the swarm parameters for electrons up to the early 1980s have been compiled by Dutton et al. [2].

6.2.1 Electron Drift Velocity

Drift velocities are usually measured in two types of experiments [1,3]. Experiments with shutter grids are of high accuracy but are limited to low E/N. The pulsed Townsend experiments are made between two parallel electrodes with a swarm of electrons produced by a photo-emission. Their accuracy is somewhat smaller, but it may still be possible to achieve an uncertainty of less than ±3%. The measurements may be extended to higher E/N. There is a class of the drift velocity experiment measuring the optical emission from short-lived excited species in the electron swarm that can be performed only at higher E/N.

In Figure 6.1 we show the typical characteristics of the drift velocity of electrons in gases as a function of DC-E/N. At sufficiently low E/N, where an electron loses all the kinetic energy equal to the gain from the field E at one elastic collision, the drift velocity is proportional to E/N (region I). In region I the mean energy of electrons is close to the thermal energy $3kT_g/2$ and the velocity distribution is approximated by the Maxwellian. A finite kinetic energy of electrons is steadily maintained when the energy gain from the field is balanced with the energy loss by elastic collisions during a unit time. Then, region II appears in Figure 6.1. Even at low E/N, polyatomic molecules have a probability of vibrational (and rotational) excitations with the electron as well as the elastic scattering. This means that there is also a region III that

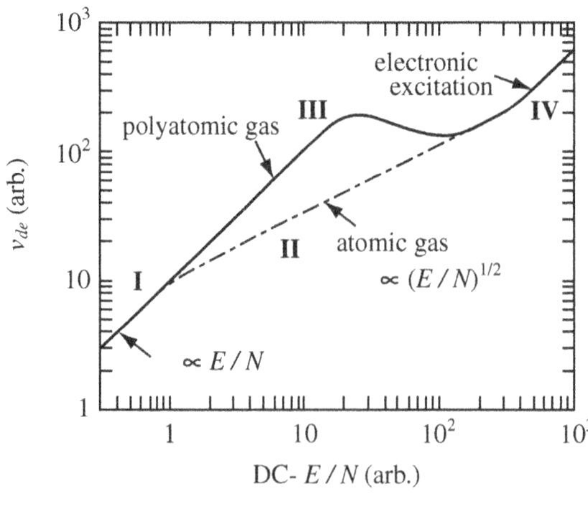

FIGURE 6.1 Typical characteristics of electron drift velocity as a function of DC-E/N.

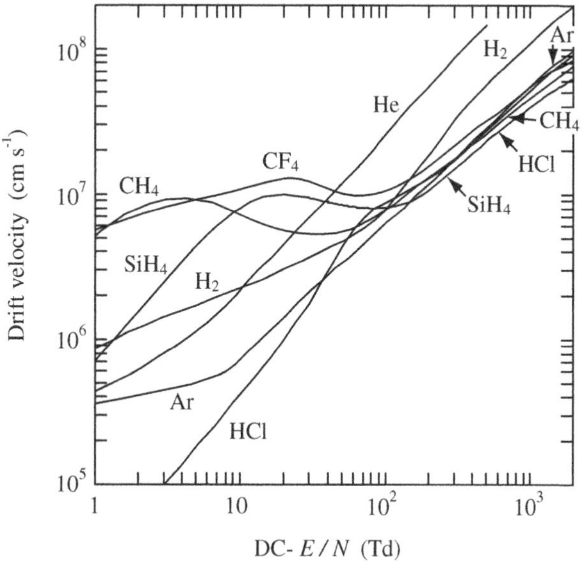

FIGURE 6.2 Electron drift velocities in various gases as a function of DC-E/N.

mixes II with I as shown in Figure 6.1. The magnitude of the drift velocity in the mixed region strongly depends on the magnitude of the vibrational cross section. A rapid increase in the drift velocity at high E/N (region IV) is caused by electronic excitations (including ionization) with a high energy loss from several eV to 20 eV [1].

The E/N dependence of the drift velocity of electrons in He, Ar, H_2, HCl, CH_4, CF_4, and SiH_4 is shown in Figure 6.2. As qualitatively shown in Figure 6.1, CF_4, SiH_4 and CH_4 have more distinct characteristics than He and Ar in the region of moderate E/N. That is, the drift velocity has a peak in region III in CF_4, SiH_4, and CH_4. The first significant increase is often known as enhanced mobility, and the decrease of the drift velocity is known as negative differential conductivity (NDC). The name of the conductivity originated from the fact that in a weakly ionized plasma the conductivity σ is defined as

$$\boldsymbol{j}_e = en_e\boldsymbol{v}_d = \sigma\boldsymbol{E}.$$

Observe, however, that the conductivity is proportional to both the drift velocity $\boldsymbol{v}_d$ and the number density n_e, and thus the NDC will also occur if the electron number density changes with E/N, which will be possible in the case of the nonconservative collisions, that is, under the electron attachment or ionization. The most interesting phenomena associated with the NDC arise from the negative slope of v_d with respect to E/N, and we use the term NDC just to describe the negative slope of the drift velocity. Typical gases that have NDC are CF_4, SiH_4, CH_4, and mixtures of Ar with molecular gases such as N_2.

Exercise 6.2.1

Derive the conditions for NDC based on the momentum and energy balance of electrons. If we start from the energy balance between the energy gain from the field E and dissipation in inelastic collisions with a threshold energy ε_j and rate R_j in a steady state with uniform n_e (here, we neglect the energy loss in elastic collisions)

$$\frac{d}{dt} < \varepsilon_m >= eEv_d(E/N) - \varepsilon_j R_j(E/N) = 0.$$

The momentum balance with the momentum transfer collision rate R_m is written in the form

$$\frac{d}{dt}(mv_d(E/N)) = eE - mv_d R_j(E/N) = 0.$$

After eliminating E from both of the balanced equations, we obtain the following condition for NDC:

$$\frac{\partial}{\partial E/N} v_d(E/N) = \frac{e\varepsilon_j}{2mv_d} \frac{R_j}{R_m} \left(\frac{1}{R_j} \frac{\partial R_j}{\partial E/N} - \frac{1}{R_m} \frac{\partial R_m}{\partial E/N} \right).$$

Here, we observe directly that if we want the slope of the drift velocity to be negative, then the collision rate for inelastic scattering R_j should decrease as a function of E/N and the collision rate for the elastic momentum transfer should increase. Although the above relation is derived for very simple assumptions, the conclusions are generally applied for the NDC.

The magnitude of the drift velocity is a strong indicator of the anisotropy of the velocity distribution $g^0(\boldsymbol{v})$ in a uniform density of electrons. The effect of enhanced mobility due to inelastic processes, for example, in SiH_4, can be observed from the plots of the velocity distribution $g^0(\boldsymbol{v})$ in Figure 6.3 at DC-E/N of 20 Td, 100 Td, and 500 Td corresponding to the maximum, minimum, and much higher value of the drift velocity. The asymmetric behavior of g^0 at $\theta = 0$ axis is caused such that the part of the distribution in the direction of acceleration is continuously supplied by the effect of the electric field, but the relative shrinking of the distribution in other directions is directly the result of inelastic collisions.

6.2.2 Diffusion Coefficients

The diffusion coefficients ND_T and ND_L are directly used in plasma modeling, though the anisotropy of the diffusion tensor is often neglected. In Chapter 2 we discussed the reasons for the anisotropy of the diffusion tensor. In Equation 2.12 of the diffusion coefficient, ξ is the logarithmic derivative of the

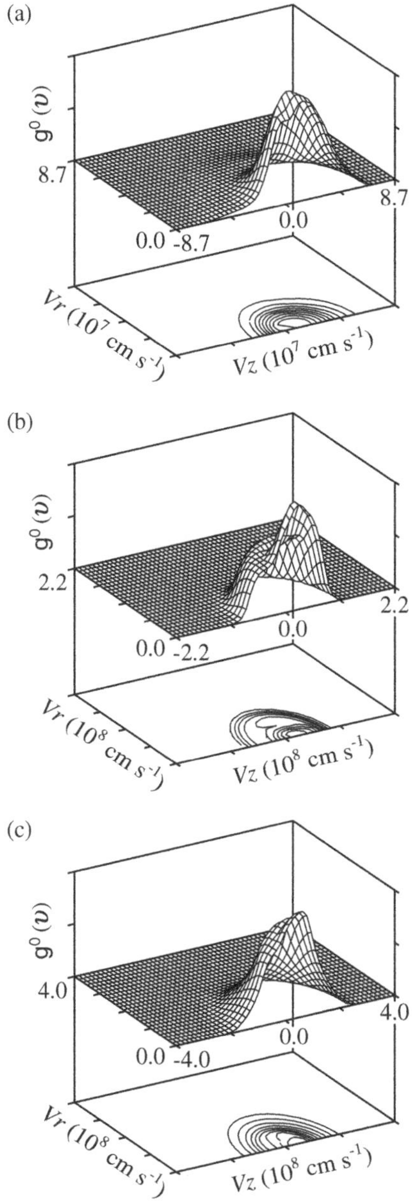

FIGURE 6.3 Velocity distribution $g^0(\boldsymbol{v})$ as a function of DC-E/N in pure SiH_4: (a) 20 Td, (b) 100 Td, and (c) 500 Td.

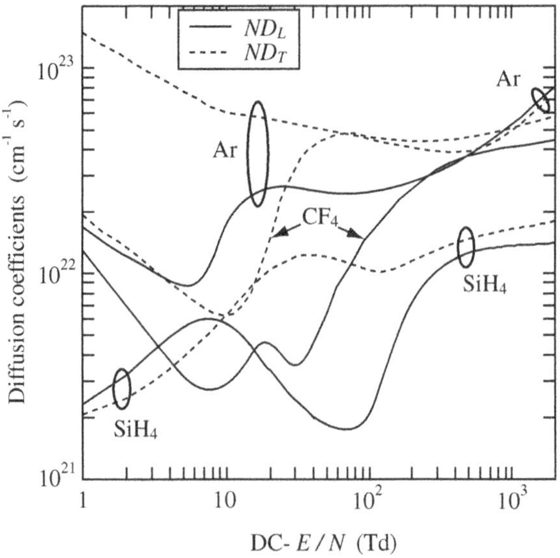

FIGURE 6.4 Diffusion coefficients of electrons in various gases as a function of DC-E/N. Transverse diffusion coefficient $(\cdots)$, longitudinal diffusion coefficient (—).

momentum transfer collision rate R_m on mean energy. When R_m is independent of the energy, the diffusion tensor is isotropic. If $\xi \gg 1$, it follows that $D_L = D_T/2$. This is the case in many molecular gases for the region of moderate E/N, which overlaps the typical applied conditions in a nonequilibrium and low-temperature collision-dominated plasma. At higher mean electron energies where the collision rate starts to decay, the dependence of the collision rate becomes close to a constant, and thus at very high E/N the components of the diffusion tensor again become similar in magnitude. In gases that have a chance for a rapid reduction of the collision rate (e.g., due to the Ramsauer–Townsend minimum), there is a chance that D_L will be greater than D_T at very low E/N. In Figure 6.4 we show data for the components of the diffusion tensor in Ar, CF_4, and SiH_4.

The variations of the D_T/D_L ratio attain a magnitude as large as 10 in gases such as Ar, CH_4, CF_4, or SiH_4. In particular, the most striking feature is the large difference between the two components of the diffusion tensor between 10 Td and 100 Td, which is almost one order of magnitude. This feature is a direct result of the very strong Ramsauer–Townsend minimum in the elastic scattering cross section.

Problem 6.2.1

Discuss the numerical algorithm to calculate $g^0(\boldsymbol{v})$ as a boundary condition problem.

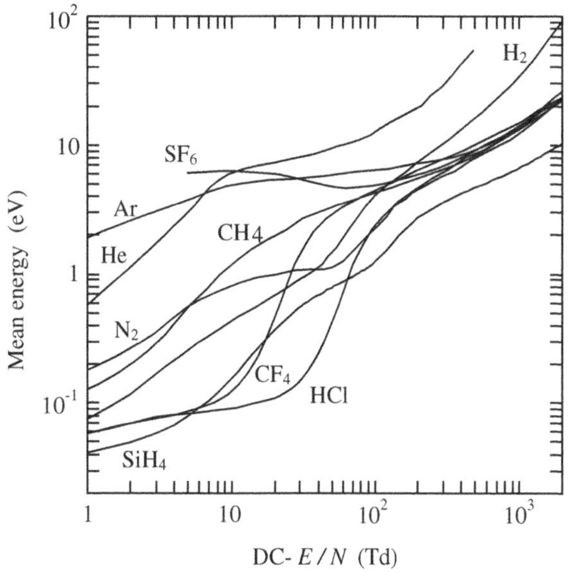

FIGURE 6.5 Electron mean energy in gases as a function of DC-E/N.

6.2.3 Mean Energy of Electrons

Figure 6.5 shows the characteristics of the electron mean energy in He, Ar, H_2, N_2, CF_4, HCl, SiH_4, SF_6, and CH_4 in a pure DC electric field. The energy is determined from the balance between the gain from the field and the loss by collisions with the molecule. The electron mean energy is the dominant one when we consider the electron swarm in gases, though it does not appear in the continuity equation of the number density, unlike other transport coefficients such as the drift velocity, diffusion coefficients, and collision rates. Usually we have no way to measure the mean energy with high accuracy.

6.2.4 Excitation, Ionization, and Electron Attachment Rates

The rates of inelastic processes (and of elastic processes) are represented as

1. Collision frequency (number of events per unit time, per electron with enegy ε, and per gas molecule : $\nu_j(\varepsilon)[\mathrm{cm}^3\,\mathrm{s}^{-1}]$);

2. Rate constant (number of events per unit time per electron and per molecule: $k_j[\mathrm{cm}^3\,\mathrm{s}^{-1}]$);

3. Rate (number of events per unit time per electron $R_j[\mathrm{s}^{-1}]$);

4. Net rate (number of events per unit time: $\Lambda_j[\mathrm{cm}^{-3}\,\mathrm{s}^{-1}]$); and

5. Townsend's spatial coefficients (number of events per unit distance traveled by an electron: $\alpha_j[\mathrm{cm}^{-1}]$).

Excitation processes that include dissociation are a special group of transport properties of electron swarms. Excitation rates are used to determine the absolute intensity of emission and dissociation and the density of short-lived species and long-lived radicals and to normalize the collision cross sections. There is a particular need for excitation rates for optimization of plasma displays, light sources, and particle detectors and for diagnostics of plasma structure and characteristics.

Ionization and electron attachment rates enter the transport equations as zeroth-order transport coefficients describing the change in the number density. All these collision processes have common characteristics having a threshold energy as a minimum energy required for the process to occur. Thus only a part of the energy distribution is convoluted to give the rates or net rates for those processes, normally the higher energy tail of the distribution. In Figure 6.6 we show some of the excitation rates with ionization and electron attachment rates in Ar and CF_4. One can see that for a range of E/N there is a negligible rate, and then at some point the rate starts to increase very rapidly. The point where this occurs is related to the magnitude of the energy loss for that process, and the rapid increase marks the overlap of the high-energy tail of the energy distribution with the cross section for that process. At some point the increase slows down until a maximum is reached, and eventually the rate starts to decay. The rate (constant) is a key to calculate the kinetics of various plasma chemical processes and densities of excited states and is thus used directly in plasma modeling.

The ionization rates are significant for relatively high mean energies (i.e., high E/N) and are essential for establishing the accuracy of the energy distribution of electrons and checking the cross section data at higher energies (see Figure 6.7). Dissociative electron attachment also has a threshold energy at moderate energy. The electron attachment to a parent gas at very low thermal energy may occur. Thus sometimes the electron attachment rate will have a maximum at thermal energy. Attachment has a threshold and peak at lower E/N than ionization, and thus there is an E/N where ionization catches up and becomes equal to attachment, as shown in Figure 6.7. Slightly above that point is the usual electrical breakdown point in electronegative gases. Thus in electronegative gases, the E/N that satisfies $R_i = R_a$ will give an ideal bulk electric field in low-temperature plasmas.

6.3 ELECTRON TRANSPORT IN RADIO-FREQUENCY ELECTRIC FIELDS

For time-varying electric fields we consider the electron transport at all frequencies. However, for low frequencies, quasi-DC behavior is achieved at each of the phases of a time-varying rf field, because the electron energy distribution

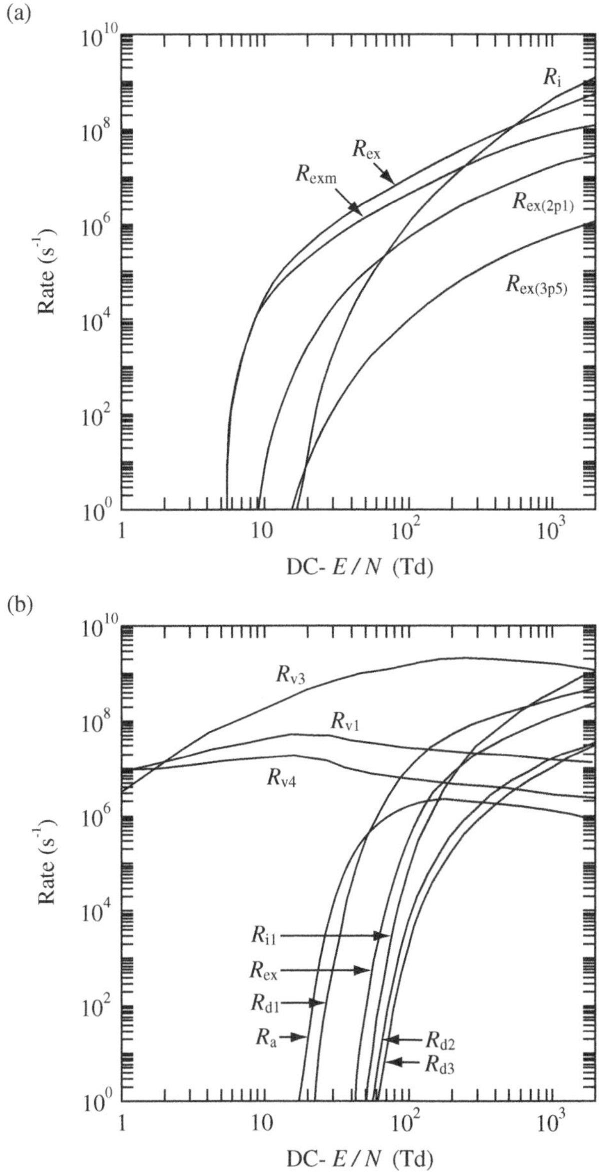

FIGURE 6.6 Excitation rates for electrons in Ar (a) and CF_4 (b) as a function of DC-E/N.

has enough time to relax to the profile at the instantaneous field by collision with neutral molecules. The most interesting phenomena under general circumstances will occur for radio frequencies at which electrons do not have enough time to relax their energy or momentum on the time scale of the

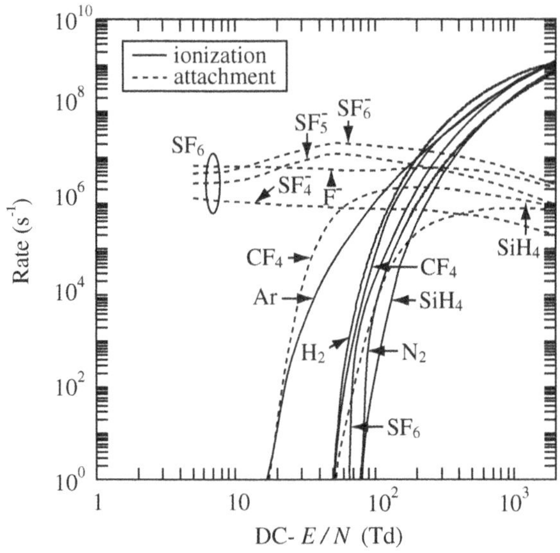

FIGURE 6.7 Ionization and attachment rates of electrons in Ar, H_2, N_2, CF_4, SF_6, SiH_4, and CH_4 as a function of DC-E/N.

changes of the field. For extremely high frequencies, electrons are not able to react to the electric field on the time scale of the period of oscillations and behave as a DC-like swarm. Ions will fail to relax their properties even at much lower frequencies due to the large mass in comparison with electrons. Here, we use the term "transport in rf fields" for all frequencies used in low-temperature plasma applications.

We consider the rf electron transport in an electric field

$$\boldsymbol{E}(t) = \boldsymbol{E}_0 \cos \omega t = \mathrm{Re}[\boldsymbol{E}_0 \exp(i\omega t)], \tag{6.1}$$

where ω and E_0 are the angular frequency and the amplitude, respectively, and $\boldsymbol{E}_0$ is usually taken to be $-E_0\boldsymbol{k}$. The rf transport is characterized as a function of E_0/N and ω/N in a gas with number density N, whereas the DC-transport is only a function of E_0/N.

6.3.1 Relaxation Time Constants

We start from the velocity distribution $g^0(\boldsymbol{v}, t)$ of electrons with a uniform density in an rf field 6.1. Usually we have time constants for momentum and energy of electrons migrating in the field under collisions with surrounding molecules in a gas or in a collision-dominated plasma. These two time constants of the electron are related to the first two-term expansion of the

Boltzmann equation (see Section 5.6.2). We have the relaxation equation for energy,

$$\frac{\partial}{\partial t} g_0^0(v,t) + \zeta \left[E_0, \omega, g_1^0(v,t)\right] = -\frac{1}{\tau_e(v)} g_0^0(v,t), \tag{6.2}$$

and for momentum relaxation

$$\frac{\partial}{\partial t} g_1^0(v,t) + \xi \left[E_0, \omega, g_0^0(v,t)\right] = -\frac{1}{\tau_m(v)} g_1^0(v,t), \tag{6.3}$$

where $g_0^0(v,t)$ and $g_1^0(v,t)$ are the isotropic and directional parts of the velocity distribution $g^0(\boldsymbol{v},t)$ expressed by the two-term approximation,

$$g^0(\boldsymbol{v},t) = g_0^0(v,t) + g_1^0(v,t)\cos\theta,$$

where θ is the angle between $\boldsymbol{v}$ and $-\boldsymbol{E}$; that is, $\theta = -(\boldsymbol{v} \cdot \boldsymbol{E})/vE$. The two time constants $\tau_e(v)$ and $\tau_m(v)$ are defined, respectively, as

$$\tau(v)^{-1} \sim Nv\Big[\frac{2m}{M} Q_m(v) + \sum \left(\frac{\epsilon_v}{\epsilon}\right)^{1/2} \frac{d}{dv}\Big(Q_v(v)v\Big) \\ + \Big(\sum Q_j(v)\frac{\epsilon_j}{\epsilon} + Q_i(v)\frac{\epsilon_i}{\epsilon} + Q_a\Big)\Big] \tag{6.4}$$

$$\tau_m(v)^{-1} = Nv\left[Q_m(v) + \sum Q_v(v) + \sum Q_j(v) + Q_i(v) + Q_a(v)\right]. \tag{6.5}$$

Typically, at 1 Torr in a low-temperature plasma the mean energy of electrons is on the order of 2 eV $\sim$ 6 eV and then $\tau_m \cong 10^{-10}$ s, whereas $\tau_e \cong 10^{-5}$ s in atomic gases or $\tau_e \cong 10^{-6}$ s - $\tau_e \cong 10^{-8}$ s in molecular gases. In Figure 6.8 we show the collisional relaxation times in Ar at 1 Torr together with the cross section Q_j and collision frequency ν_j. We see that at low energies ($\varepsilon < 11.55$ eV) the energy relaxation is slow as it is conducted purely by elastic collisions, whereas at higher energies first electronic excitation and then ionization set in and shorten τ_e by several orders of magnitude. We should recognize that the relaxation of the high-energy tail of the electron energy distribution will be much faster than that of the bulk of the distribution. The relaxation of the electron transport in rf fields can be determined directly by the solution to the Boltzmann equation or by Monte Carlo simulation. In Figure 6.9 we show a relaxation of the mean energy $< \varepsilon(t) >$ of electrons in rf fields for 1 GHz, 10 MHz, and 100 KHz. The transition region may be easily recognized, and then quasi-stationary conditions develop where the same periodic function is repeated even though the electrons may not be fully relaxed to the properties they would have for the instantaneous electric field. Nevertheless, a periodic steady state or quasi-stationary distribution develops.

The characteristics of the electron transport can be divided into three distinct regions depending on the relative magnitudes of the relaxation time constants and period of oscillations of the rf field.

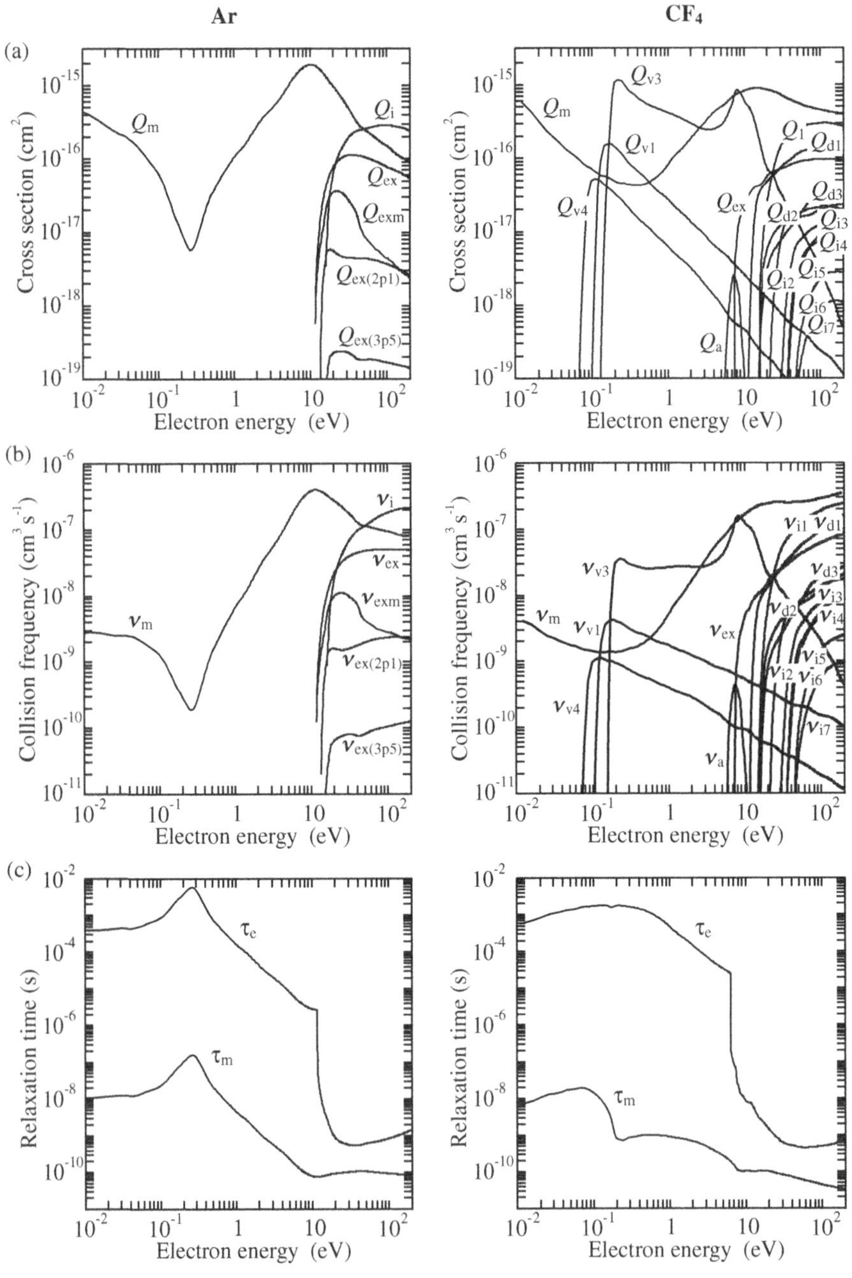

FIGURE 6.8 A set of cross sections of electrons in pure Ar and CF_4: (a) cross section, (b) collision frequency, and (c) collisional relaxation time at 1 Torr.

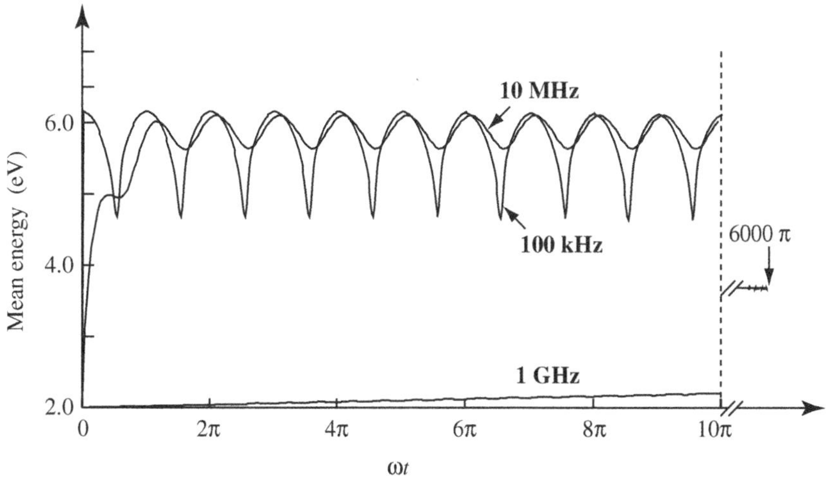

FIGURE 6.9 Examples of the relaxation of the mean energy of electrons initiated at $g_M(\boldsymbol{v})$ at 2 eV at $E_0/N = 50\sqrt{2}$ Td and 1 Torr in Ar: (a) 100 kHz, (b) 10 MHz, and (c) 1 GHz.

1. $\langle\tau_m(v)\rangle \ll \langle\tau_e(v)\rangle \ll \omega^{-1}$: Low frequencies.
   Under these circumstances, at all times the electron swarm and the velocity distribution are relaxed to the DC field conditions corresponding to the instantaneous electric field $E(t)$. Thus this is usually labeled the regime of "instantaneous field approximation."

2. $\langle\tau_m(v)\rangle \ll \omega^{-1} \ll \langle\tau_e(v)\rangle$: High frequencies.
   Under these conditions the momentum will mostly be relaxed at all times, but the energy of the electron swarm will not be relaxed and this will lead to numerous phenomena that may not be explained easily on the basis of DC-field transport. In all cases there will be a delay introduced between the field and the transport coefficients.

3. $\omega^{-1} \ll \langle\tau_m(v)\rangle \ll \langle\tau_e(v)$: Very (ultra) high frequencies.
   For this set of circumstances both the energy and momentum will be delayed in response to the electric field, and therefore all transport coefficients will have a considerable phase shift to the electric field. The ratio of the alternating-current component to the DC component of the coefficient will decrease with the increase of ω of $E(t)$, and the DC component is caused by the superimposition of the property during the one period.

The properties in the second and third region are observed in the ionization rate R_i in Ar in Figure 6.10. Periodic steady-state characteristics with twice the frequency of the external field are realized as a function of ωt. The time-averaged ionization rate is kept at an almost constant value up to 100 MHz.

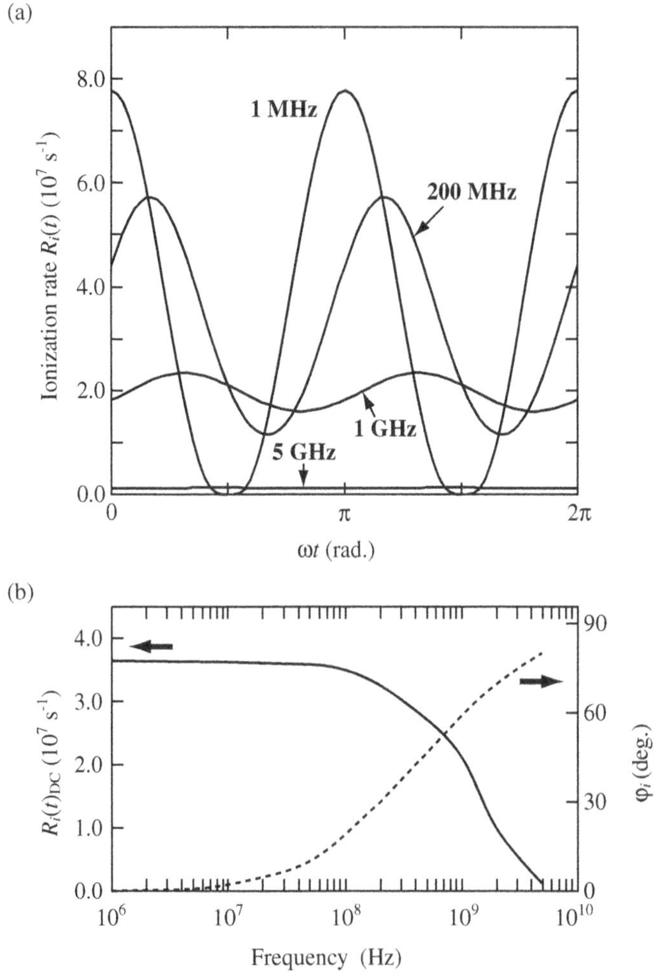

FIGURE 6.10 Ionization rate $R_i(t)$ at $E_0/N = 300\sqrt{2}$ Td at 1 Torr in Ar as a function of external frequency: (a) time-modulation of $R_i(t)$, (b) DC-component and phase delay.

At frequencies greater than 100 MHz, the DC component rapidly decreases, because electrons do not have enough time to get energy from the external field before its direction changes. That is, electrons fall into a spatial trapping as is shown in the term of the phase delay with respect to the external $\boldsymbol{E}(t)$.

Exercise 6.3.1

By considering the time evolution of the electron number density $n_e(t)$ due to ionization and electron attachment, estimate the form of $n_e(t)$ under a periodic steady-state condition.

The time-dependent distribution function $G^0(\boldsymbol{v},t)$ may be separated into the product of a time-varying number density $n_e(t)$ and a velocity distribution function $g^0(\boldsymbol{v},t)$ normalized to unity under a periodic steady state

$$G^0(\boldsymbol{v},t) = n_e(t) g^0(\boldsymbol{v},t) \quad \text{where} \quad \int_{\boldsymbol{v}} g^0(\boldsymbol{v},t) d\boldsymbol{v} = 1. \tag{6.6}$$

$n_e(t)$ and $g^0(\boldsymbol{v},t)$ are separated using a separation function $R_T(t)$:

$$\frac{\partial}{\partial t} g^0(\boldsymbol{v},t) + \frac{e\boldsymbol{E}(t)}{m} \cdot \frac{\partial}{\partial \boldsymbol{v}} g^0(\boldsymbol{v},t) = J(G,F) - R_T(t) g^0(\boldsymbol{v},t), \tag{6.7}$$

where $R_T(t)$ represents the difference between the ionization and attachment rates, $R_i(t) - R_a(t)$, and is related to the electron number density by

$$n_e(t) = n_o \exp\left(\int_0^t R_T(t) dt \right), \tag{6.8}$$

where n_o is the density at $t = 0$.

6.3.2 Effective Field Approximation

In atomic gases such as He and Ar, most of the energy loss of electrons is due to elastic collisions with neutral atoms up to moderate $E(t)/N$. Under these conditions, the energy relaxation time τ_e will be much longer than the period of the field, $\tau_e \gg \omega^{-1}$, whereas τ_m is much shorter (see for example Figures 6.11 and 6.12). Then, the time dependence of the velocity distribution $g^0(\boldsymbol{v},t)$ is expressed as

$$g^0(\boldsymbol{v},t) = g_0^0(v) + g_1^0(v) \cos\theta \cos\omega t$$

in terms of an effective DC-field $E_{eff}(v)$,

$$E_{eff}(v) = \frac{E_0}{\sqrt{2}} \frac{1}{\left[1 + (\omega \tau_m(\boldsymbol{v}))^2\right]^{1/2}} = \frac{E_0}{\sqrt{2}} \frac{1}{\left[1 + \left(\frac{\omega}{R_m(v)}\right)^2\right]^{1/2}} \tag{6.9}$$

The method of using the effective DC-field instead of a time-varying field 6.1 is very simple and has been widely used to study swarm transport. With this procedure, the transport coefficients defined in terms of $g_0^0(v)$, such as $< \varepsilon >$ and R_i, are constant in time, though the drift velocity given by $g_1^0(v,t)$ has a periodic time response. For molecular gases with large vibrational excitations from low energy on the order of 0.1 eV to several eV, the situation is not so simple even at high frequency. In that case, the isotropic part of the distribution $g_0^0(v,t)$ as well as $g_1^0(v,t)$ is modulated in time owing to $\tau_e < \omega^{-1}$ even at high frequency.

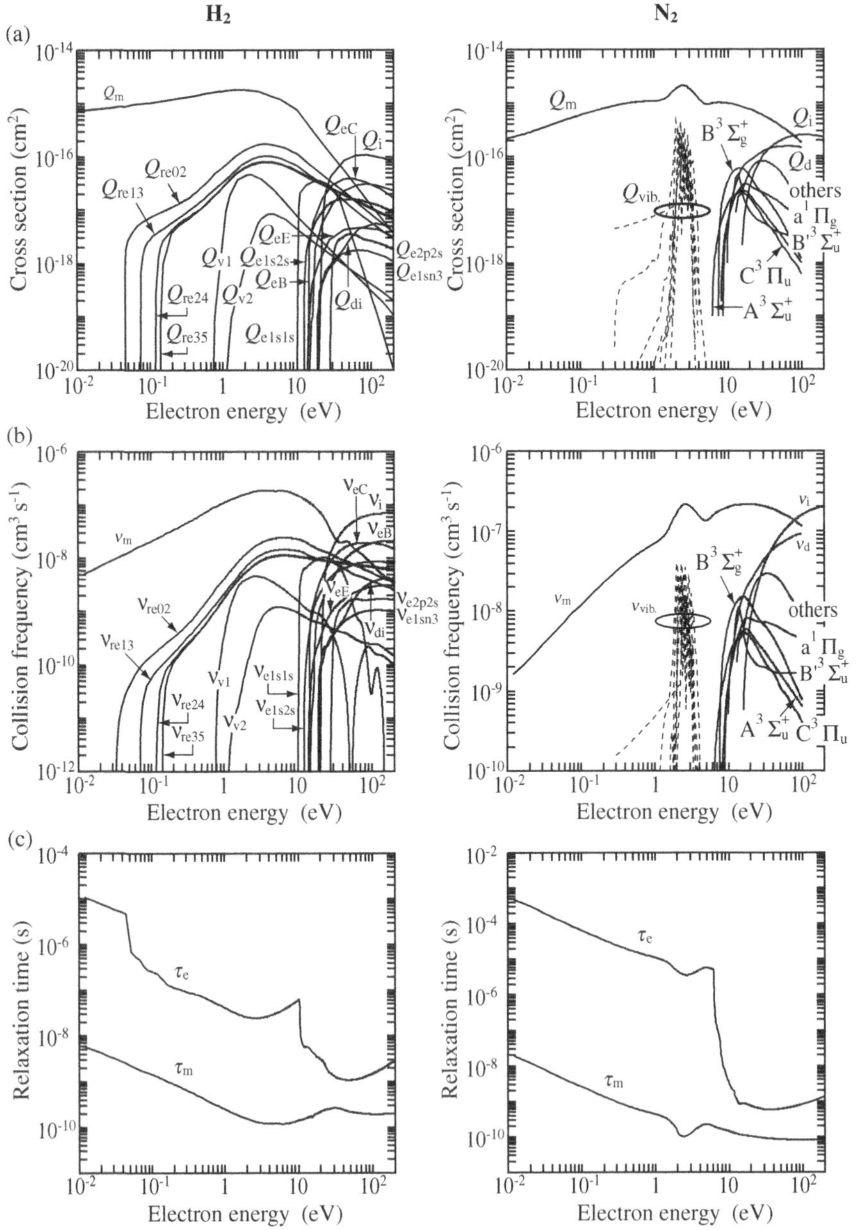

FIGURE 6.11 A set of cross sections of electrons in H_2 and N_2: (a) cross section, (b) collision frequency, and (c) collisional relaxation time at 1 Torr.

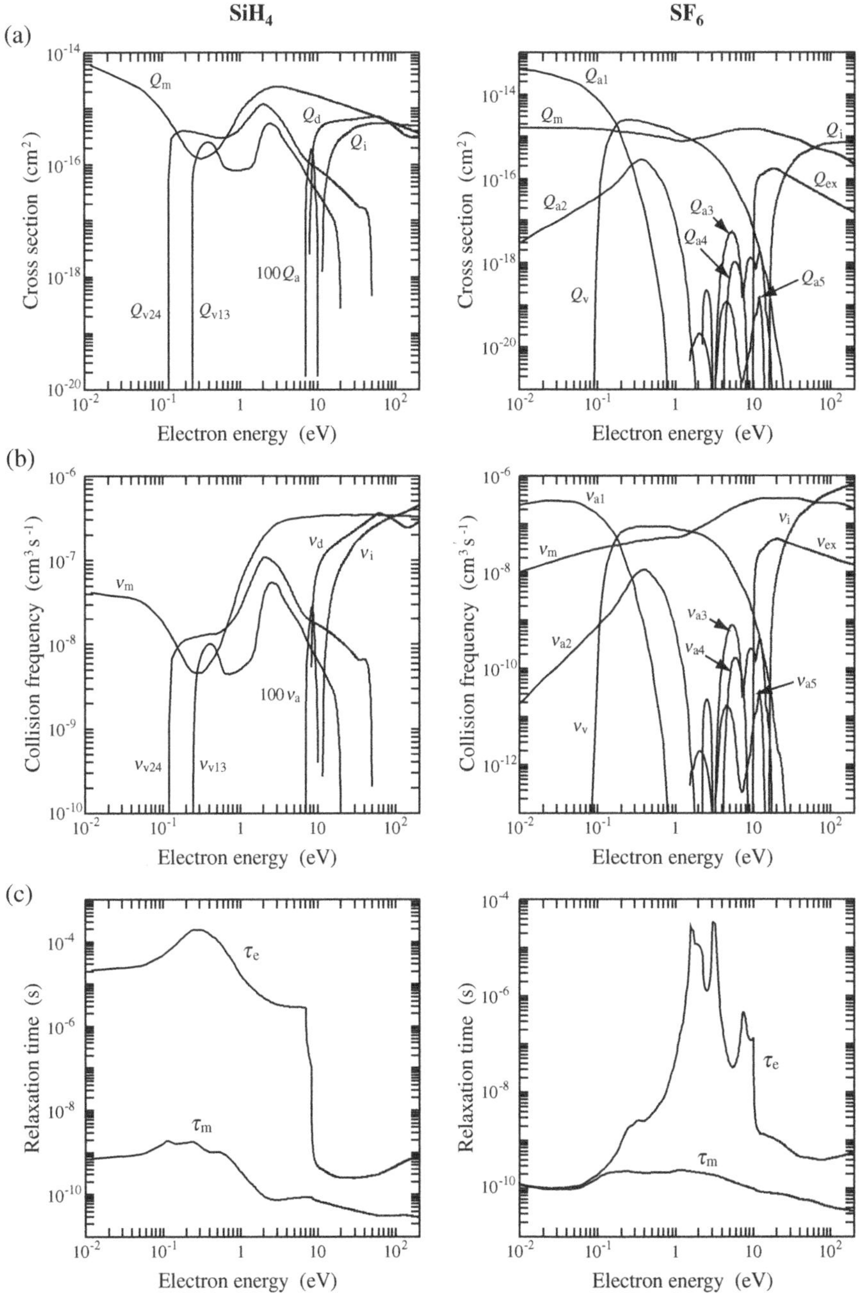

FIGURE 6.12 A set of cross sections of electrons in SiH_4 and SF_6: (a) cross section, (b) collision frequency, and (c) collisional relaxation time at 1 Torr.

Exercise 6.3.2

Derive the effective field approximation 6.9 by using the momentum balance equation of electrons.

The momentum balance of electrons with a uniform density in gases is given by

$$m\frac{\mathrm{d}}{\mathrm{d}t}\boldsymbol{v}_d(t) = -e\boldsymbol{E}(t) - R_m m\boldsymbol{v}_d(t),$$

where $\boldsymbol{v}_d$ is the drift velocity and R_m is the momentum transfer rate. The solution is

$$\boldsymbol{v}_d(t) = -\frac{e\boldsymbol{E}_0}{mR_m}\frac{1}{[1+(\frac{\omega}{R_m})^2]^{1/2}}\cos(\omega t+\phi) \quad \text{where} \quad \phi=\tan^{-1}\left(\frac{\omega}{R_m}\right). \tag{6.10}$$

In the limit of low frequency ($(\omega/R_m)^2 \ll 1$), the drift velocity converges to an instantaneous DC-like solution, $v_d(t) = -eE_0/mR_m \cos\omega t$, and for the high frequency limit $(\omega/R_m)^2 \gg 1$) the solution is

$$\boldsymbol{v}_d(t) = -\frac{e\boldsymbol{E}_0}{m\omega}\cos(\omega t + \phi). \tag{6.11}$$

We can see that in the limit of $(R_m/\omega)^2 \to 0$ the electron motion is always in the field direction and thus the electron cannot gain any energy from the field. If there are collisions (represented by R_m), electrons are able to attain some mean energy. Thus collisions with gas or any other dissipation of momentum will lead to a net gain of energy for electrons in rf fields. Next, the power absorption of electrons is achieved by the drift in an rf field:

$$\begin{aligned} P_W &= \int_0^\pi -e\boldsymbol{E}(t)\cdot\boldsymbol{v}_d(t)d\omega t \Big/ \int_0^\pi d\omega t \\ &= \frac{1}{mR_m}\left(\frac{eE_0}{\sqrt{2}(1+(\omega/R_m)^2)^{1/2}}\right)^2 = \frac{e^2E_{eff}^2}{mR_m}. \end{aligned} \tag{6.12}$$

Here, E_{eff} is equal to the effective field in Equation 6.9.

Although the effective field approximation may give averaged properties of electrons in rf fields, it will not be so accurate in the most interesting frequency region, and it certainly cannot provide time-dependent results. We proceed by discussing in greater detail the time-dependence of the transport data for quasi-stationary (relaxed) conditions.

6.3.3 Expansion Procedure

A finite phase delay with respect to an external field, Equation 6.1 will produce higher-order harmonics of the fundamental wave with frequency ω in the velocity distribution of electrons. We consider the temporal harmonic behavior of electron transport under a purely sinusoidal field 6.1. Inasmuch as the velocity distribution $g^0(\boldsymbol{v},t)$ is symmetric with respect to the field direction, the velocity distribution can be expanded in terms of spherical harmonics, and the temporal behavior is determined by the sum of the higher-order harmonics of the fundamental wave with frequency ω. Expanding $g^0(\boldsymbol{v},t)$ in Legendre polynomials in velocity space and in a Fourier series in time, we obtain

$$g^0(\boldsymbol{v},t) = \mathrm{Re}\left[\sum_k \sum_s g_s^{0k}(v) P_s(\cos\theta)\exp(ik\omega t)\right], \tag{6.13}$$

where θ is the angle between $-\boldsymbol{E}(t)$ and $\boldsymbol{v}$ is expressed as $\theta = \cos^{-1}[-(\boldsymbol{v} \cdot \boldsymbol{E})/vE$. $g_s^{0k}(v)$ is real for $k = 0$ and complex for $k > 0$, expressing the phase lag of electron transport with respect to the applied rf field 6.1. It should be noted that the isotropic part of the distribution $g_o^{0k}(v)$ possesses only even harmonics in time, whereas the directional-drift part, $g_1^{0k}(v)$, possesses only odd harmonics. Physically, this is because under a pure sinusoidal field 6.1 the energy gain described by the isotropic velocity component is proportional to $\boldsymbol{E}(t)^2$, that is, even time harmonics. The directional-drift component is a function of $\boldsymbol{E}(t)$, that is, odd harmonics.

It is convenient to define an energy distribution $f_s^{0k}(\varepsilon)$ according to the relation

$$f_s^{0k}(\varepsilon) = (4\pi/m) v g_s^{0k}(v); \quad \varepsilon = mv^2/2, \tag{6.14}$$

where $f_s^{0k}(\varepsilon)$ is normalized to $\int_0^\infty f_0^0(\varepsilon)d\varepsilon = 1$. In a pure sinusoidal field, symmetry considerations show that only components $f_s^{0k}(\varepsilon)$ where $s+k$ is an even number can exist; that is, $s+k = 2\beta$, $\beta = 1, 2, \ldots$. In that case, insertion of expansion 6.13 into the Boltzmann equation without a density gradient 5.85 gives a set of coupled differential/difference equations of the form [5]:

$$\begin{aligned}
&ik\omega(m/2\epsilon)^{1/2} f_s^k(\epsilon) \\
&+ \frac{eE_R}{\sqrt{2}} \frac{s}{2s-1}\left(\frac{d}{d\epsilon}[\lambda_1 f_{s-1}^{k-1}(\epsilon) + f_{s-1}^{k+1}(\epsilon)] - \frac{s}{2\epsilon}[\lambda_1 f_{s-1}^{k-1}(\epsilon) + f_{s-1}^{k+1}(\epsilon)]\right) \\
&+ \frac{eE_R}{\sqrt{2}} \frac{s+1}{2s+3}\left(\frac{d}{d\epsilon}[\lambda_1 f_{s+1}^{k-1}(\epsilon) + f_{s+1}^{k+1}(\epsilon)] + \frac{s+1}{2\epsilon}[\lambda_1 f_{s+1}^{k-1}(\epsilon) + f_{s+1}^{k+1}(\epsilon)]\right) \\
&= \left(\frac{m}{2\epsilon}\right)^{1/2} I_s^k(\epsilon) \left| \begin{array}{ll} +J[f_0^k(\epsilon), T_g] & s = 0 \\ -N[Q_m(\epsilon) + \sum Q_j(\epsilon) + Q_i(\epsilon) + Q_a(\epsilon)] f_s^k(\epsilon) & s \neq 0, \end{array}\right.
\end{aligned} \tag{6.15}$$

where $\lambda_1 = 1$ for all $k \neq 1$, and $\lambda_1 = 2$ for $k = 1$. T_g is the gas temperature, and $I_s^k(\epsilon)$ satisfies the relation

$$R_T(t)\mathrm{Re}\left[\sum_k f_s^k(\epsilon)\exp(ik\omega t)\right] = \mathrm{Re}\left[\sum_k I_s^k(\epsilon)\exp(ik\omega t)\right]$$

and can be expressed as

$$I_s^k(\epsilon) = \frac{\lambda_o}{2}\sum_{k'=0}^{k} R_T^{k'} f_s^{k-k'}(\epsilon) + \frac{1}{2}\sum_{k'=0}^{\infty}[R_T^{k+k'} f_s^{k'*}(\epsilon) + R_T^{k'*} f_s^{k+k'}(\epsilon)], \quad (6.16)$$

where $\lambda_o = 1$ for all $k \neq 0$, and $\lambda_o = 2$ for $k = 0$. R_T^k is the kth component in the Fourier series of $R_T(t)$, and the symbol $*$ denotes the complex conjugate. The collision term in Equation 6.15 is expressed as follows:

$$\begin{aligned} J[f_0^k(\epsilon), T_g] &= \sqrt{\epsilon}\frac{2m}{M}\frac{d}{d\epsilon}\left(NQ_m(\epsilon)\epsilon^{3/2} f_o^k(\epsilon) + k_B T_g N Q_m(\epsilon)\epsilon^2 \frac{d}{d\epsilon}[\epsilon^{-1/2} f_o^k(\epsilon)]\right) \\ &+ \epsilon^{-1/2}\sum_j \frac{d}{d\epsilon}\int_{\epsilon}^{\epsilon+\epsilon_j} \epsilon^{1/2} N Q_j(\epsilon) f_o^k(\epsilon) d\epsilon \\ &+ \epsilon^{-1/2}\frac{d}{d\epsilon}\int_{\epsilon_a}^{o} \epsilon^{1/2} N Q_a(\epsilon) f_o^k(\epsilon) d\epsilon \\ &+ \epsilon^{-1/2}\frac{d}{d\epsilon}\left[\left(\int_{\epsilon}^{\epsilon/\delta+\epsilon_i} + \int_{o}^{\epsilon/(1-\delta)+\epsilon_i}\right)\epsilon^{1/2} N Q_i(\epsilon) f_o^k(\epsilon) d\epsilon\right], \end{aligned} \quad (6.17)$$

where k_B is the Boltzmann constant. ε_j, ε_i, and ε_a are the threshold energies for excitation, ionization, and electron attachment, respectively. δ:$(1-\delta)$ is the energy partition ratio between two ejected electrons after ionization.

In particular, at ultra-high frequency, most of the component $f_0^{00}(\varepsilon)$ lies in an energy range less than the electronic excitation threshold ε_j, and collisions are almost completely limited to elastic, rotational, and vibrational scatterings. Under these circumstances, $f_0^{00}(\varepsilon)$ can be expressed in an analytical form from the Boltzmann Equation 6.15 as

$$f_0^{00}(\epsilon) = A\sqrt{\epsilon}\exp\left(-\int_0^{\epsilon}\frac{1+\sum[Q_r(\epsilon)\epsilon_r + Q_v(\epsilon)\epsilon_v]/(2m/M)\epsilon Q_m(\epsilon)}{k_B T_g + \frac{(eE_R/N)^2}{1+[\omega/\sqrt{2\epsilon/m}N\sum Q(\epsilon)]^2}/3\frac{2m}{M}\epsilon Q_m(\epsilon)\sum Q(\epsilon)} d\epsilon\right) \quad (6.18)$$

This is identical to the expression of Margenau and Hartman [6], except for the presence of rotational and vibrational collisions. In atomic gases subject only to elastic scattering, the distribution $f_0^{00}(\varepsilon)$ in the limit, $\omega^{-1} \ll \tau_m(\varepsilon)$, has a Maxwellian form with an effective gas temperature

$$T_{eff} = T_g\left[1 + \left(\frac{eE_R}{m\omega}\right)^2 / \left(\frac{3k_B T_g}{M}\right)\right]. \quad (6.19)$$

The presence of rotational and vibrational collisions with low threshold energies in molecular gases changes the shape of the distribution substantially in comparison with that obtained in atomic gases. That is, $f_0^{00}(\varepsilon)$ can be expressed in the asymptotic form of Equation 6.19 as

$$f_0^{00}(\epsilon) = A\sqrt{\epsilon}\exp\left\{-\frac{1}{k_B T_{eff}}\int_0^{\epsilon}\left[1+\frac{\sum[Q_r(\epsilon)\epsilon_r + Q_v(\epsilon)\epsilon_v]}{(2m/M)\epsilon Q_m(\epsilon)}\right]d\epsilon\right\}. \quad (6.20)$$

The second term of the integral in Equation 6.20 shows the ratio of the collisional energy loss between inelastic and elastic scattering. Recalling the fact that the higher the threshold energy, the faster the energy relaxation time, one would expect that $\varepsilon/\varepsilon_j$ should be weighted by the energy relaxation time for inelastic scattering and that the effective relaxation time is expressed as

$$\tau_e^{eff}(\varepsilon)^{-1} = N[(2m/M)Q_m(\varepsilon) + \sum Q_j(\varepsilon)\varepsilon_j/\varepsilon]v \quad (6.21)$$

when referring to inelastic collision processes. The electrons with energy ε follow the time-varying field closely when $\omega^{-1} \gg \tau_e^{eff}(\varepsilon)$.

The macroscopic behavior of the swarm is described in terms of swarm parameters (transport parameters), which are ensemble averages of quantities with respect to the time-dependent velocity (energy) distribution.

The mean energy, $<\varepsilon(t)>$, is expressed as

$$\begin{aligned} <\varepsilon(t)> &= \mathrm{Re}\left[\frac{1}{2}\int_{\boldsymbol{v}} mv^2 g^0(\boldsymbol{v},t)d\boldsymbol{v}\right] \\ &= \mathrm{Re}\left[\sum_k 2\pi e^{2ik\omega t}\int_0^{\infty} mv^4 g_o^{2k}(v)dv\right] \\ &= \mathrm{Re}\left[\sum_k e^{2ik\omega t}\int_0^{\infty} \epsilon f_0^{2k}(\epsilon)d\epsilon\right]; \end{aligned} \quad (6.22)$$

the ionization rate, $R_i(t)$, as

$$\begin{aligned} R_i(t) &= \mathrm{Re}\left[\int_{\boldsymbol{v}} NQ_i(v)vg^0(\boldsymbol{v},t)d\boldsymbol{v}\right] \\ &= \mathrm{Re}\left[\sum_k 4\pi e^{2ik\omega t}\int_0^{\infty} NQ_i(v)v^3 g_o^{2k}(v)dv\right] \\ &= \mathrm{Re}\left[\sum_k (2/m)^{1/2} e^{2ik\omega t}\int_{\epsilon_i}^{\infty} NQ_i(\epsilon)\epsilon^{1/2} f_o^{2k}(\epsilon)d\epsilon\right]; \end{aligned} \quad (6.23)$$

and the drift velocity as

$$\begin{aligned} V_d(t) &= \mathrm{Re}\left[\int_{\boldsymbol{v}} \boldsymbol{v}g^0(\boldsymbol{v},t)d\boldsymbol{v}\right] \\ &= \mathrm{Re}\left[\sum_k \frac{4\pi}{3} e^{i(2k+1)\omega t}\int_0^{\infty} v^3 g_1^{2k+1}(v)dv\right] \\ &= \mathrm{Re}\left[\sum_k \frac{1}{3}(2/m)^{1/2} e^{i(2k+1)\omega t}\int_0^{\infty} \epsilon^{1/2} f_1^{2k+1}(\epsilon)d\epsilon\right]. \end{aligned} \quad (6.24)$$

6.3.4 Direct Numerical Procedure

It is sometimes difficult to obtain the time-dependent energy distributions from the set of Equations 6.15 by using the expansion procedure both in velocity space and in time as described in the preceding section. Developments in large memory computing hardware, however, permit the application of direct numerical methods. We describe a direct numerical procedure (DNP) for solution of the Boltzmann equation in an rf field [7], and the results are shown to illustrate the usefulness of the procedure. Because the spatially homogeneous Boltzmann equation represents the time evolution of the velocity distribution only in velocity space, the Boltzmann Equation 5.8 is completely equivalent to the expression

$$g(\boldsymbol{v} + \Delta\boldsymbol{v}, t + \Delta t) = g(\boldsymbol{v}, t) + J(g, F)\Delta t, \tag{6.25}$$

when Δt approaches 0. The velocity increment $\Delta\boldsymbol{v}$ is related to Δt as follows,

$$\Delta\boldsymbol{v} = [e\boldsymbol{E}(t)/m]\Delta t. \tag{6.26}$$

Equation 6.25 suggests that it should be possible to calculate the time evolution from an arbitrary initial distribution to a final periodic steady state under the rf field by a differential method. It is essential to consider the change both of the incident angle and of the energy at collision in terms of a deterministic finite probability for each type of collision. For this purpose, it is convenient to evaluate the collisional scattering in spherical coordinates. The electron velocity also changes, however, due to acceleration or deceleration in the electric field, and Cartesian coordinates are more suitable for this case. In fact, the transformation of quantities between the two coordinate systems is a serious practical issue, and, therefore, a simulation in velocity space instead of energy space may be essential.

The collision term is expressed in the case of a cold gas, $T_g = 0$, as

$$\begin{aligned} J(g, F) &= N\int_\Omega \frac{v_1'^3}{v^3} g(\boldsymbol{v}_1', t)v_1'\sigma_{el}(v_1', \theta)d\Omega \\ &\quad + N\sum_j \int_\Omega \frac{v_2'}{v} g(\boldsymbol{v}_2', t)v_2'\sigma_j(v_2', \theta)d\Omega \\ &\quad + N\frac{1}{(1-\delta)}\int_\Omega \frac{v_3'}{v} g(\boldsymbol{v}_3', t)v_3'\sigma_i(v_3', \theta)d\Omega \\ &\quad + N\frac{1}{\delta}\int_\Omega \frac{v_4'}{v} g(\boldsymbol{v}_4', t)v_4'\sigma_i(v_4', \theta)d\Omega \\ &\quad - N\int_\Omega g(\boldsymbol{v}, t)v[\sigma_{el}(v, \vartheta) + \Sigma\sigma_j(v, \theta) + \sigma_a(v, \vartheta) + \sigma_i(v, \theta)]d\Omega. \end{aligned} \tag{6.27}$$

Here, $\sigma(v, \theta)$ is a differential cross section, defined as the number of electrons scattered into a solid angle $d\Omega = v^2 \sin\theta d\theta d\phi dv$ at scattering angle θ. The

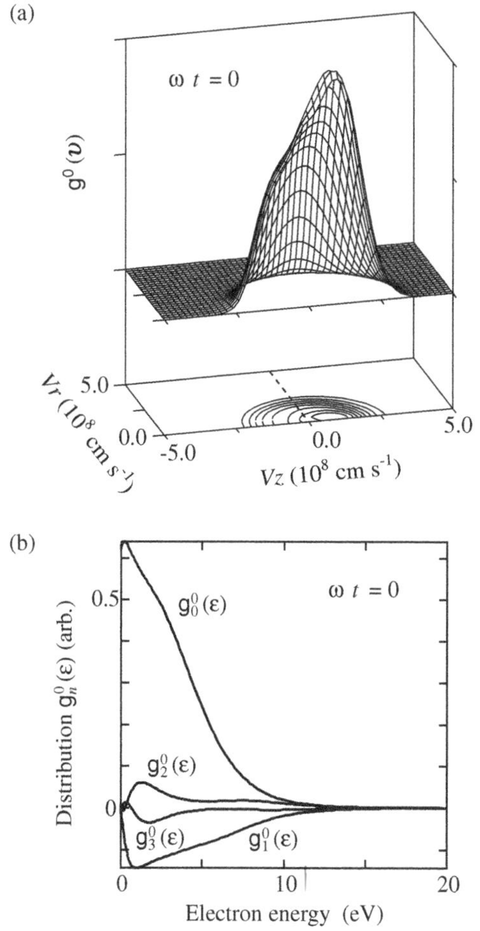

FIGURE 6.13 Electron velocity distribution $g^0(\boldsymbol{v}, \omega t = 0)$ (a) and each component of Legendre polynomials $g_n^0(\epsilon)(n = 0, 1, 2, 3)$ (b) at $E_0/N = 300\sqrt{2}$ Td and 1 MHz at 1 Torr in Ar.

subscripts of $\sigma(v, \theta)$, el, j, a, and i denote elastic, excitation, attachment, and ionization collisions, respectively. The velocity $\boldsymbol{v}_1'$ of the incoming flux prior to elastic scattering is related to the velocity of the outgoing flux $\boldsymbol{v}$ as

$$v_1' = \frac{v}{1 - (m/M)(1 - \cos\vartheta)}.$$

For excitation, v_2' is expressed as $v_2' = (v^2 + 2\varepsilon_j/m)^{1/2}$. Similarly, $v_3' = (v^2/\delta + 2\varepsilon_i/m)^{1/2}$ and $v_4' = (v^2/(1-\delta) + 2\varepsilon_i/m)^{1/2}$ are the velocity relations for two electrons after ionization. When the scattering is not strongly dependent on angle θ, we may approximate it by isotropic scattering, which means that the differential cross section σ is independent of θ. For isotropic scattering, $\sigma(v,\theta)$ is replaced by $Q(v)/4\pi$, and Equation 6.25 can be rewritten in a form for numerical calculation of the time evolution:

$$g(v_x, v_y, v_z+(eE(t)/m)\Delta t, t+\Delta t) = g(v_x, v_y, v_z, t)+J(v_x, v_y, v_z, t)\Delta t, \quad (6.28)$$

where

$$\begin{aligned} J(v_x, v_y, v_z, t) \quad = \quad & \frac{NQ_m(v_1')v_1'^4}{4\pi v^3} \int_\Omega g(\boldsymbol{v}_1', t)d\Omega \\ & +N\frac{\Sigma Q_j(v_2')v_2'^2}{4\pi v} \int g(\boldsymbol{v}_2', t)d\Omega \\ & +\frac{NQ_i(v_3')v_3'^2}{4\pi v(1-\delta)} \int g(\boldsymbol{v}_3', t)d\Omega \\ & +\frac{NQ_i(v_4')v_4'^2}{4\pi v\delta} \int g(\boldsymbol{v}_4', t)d\Omega \\ & -N[Q_m(v)+\sum Q_j(v)+Q_a(v) + Q_i(v)]vg(\boldsymbol{v}, t). \quad (6.29) \end{aligned}$$

Because there is axial symmetry around the z-axis, g is stored only as a function of v_z and v_x (or v_y) in such a way that the velocity increment Δv_z satisfies the relation $\Delta v_z = (eE_z(t)/m)\Delta t$. By so doing, evaluation of g in Equation 6.28 merely involves a shifting of the two-dimensional array $g(v_x, v_z)$ along v_z at each time interval Δt and addition of the corresponding collision term $J(v_x, v_y, v_z, t)\Delta t$ to each array element. The collision integrals for each v_x and v_z at time t are calculated by integrating the distribution of v and θ over both the velocity v and the polar angle θ. The time step Δt in an explicit method is limited to a very short time in order to satisfy the Courant–Friedrichs–Lewy condition relating to the velocity cell size, $\Delta \boldsymbol{v}$,

$$[e\boldsymbol{E}(t)/m]\Delta t \leq \Delta \boldsymbol{v}, \quad (6.30)$$

and to satisfy the differential expression

$$\mathrm{Max}[\tau_{\mathrm{m}}(\mathrm{v})^{-1}] \times \Delta \mathrm{t} << 1. \quad (6.31)$$

A method not restricted by the relation in Equation 6.30 has been proposed [8].

In Figure 6.14 we show the time modulation of the velocity distributions $g^0(v_z, t)$ in Ar at $E_0/N = 50\sqrt{2}$ Td and 13.56 MHz in a periodic steady state. The distribution at each phase is not symmetric, leading to a directional velocity component in a direction opposite that of the field and the asymmetry changes with the field, $E_0 \cos\omega t$. At low energy of less than 2 eV, especially in the Ramsauer–Townsend minimum, the collisional momentum relaxation is not complete. The energy relaxation time in Ar is much longer

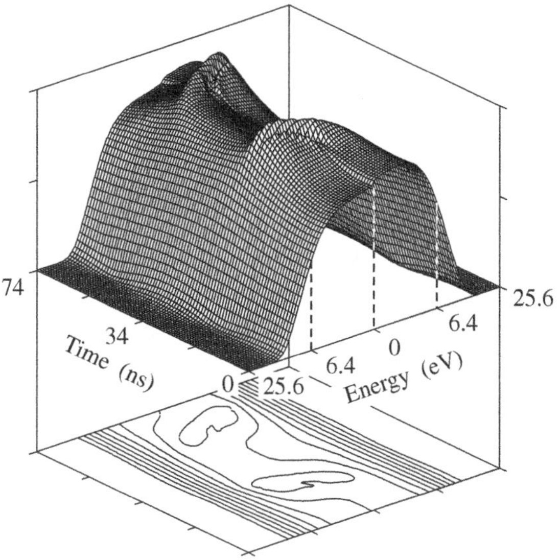

FIGURE 6.14 The modulation of the electron velocity distribution $g^0(v_z, t)$ at $E/N = 50\sqrt{2}\cos\omega t$ Td and 13.56 MHz at 1 Torr in Ar.

than the one period, 74 ns, except energies greater than the excitation threshold 11.56 eV. As a result, $g^0(v_z, t)$ exhibits temporal characteristics peculiar to the rf field. Most important, the effects of the limited times of relaxation become observable for some parts of the periods and also for some parts of the energy distribution, and the time behavior of the transport coefficients become quite complex. This leads to general kinetic phenomena, and here we describe two such processes: the negative differential conductivity and anomalous anisotropic diffusion intrinsic to the rf field.

Problem 6.3.1

In a high density plasma at high pressure, long-lived metastable particles will be accumulated and contribute to the plasma production by the metastable pooling. For example, in pure Ar plasma

$$Ar(1s_5) + Ar(1s_5) \rightarrow e + Ar^+ + Ar(^1S_0).$$

Then, the collision term in Equation 6.29 has to include a new source term caused by the metastable pooling. That is,

$$\frac{\Lambda_{mp}}{n_e(t)4\pi v^2 dv}\delta\left(\frac{2(2\epsilon_j - \epsilon_i)}{m}\right)^{1/2}.$$

Here, Λ_{mp} is the net ionization rate by the metastable pooling in an electron density $n_e(t)$. Derive the above source term.

6.3.5 Time-Varying Swarm Parameters

The time dependence of most transport coefficients of electron swarms in rf fields is explained by starting from the instantaneous field approximation and adding temporal relaxation for momentum and energy. Figure 6.15 exhibits the time modulation of the electron swarm parameters in SiH_4 at $E_0/N = 160\sqrt{2}$ Td as a function of applied frequency $f(= \omega/2\pi)$. The ensemble average of the energy $< \varepsilon(t) >$ has several interesting characteristics. One is its time modulation with a frequency of 2ω. Another is a gradual increase of the minimum value due to failure to achieve full relaxation of the mean energy. A third is a decrease of the maximum due to the lack of energy gain from the field during a half period. As a result, the fourth is the deduction of the amplitude of the modulation and of the mean value. In addition, we observe a phase delay with respect to the waveform of the external field $E_0 \cos \omega t$ by lack of collisions for momentum transfer during a half period with increasing frequency $f(\omega/2\pi)$.

The drift velocity $v_d(t)$ in Figure 6.15 changes its shape and the amplitude as a function of frequency. At low frequency, we observe an NDC similar to that in a DC field, for example, at the first and third maximum at 100 kHz. From the DC E/N dependence of the drift velocity and assuming instantaneous relaxation, one will expect symmetric nonsinusoidal profiles of the drift velocity with two or even three maxima (and minima) depending on the maximum value of E/N, which is confirmed at a low frequency of 100 kHz in Figure 6.15. With increasing frequency, the third peak disappears. The first peak is skewed and then disappears at a much higher frequency. As a result, a sinusoidal shape with one peak will appear at microwave frequency. The temporal drift velocity profiles at low and high frequency suggest that the power deposition to electrons, given by the convolution between the electric field and the drift velocity, will be quite different from the expected sinusoidal profile. As the frequency (in detail ω/N) increases, a phase lag appears in the drift velocity. This coincides with the beginning of electron trapping. Finally, collisionless phenomena occur between electrons and molecules during one cycle of the driving field. The electron swarm behaves as a group vibrating in a vacuum under the effect of the field $\boldsymbol{E}_0 \cos \omega t$. The instantaneous group velocity has the form

$$\boldsymbol{v}_d(t) = -\mathrm{Re}[(e\boldsymbol{E}_o/m\omega)\cos(\omega t + \pi/2), \tag{6.32}$$

and its magnitude decreases with ω.

Problem 6.3.2

Discuss how to calculate the velocity distribution $g^0(\boldsymbol{v}, t)$ of electrons in a periodic steady state by using a time evolution algorithm of the DNP.

The second kinetic phenomenon observed only in rf fields is the anomalous behavior of the longitudinal component of the diffusion tensor [7, 9], and the example is shown in SiH_4 in Figure 6.15. The effect is observed in the time dependence of the transverse and longitudinal diffusion coefficients. The instantaneous field approximation gives minima equal to the thermal values

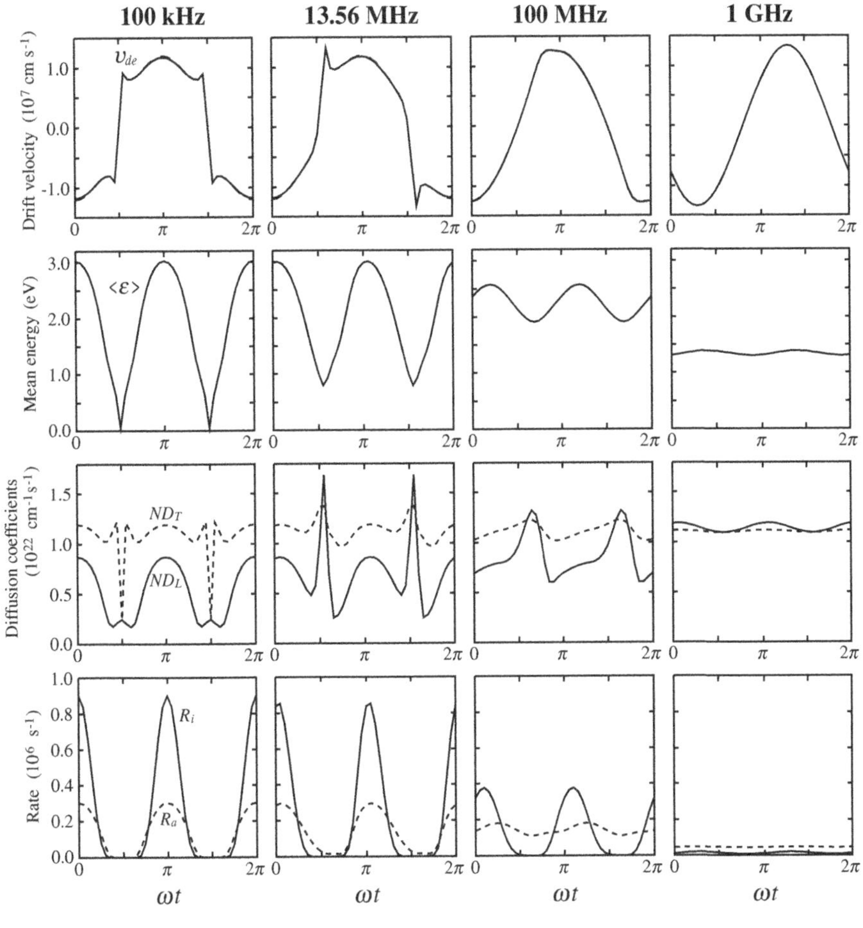

FIGURE 6.15 Time modulation of the swarm parameters in SiH_4 at $E_0/N = 160\sqrt{2}$ Td frequencies of 100 kHz, 13.56 MHz, 100 MHz, and 1 GHz.

for both components of the diffusion tensor. The transverse component follows the expected behavior with a somewhat delayed minimum that is not as deep as predicted due to a failure to relax to the instantaneous field at this very high frequency. However, the longitudinal component shows two unexpected aspects. When the electric field changes its sign, a peak, rather than a minimum, occurs in the longitudinal diffusion coefficient. Sometimes that peak even exceeds the value of the transverse diffusion coefficient, which is also unexpected, because we expect the transverse component to be larger for the cross sections that were used for all mean energies except the thermal energy (see D_L at 13.56 MHz in Figure 6.15). The importance of the

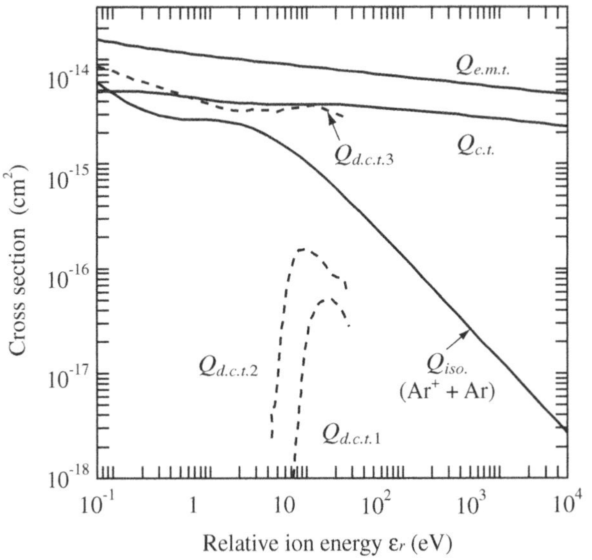

FIGURE 6.16 Cross sections for Ar^+ in Ar (—) and CF_4(· · ·).

effect of anomalous diffusion is that it is a kinetic effect occurring when the field changes sign, which cannot be predicted on the basis of any effect for DC fields. This effect may be relevant for the calculation of power input into electrons in capacitively and inductively coupled plasmas.

6.4 ION TRANSPORT IN DC-ELECTRIC FIELDS

Extending the discussion to all aspects of ion transport in neutral gases would require another volume [11–13]. Here, we briefly mention some basic properties of ion transport, avoiding the details of the ion molecule reactions and interaction potentials as well as other details. The data on the swarm parameters in gases up to the early 1980s have been compiled by Mason et al. [14].

A typical example of the cross section of ions in gas is shown in Figure 6.16. Here, the total momentum transfer cross section is shown together with the cross section for isotropic-elastic scattering and backward scattering equivalent to the symmetric charge exchange. In principle, ions exchange a lot of energy in elastic collisions and therefore require a much higher E/N to reach nonthermal conditions of energy as compared with electrons. The dependence of the energy distribution of ions on E/N is shown in O^- ions in O_2 in Figure 2.5 in Chapter 2. The energy distributions give general characteristics as a function of relative energy between an ion and molecule. That is, the relative energy distribution is Maxwellian at a low E/N of 20 Td and has a broad profile at a moderate E/N where the attractive and repulsive forces are balanced between the ion and molecule. At higher E/N where a repulsive force is predominant, the energy distribution will shrink in width.

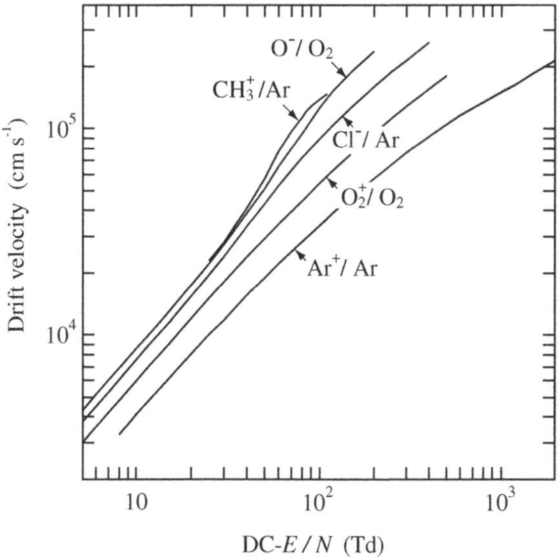

FIGURE 6.17 Ion drift velocities in gases as a function of DC-E/N.

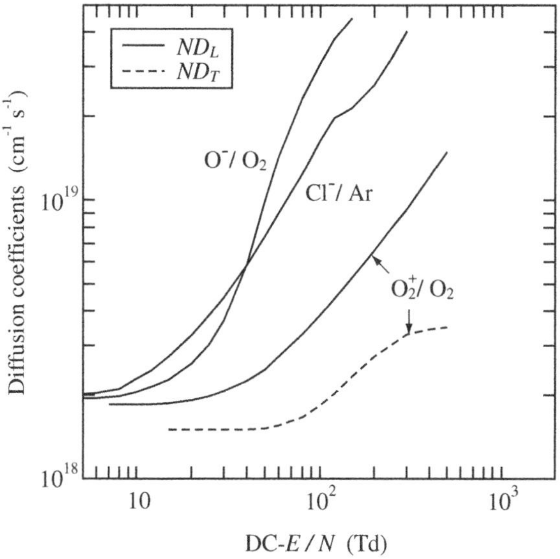

FIGURE 6.18 Diffusion coefficients of ions in gases as a function of DC-E/N. Transverse diffusion coefficient (· · ·) and longitudinal diffusion coefficient (—).

Examples of the drift velocity of ions are shown in Figure 6.17. The drift velocity is almost proportional to E/N at $E/N < 100$ Td. In addition to the characteristics dominated by the charge transfer, there are many other features that are possible for different ions. For example, ions that are not in their parent gas (such as N^+ in N_2 or Cl^- in Ar) may not have such a large charge transfer cross section at high energies. This leads to a situation in which the high energy ions may have a small cross section and may increase their energy continuously, never coming to quasi-equilibrium with the external field (i.e., runaway). Fast neutrals produced in charge transfer collisions should be followed separately, as they may participate in a number of processes, including gas phase excitation (or even ionization), and secondary electron production and sputtering and etching on a surface. Diffusion coefficients are classified into longitudinal and transverse coefficients with respect to the direction of the external electric field. Some diffusion coefficients are shown in Figure 6.18 as a function of E/N.

References

[1] Huxley, L.G.H. and Crompton, R.W. 1973. *The Drift and Diffusion of Electrons in Gases.* New York: Wiley Interscience.

[2] Dutton, J. 1983. *J. Phys. Chem. Ref. Data* 4: 577; Gallagher, J.W., Beaty, E.C., Dutton, J., and Pitchford, L.C. *J. Phys. Chem. Ref. Data* 12:109.

[3] Tagashira, H., Sakai, Y., and Sakamoto, S. 1977. *J. Phys. D* 10:1051; Taniguchi, T., Tagashira, H., and Sakai, Y. *J. Phys. D* 10:2301.

[4] von Engel, A. 1983. *Electric Plasmas: Their Nature and Uses.* London: Taylor & Francis.

[5] Goto, N. and Makabe, T. 1990. *J. Phys. D* 23:686.

[6] Margenau, H. and Hartman, L.M. 1948. *Phys. Rev.* 73:309.

[7] Maeda, K., Makabe, T., Nakano, N., Bzenic, S., and Petrovic, Z.Lj. 1997, 1994. *Phys. Rev. E* 55:5901; Maeda, K. and Makabe, T. *Physica Scripta* T53:61; Maeda, K. and Makabe, T. *Europhys. Conf. Abstr.* 18E:151.

[8] Matsui, J., Shibata, M., Nakano, N., and Makabe, T. 1998. *J. Vac. Sci. Technol. A* 16:294.

[9] White, R.D., Robson, R.E., and Ness, K.F. 1995. *Aust. J. Phys.* 48:925.

[10] McDaniel, E.W. and Mason, E.A. 1973. *The Mobility and Diffusion of Ions in Gases.* New York: John Wiley & Sons.

[11] Mason, E.A. and McDaniel, E.W. 1988. *Transport Properties of Ions in Gases.* New York: John Wiley & Sons.

[12] Farrar, J.M. and Saunders, W.H. Jr. Ed. 1988. *Techniques for the Study of Ion-Molecule Reactions.* New York: John Wiley & Sons.

[13] Lindinger, W., Mark, T.D., and Howorka, F. Ed. 1984. *Swarms of Ions and Electrons in Gases.* Wien: Springer Verlag.

[14] Mason, E.H. and Viehland, L.A. 1976, 1978, 1984. *Atomic Data and Nuclear Data Tables*, 17: 177; Ellis, H.W., McDaniel, E.W., Albritton, D.A., Viehland, L.A., Lin, S.L., and Mason, E.A., *ibid.* 22:179; Ellis, H.W., Thackston, G., McDaniel, E.W., and Mason, E.A. *ibid.* 31:113.

CHAPTER 7

Modeling of Nonequilibrium (Low-Temperature) Plasmas

7.1 INTRODUCTION

Early attempts to model plasmas were based on zero-dimensional and phenomenological models. An example of this approach is the circuit model, which represents plasma behavior using passive circuit elements, resistance, capacitance, and inductance. Although such an approach is useful for representing specific plasmas and studying how they behave once connected to the external circuit, it is beyond the scope of this chapter to specify plasma characteristics intrinsic in a feed gas molecule. There is a rate equation model describing zero-dimensional plasma, which has proven quite useful for the kinetic optimization of a gas laser, mainly by utilizing a bulk plasma.

During the 1980s and early 1990s, it was recognized that plasma-enhanced chemical vapor deposition of amorphous silicon and plasma etching for integrated circuits could not be improved further by a trial-and-error approach, and that a complete understanding of these processes was required for a new generation of technologies. This was partially motivated by the increasing cost of the development of the equipment and complexity of reaction chambers.

The need for radio frequency (rf) plasmas arose from the fact that, in these technologies, one needs also to treat metals, semiconductors, and dielectrics at the same time. There is an important difference between modeling of direct current (DC) and alternating current (AC) discharge plasmas, especially those operating in an rf range. Rfs have periods on the order of nanoseconds, much shorter than the characteristic lifetimes of many of the gas-phase processes, and yet the discharge clearly develops and exhibits complex behavior on such short time scales, so the characteristic time step of the numerical simulation must be significantly shorter than the period. At the same time, one must take into account slow processes such as the kinetics of chemically active neutrals and the flow of gas through the system that will occur on time scales from milliseconds to seconds. As a result of this vast difference in time scales, we need to cover as many as 10 orders of magnitude in time development. Thus the rf plasma for material processings belongs to the class of the stiff system, and special attention should be paid to the numerical techniques or physical approximations. Under these circumstances, the basic numerical difficulty that sometimes limits the stability and accuracy of computer simulation will arise from the orders-of-magnitude difference among the time constants (relaxation times) associated with the governing differential equations. The governing equation system is classified as a set of stiff differential equations [2].

Another issue associated with the use of rf plasma is that, due to the change of the amplitude and the direction of the local field, the properties of electrons and ions will not be relaxed to the instantaneous electric field, or, at the very least, temporal relaxation will take up a considerable part of the period. Both DC and rf discharge plasmas, however, engender the problems with the spatial relaxation of the properties of charged particles, particularly electrons.

As a result of an extensive scientific effort, a number of low-temperature plasma models were developed in the 1980s and early 1990s. The general characteristics of these models are that they are predominantly numeric and that they (to differing degrees) try to represent the entire plasma, often including external circuits. In this chapter, we present several techniques for modeling low-temperature (nonequilibrium) plasmas [2, 3]. See Figure 7.1.

7.2 CONTINUUM MODELS

Continuum models are based on the moments of the Boltzmann equation for both charged particles and neutral molecules. As outlined in Chapter 5, these moments are obtained by integrating the Boltzmann equation with various powers of velocity (see Section 5.4). The corresponding conservation laws are

$$\text{particle number density} \quad \Leftarrow \quad \int (\text{Boltzmann Eq.}) \; d\boldsymbol{v} \qquad (7.1)$$

$$\text{momentum balance} \quad \Leftarrow \quad \int (\text{Boltzmann Eq.}) \; \boldsymbol{v} d\boldsymbol{v} \qquad (7.2)$$

$$\text{energy balance} \quad \Leftarrow \quad \int (\text{Boltzmann Eq.}) \; \boldsymbol{v}^2 d\boldsymbol{v}. \qquad (7.3)$$

These laws should be written for all particles involved and should be completed by a set of Maxwell's equations. This model treats particles as fluids. Continuum models havea key and practical advantage of having a short

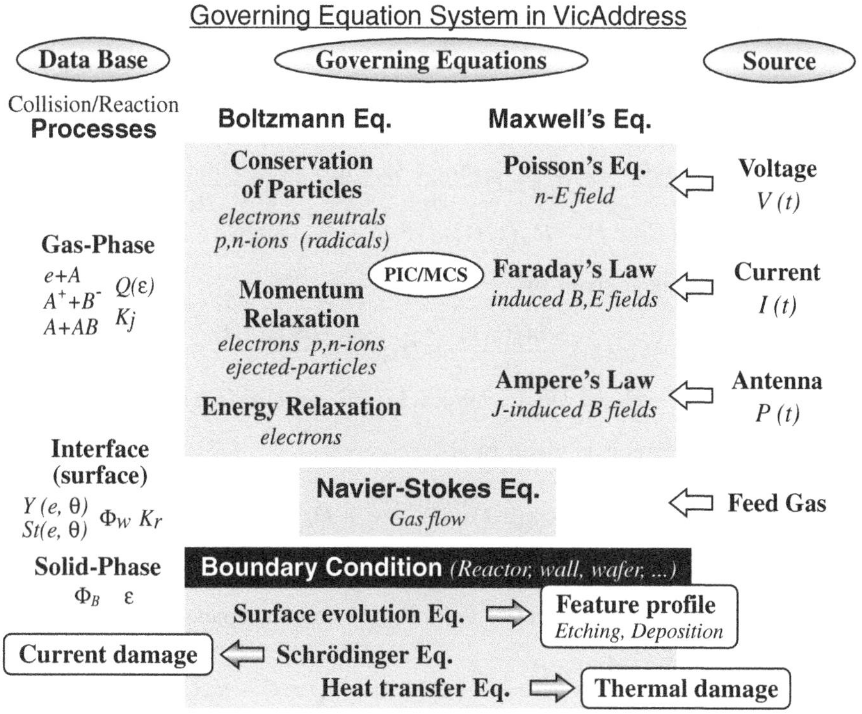

FIGURE 7.1 Governing equation system of a low-temperature plasma for material processing.

computation time, and the validity is kept under conditions such that the mean free path of charged particles, especially electrons, is shorter than or equivalent to the characteristic length of a plasma reactor. In principle, these Equations 7.1 to 7.3 are not closed, so additional equations should be added. For example, in the case of thermal equilibrium, the equation of state is used to complete the system. In nonequilibrium plasmas, the electron energy distribution, which is dominated by the short-range electron–molecule collisions, is non-Maxwellian, and we need a procedure to find solutions to the corresponding transport (Boltzmann) equation.

Typically, nonequilibrium rf plasma consists of both a bulk plasma with quasi-neutrality between electrons and positive ions and of ion sheaths. Both regions are periodically modulated in time even when quasi-stationary conditions are achieved. In order to deal with such a complex system, we need to establish a modeling hierarchy that involves several different models.

7.2.1 Governing Equations of a Continuum Model

The minimum set of equations required to describe nonequilibrium plasmas in time and configuration space, as mentioned above, consists of continuity equations (i.e., conservation of number density) for electrons, of positive and negative ions, and of Poisson's equation.

In the one-dimensional space along the axial field direction (z-axis), the continuity equation for electrons has the form

$$\frac{\partial n_e(z,t)}{\partial t} = -V_{de}(z,t)\frac{\partial n_e(z,t)}{\partial z} + D_{Le}(z,t)\frac{\partial^2 n_e(z,t)}{\partial z^2}$$
$$+\{R_i(z,t) - R_a(z,t)\}n_e(z,t) - R_{re}(z,t)n_e(z,t)n_p(z,t). \quad (7.4)$$

For positive ions it is given by

$$\frac{\partial n_p(z,t)}{\partial t} = -V_{dp}(z,t)\frac{\partial n_p(z,t)}{\partial z} + D_{Lp}(z,t)\frac{\partial^2 n_p(z,t)}{\partial z^2} + R_i(z,t)n_e(z,t)$$
$$- R_{re}(z,t)n_e(z,t)n_p(z,t) - R_{ri}(z,t)n_n(z,t)n_p(z,t), \quad (7.5)$$

and for negative ions it is given by

$$\frac{\partial n_n(z,t)}{\partial t} = -V_{dn}(z,t)\frac{\partial n_n(z,t)}{\partial z} + D_{Ln}(z,t)\frac{\partial^2 n_n(z,t)}{\partial z^2}$$
$$+R_a(z,t)n_e(z,t) - R_{ri}(z,t)n_n(z,t)n_p(z,t). \quad (7.6)$$

Poisson's equation for the electric field $E(z,t)$ and the space potential $V(z,t)$ is

$$\frac{\partial E(z,t)}{\partial z} = -\frac{\partial^2 V(z,t)}{\partial z^2} = \frac{e}{\varepsilon_0}\{n_p(z,t) - n_e(z,t) - n_n(z,t)\}. \quad (7.7)$$

Here, the number densities of electrons, positive ions, and negative ions are $n_e(z,t)$, $n_p(z,t)$, and $n_n(z,t)$, respectively; V_{dj} and D_{Lj} are the drift velocity and the longitudinal component of the diffusion tensor of the jth particles (e, p, n); R_i and R_a are the ionization and electron attachment rates; and R_{re} and R_{ri} are rates of recombination in electron–ion and ion–ion collisions. The data for transport coefficients V_{dj}, D_{Lj}, R_i, and so on, which are taken from a database of experimental or theoretical data, are provided for the local and instantaneous electric field $E(z,t)$. One example of a set of data for the collision rate for O_2 is shown in Figure 7.2. In Equation 7.7, $e(> 0)$ is the elementary charge and ε_0 is the permittivity (dielectric constant) of the vacuum.

The set of Equations 7.4 to 7.7 is numerically calculated for a time-varying one-dimensional capacitively coupled plasma (CCP) with infinite parallel plates. In a finite plasma source or axisymmetric plasma, these equations are developed in the two- or three-dimensional configuration space.

Problem 7.2.1

Derive the continuity of the current density in the one-dimensional position space

$$\frac{\partial J_T}{\partial z} = \frac{\partial}{\partial z}\left(J_e(z,t) + J_p(z,t) + J_n(z,t) + J_{Dis}(z,t)\right) = 0 \quad (7.8)$$

from the Equations 7.4 to 7.7. Here J_e, J_p, and J_n are the current density of electrons and of positive and negative ions, respectively. J_{Dis} is the displacement current density.

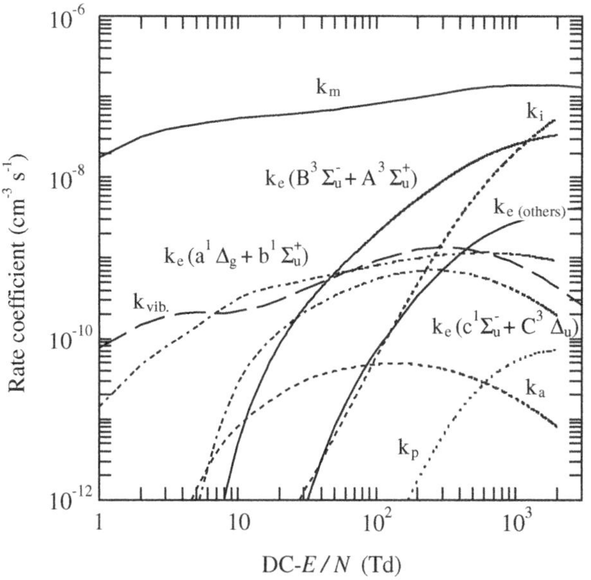

FIGURE 7.2 Collision rates of the electron in pure oxygen as a function of DC-E/N.

We need information on the induced magnetic and electric fields, $B(t)$ and $E(t)$, in addition to the static Poisson field in the case of inductively coupled plasmas (ICPs), and for that purpose we must solve Maxwell's equations. In the system it is necessary to extend the calculation to the two- or three-dimensional position space. For cylindrical coordinate (r, θ, z), the Equation 7.7 is replaced by the following system of Maxwell's equations, which includes Poisson's equation:

$$
\begin{aligned}
\frac{1}{r}\frac{\partial}{\partial r}(rE_r(r,\theta,z,t)) \quad + \quad & \frac{1}{r}\frac{\partial E_\theta(r,\theta,z,t)}{\partial \theta} + \frac{\partial E_z(r,\theta,z,t)}{\partial z} \\
= \quad & -\frac{1}{r}\frac{\partial}{\partial r}\left(r\frac{\partial V}{\partial r}\right) - \frac{1}{r^2}\frac{\partial^2 V}{\partial \theta^2} - \frac{\partial^2 V}{\partial z^2} \\
= \quad & \frac{e}{\varepsilon_0}(n_p - n_e - n_n); \qquad (7.9)
\end{aligned}
$$

Faraday's law:

$$
\left| \begin{matrix} \dfrac{1}{r}\dfrac{\partial E_z}{\partial \theta} - \dfrac{\partial E_\theta}{\partial z} \\[2ex] \dfrac{\partial E_r}{\partial z} - \dfrac{\partial E_z}{\partial r} \\[2ex] \dfrac{1}{r}\dfrac{\partial}{\partial r}(rE_\theta) - \dfrac{1}{r}\dfrac{\partial E_r}{\partial \theta} \end{matrix} \right| = -\frac{\partial}{\partial t} \left| \begin{matrix} B_r \\[2ex] B_\theta \\[2ex] B_z \end{matrix} \right|; \qquad (7.10)
$$

Ampere's law:

$$\frac{1}{\mu_0}\begin{vmatrix} \dfrac{1}{r}\dfrac{\partial B_z}{\partial \theta} - \dfrac{\partial B_\theta}{\partial z} \\ \dfrac{\partial B_r}{\partial z} - \dfrac{\partial B_z}{\partial r} \\ \dfrac{1}{r}\dfrac{\partial}{\partial r}(rB_\theta) - \dfrac{1}{r}\dfrac{\partial B_r}{\partial \theta} \end{vmatrix} = -\varepsilon_0 \frac{\partial}{\partial t}\begin{vmatrix} E_r \\ E_\theta \\ E_z \end{vmatrix} + \begin{vmatrix} J_{er} \\ J_{e\theta} \\ J_{ez} \end{vmatrix} + \begin{vmatrix} J_{pr} \\ J_{p\theta} \\ J_{pz} \end{vmatrix}; \tag{7.11}$$

and Coulomb's law in magnetics:

$$\frac{1}{r}\frac{\partial}{\partial t}(rB_r) + \frac{1}{r}\frac{\partial B_\theta}{\partial \theta} + \frac{\partial B_z}{\partial z} = 0; \tag{7.12}$$

where μ_0 is the magnetic permeability of vacuum and where $\mu_0\varepsilon_0 = c^{-2}$ (c is the speed of light).

Problem 7.2.2

Derive the following continuity equation for a short-lived excited molecule with a number density $N_j(k)$

$$\frac{\partial}{\partial t}N_k(t) = \Sigma_j k_{jk} n_e N_j + \Sigma_l \Sigma_m k_{lk} N_l N_m - \Sigma_j \Sigma_l k_{kj} N_k N_l - \frac{N_k}{\tau_{rad}^k}, \tag{7.13}$$

where k_{jk} is the collisional rate constant from the state j to k. τ_{rad}^k is the radiative lifetime of the molecule in the k state. The system describing a linkage among collisional excited molecules $N_j(j = 1, 2, \ldots)$ with short lifetimes is given by the simultaneous equations similar to Equation 7.13. The model is called a collisional radiative (CR) model.

7.2.2 Local Field Approximation (LFA)

When the transport coefficients in the continuity Equations 7.4 to 7.6 are given at each time as a function of reduced field strength $E(z,t)/N$, given by the number density of the feed gas N and the local instantaneous field $E(z,t)$, we can solve the governing Equations 7.4 to 7.7 and obtain the number densities of charged particles and the space potential. This approximation, known as the local field approximation (LFA), assumes a quasi-equilibrium of charged particles at all points in space and for all times. As mentioned earlier, this assumption allows the direct and fastest possible application of swarm data through an appropriate database. However, when the external driving frequency of the plasma increases, the temporal variation of the electric field occurs over a very short period of time, that is, comparable with or even shorter than the relaxation times for charged particles in an rf plasma, and the spatial variation of the field will be less than the relaxation length. Application of LFA becomes more difficult as the gas number density N is reduced, because the collisional relaxation time of the charged particle is proportional to N^{-1}. Even for higher neutral density (pressure), the spatial variation of the field in the sheath close to the electrode is very large and the application of LFA for these regions will be inaccurate. In practical terms, the LFA algorithm provides very fast numerical calculation and thus should be the first algorithm to be attempted whenever there exists a database of the transport (swarm) parameters.

7.2.3 Quasi-Thermal Equilibrium (QTE) Model

When electrons are in thermal equilibrium in a plasma, their energy distribution is Maxwellian. Maxwellian distribution is determined by a single parameter, electron temperature T_e, which significantly simplifies the calculation of collision rates. On the other hand, the temperature will be spatially dependent, and it should certainly depend on the time in an rf plasma, that is, $T_e(z,t)$. Nevertheless, this simplifies the calculation tremendously. If we assume that the cross sections for the ionization and electron attachment are constant with threshold energies ε_i and ε_a, respectively, we obtain rates in a very simple analytic form. For the ionization rate,

$$R_i(z,t) = Nk_i \exp\left(-\frac{\varepsilon_i}{kT_e(z,t)}\right), \tag{7.14}$$

where k is the Boltzmann constant and k_i is the constant intrinsic to the feed gas. The electron temperature is obtained from the energy $3\,kT_e(z,t)/2$ conservation equation given as

$$\begin{aligned}\frac{\partial}{\partial t}\left(\frac{3}{2}n_e(z,t)kT_e(z,t)\right) &+ \frac{\partial q_e(z,t)}{\partial z} \\ &= eE(z,t)\left(-n_e(z,t)V_{de}(z,t) + D_{Le}\frac{\partial n_e(z,t)}{\partial z}\right) \\ &- \varepsilon_i n_e(z,t)Nk_i \exp\left(-\frac{\varepsilon_i}{kT_e(z,t)}\right) \\ &- \varepsilon_a n_e(z,t)Nk_a \exp\left(-\frac{\varepsilon_a}{kT_e(z,t)}\right) \\ &- \sum \varepsilon_j n_e(z,t)Nk_j \exp\left(-\frac{\varepsilon_j}{kT_e(z,t)}\right),\end{aligned} \tag{7.15}$$

where the term $q_e(z,t)$ is the enthalpy defined for electrons as

$$q_e(z,t) = \frac{5}{3}\left(n_e(z,t)V_{de}(z,t) - D_{Le}\frac{\partial n_e(z,t)}{\partial z}\right)\frac{3}{2}kT_e\,. \tag{7.16}$$

This model is based on the second-order velocity moment, that is, energy balance. It is applicable to a plasma modeling, if one can represent data for the

TABLE 7.1 Comparison between a Reactive Processing Plasma and Conventional Collisional Plasma

Item	Reactive Processing Plasma	Conventional Plasma
Elements	Electrons, P(and N)-ions, Radicals, Photons, Feed gas	Electrons, P(and N)-ions, Photons, Feed gas
Indicator	Degree of ionization & Degree of dissociation	Degree of ionization
Characteristics	Plasma & radical densities	Plasma density
Function	Ion beam and Radical flux	Ion beam, Emission
Application	Surface etching, deposition & modification	Ion & Light sources

drift velocities and diffusion coefficients V_{dj} and D_{Lj} as a function of mean electron energy or, consequently, electron temperature $T_e(z,t)$. The QTE model assumes no phase delay between the mean electron energy $\varepsilon_m(z,t)$ given by the bulk of the energy distribution and the ionization rate R_i produced by high-energy electrons to be small, whereas in reality it is large and thus the ionization in a plasma will be overestimated.

Conditions that favor applicability of QTE are those in which there is a high degree of ionization so that Coulomb forces couple charged particles. In these circumstances, electrons lose their energy through the long-range interactions in e–e scattering and in e–p scattering, which can be represented as a large number of momentum transfers with small scattering angles (see Chapter 4). In such collisions, electrons are thermalized and reach a velocity distribution closer to the Maxwellian. The conditions in a low-temperature plasma used in integrated circuit production or thin film preparation and coating, and so on, range mainly from a low degree of ionization in which the velocity distribution of electrons is in nonthermal equilibrium to a high degree in which it is Maxwellian. Thus one must pay attention to which circumstances apply.

The shape of the Maxwellian energy distribution does not depend directly on the cross sections of the feed gas (see Section 2.3.2), where the Maxwellian is a logarithmically straight line depending only on the electron temperature T_e. By contrast, nonequilibrium energy distribution is strongly affected by the intrinsic short-range collisions and the cross sections, and the energy tail of the distribution will drop down very rapidly and to a much greater degree than in the Maxwellian distribution in the range greater than the threshold energy of the electronic excitation collisions. See Table 7.1.

7.2.4 Relaxation Continuum (RCT) Model

As mentioned earlier, an rf plasma maintained externally by an rf power source has two distinct regions: a positive ion sheath and a quasi-neutral bulk plasma.

TABLE 7.2 Low-Temperature Plasma Model and the Variable

Model	Variable	References
LFA	$E(\mathbf{r},t)/N$	Boeuf [4]
QTE	$< \varepsilon_e(\mathbf{r},t) >$	Graves [5]
RCT	$E^m_{eff}(\mathbf{r},t)/N$ for momentum $E^e_{eff}(\mathbf{r},t)/N$ for energy	Makabe [6]
Phase space	$\varepsilon_e(\mathbf{r},t)/N$	Sommerer [7]
Particle	$\varepsilon_e(\mathbf{r},t)$, $(\varepsilon_p(\mathbf{r},t))$	Birdsall [8]
Hybrid	—	Kushner [9]
Circuit	$V(t)$ or $I(t)$	—

In the sheath region, the electric field is very strong and changes rapidly both in space and time. It is usually impossible for charged particles to come to equilibrium under such conditions. In electropositive gases, the electron relaxation process is the most important in sustaining the rf plasma, whereas in electronegative gas both the massive positive ions and the negative ions should be considered in addition to the light electrons.

The relaxation continuum (RCT) model considers the momentum and energy relaxations in a simple and physically natural way. First we consider the momentum relaxation. From Equation 7.2, we write the momentum conservation in the form of a relaxation equation:

$$\frac{\partial}{\partial t}\Big(n_e(z,t)mV_{de}(z,t)\Big) = eE(z,t)n_e(z,t)$$
$$-\frac{n_e(z,t)mV_{de}(z,t)}{\langle\tau_m\rangle} - n_e(z,t)mV_{de}(z,t)\frac{\partial}{\partial z}V_{de}(z,t), \tag{7.17}$$

where $\langle\tau_m\rangle$ is the effective time constant for momentum transfer (the inverse of the momentum transfer collision rate and approximately equal to mV^0_{de}/eE by use of the DC-drift velocity V^0_{de}). The first term on the right-hand side is the momentum gain due to electric field acceleration, the second term is the loss due to momentum transfer collisions, and the third term describes the effect of the spatial gradient of drift velocity on relaxation.

The relaxation of the mean energy of electrons, $\varepsilon_m(z,t)$, is described by the conservation of energy,

$$\frac{\partial}{\partial t}\Big(n_e(z,t)\varepsilon_m(z,t)\Big) = eE(z,t)\Big(-n_e(z,t)V_{de}(z,t) + D_{Le}\frac{\partial n_e(z,t)}{\partial z}\Big)$$
$$-\Big(\frac{2m}{M}R_m\varepsilon_m(z,t) + \sum R_j\varepsilon_j + R_i\varepsilon_i\Big)n_e(z,t)$$
$$-\frac{\partial}{\partial z}\Bigg[\Big(n_e(z,t)V_{de}(z,t) - D_{Le}\frac{\partial n_e(z,t)}{\partial z}\Big)\varepsilon_m(z,t)\Bigg]. \tag{7.18}$$

The r.h.s in Equation 7.18 is described by the terms of the energy gain due to flux of charged particles in the electric field, followed by energy losses in elastic momentum transfer, inelastic and ionization collisions, and finally the spatial gradient of the mean energy. We introduce the effective field E_{eff}, where $E(z,t)^2_{eff} = \varepsilon_m / e\mu\langle\tau_e\rangle$, which gives a field modified from the actual field $E(z,t)$ to take into account the energy relaxation from the proper transport data under quasi-equilibrium. The relaxation equation for the effective field may be written as

$$\frac{\partial}{\partial t}\left(E_{eff}(z,t)^2 n_e(z,t)\right) = \frac{\left\{E_{eff}(z,t)^2 - E(z,t)^2\right\} n_e(z,t)}{\langle\tau_e\rangle} - \frac{\partial}{\partial z} q_e(z,t), \tag{7.19}$$

where the enthalpy is defined as

$$q_e(z,t) = n_e(z,t)V_{de}(z,t)E_{eff}(z,t)^2 - D_{Le}(z,t)\frac{\partial}{\partial z}\left(n_e(z,t)E_{eff}(z,t)^2\right). \tag{7.20}$$

The temporal relaxation time for electron diffusion is of the same magnitude as that for the mean energy. However, the energy relaxation time of electrons with energy greater than the excitation threshold ϵ_j is more difficult to estimate than those for drift velocity and mean energy, which are the result of the effect of the entire energy distribution of electrons. In particular, it is important to establish the relaxation time constant of the ionization rate τ_{R_i}. The ionization is the result of only the electrons at the highest end of the energy distribution, and these have kinetics quite different from the bulk of the distribution. Therefore, for τ_{R_i} we must make an estimate from the relaxation times for the mono-energetic groups of electrons (see Figure 7.3). The RCT model is a theory that maintains all the advantages of continuum models with respect to speed, simplicity, and ease of interpretation.

Problem 7.2.3

Discuss the time behavior of the drift velocity of electrons as a function of external field $E(t) = E_0 \cos\omega t$ from the momentum relaxation Equation 7.17 in the case of $\tau_m \ll \omega^{-1}$ (a) and $\tau_m \gg \omega^{-1}$ (b). For simplicity, we assume that the electron number density and the drift velocity are distributed uniformly in space.

7.2.5 Phase Space Kinetic Model

In order to determine the transport and rate coefficients of electrons precisely in space and time, we must solve the equation for the electron transport in phase space. We have already shown methods to solve the Boltzmann equation (see Section 5.6), but the most difficult issues are the need for huge CPU time and the possibility of numerical diffusion if the resolution is reduced in

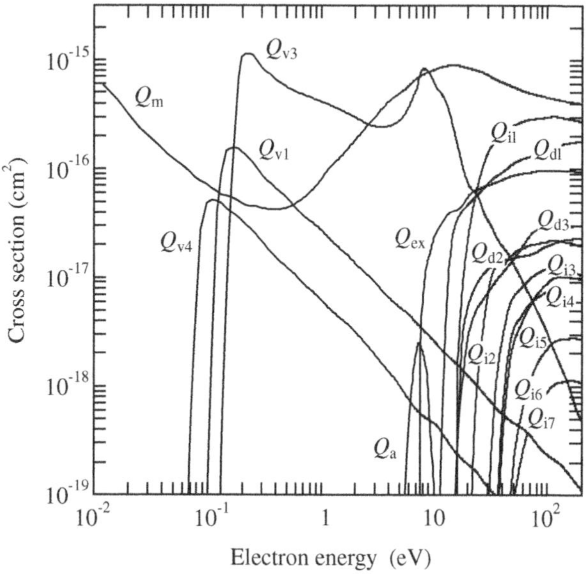

FIGURE 7.3 Set of collision cross sections in CF_4 as a function of electron energy.

order to speed up the computation. Here we discuss the two-term approximation (TTA) as a means of solving the Boltzmann equation for the time- and space-modulated electric field $E(z,t)$. The equation for the energy distribution, $f(\varepsilon, z, t)$, based on the TTA of the velocity distribution of electrons is

$$\begin{aligned}
&\frac{\partial}{\partial t} f(\varepsilon, z, t) \\
&= \sqrt{\frac{2}{m}} \frac{2m}{M} \frac{\partial}{\partial \varepsilon}\left(NQ_m(\varepsilon)\varepsilon^{3/2} f + \frac{MeE(z,t)\varepsilon}{6mNQ_T(\varepsilon)}\left(eE\frac{\partial}{\partial \varepsilon}(\varepsilon^{-1/2}f) + \frac{\partial}{\partial z}(\varepsilon^{-1/2}f)\right)\right) \\
&+ \sqrt{\frac{2}{m}} \frac{\varepsilon}{3NQ_T(\varepsilon)}\left(eE(z,t)\frac{\partial^2}{\partial \varepsilon \partial z}(\varepsilon^{-1/2}f) + \frac{\partial^2}{\partial z^2}(\varepsilon^{-1/2}f)\right) \\
&+ \sqrt{\frac{2}{m}} \sum_j \frac{\partial}{\partial \varepsilon} \int_{\varepsilon}^{\varepsilon + \varepsilon_j} \varepsilon^{1/2} NQ_j(\varepsilon) f d\varepsilon - \sqrt{\frac{2}{m}} \frac{\partial}{\partial \varepsilon} \int_0^{\varepsilon_a} \varepsilon^{1/2} NQ_a(\varepsilon) f d\varepsilon \\
&+ \sqrt{\frac{2}{m}} \frac{\partial}{\partial \varepsilon}\left(\int_{\varepsilon}^{(1+k)\varepsilon + \varepsilon_i} + \int_0^{\frac{1+k}{k}\varepsilon + \varepsilon_i}\right) \varepsilon^{1/2} NQ_i(\varepsilon) f d\varepsilon\,. \qquad (7.21)
\end{aligned}$$

Here, the energy distribution function f (actually the probability density) is normalized as $\int f(\varepsilon, z, t) d\varepsilon = 1$. The sum of the cross sections Q_m, Q_j, Q_a, and Q_i is the total cross section $Q_T = Q_m + \sum Q_j + Q_a + Q_i$. ϵ_j, ϵ_i, and ϵ_a are the threshold energies of the excitation, ionization, and electron attachment, respectively. (1:k) is the ratio of the energy partition between two electrons after ionization.

Equation 7.21 should be included within Equations 7.4 to 7.7 to make a complete set through the transport coefficients given from the integrated forms of $f(\varepsilon, z, t)$. In TTA, the longitudinal diffusion coefficient D_{L_e} is not well defined and should be estimated, as it is the most important component of the diffusion tensor in plasma modeling. However, the problem is also to properly represent the boundary conditions. In addition, in the region between the sheath and bulk plasma, where the net ionization rate is at a maximum, the velocity distribution of electrons is far from spherical and the accuracy of the TTA will go down.

We have already shown the (DNP) numerical procedure to solve the Boltzmann equation (see Sections 5.6 and 6.3). One of the procedures to overcome the appearance of the anisotropic velocity distribution in a high field and low-pressure region is the use of the DNP of the Boltzmann equation (see Section 6.3.4). In these circumstances, the most difficult issues are the need for huge CPU time and the possibility of numerical diffusion at low resolution.

7.3 PARTICLE MODELS

At low pressures the mean free path of electrons $\langle\lambda_e\rangle$ becomes comparable with or even larger than the characteristic dimension of the reactor d,

$$\langle\lambda_e\rangle \geq d, \tag{7.22}$$

and then the characteristics of plasma depend greatly on particle–surface interaction. It is difficult for electrons to relax their energy and momentum by binary collisions with the feed gas molecules. Thus, in the gas phase, electron transport strongly depends on previous history (non-Markovian process, or processes with memory), that is, on when and where the electron was produced. Under such conditions it is difficult to treat plasmas as fluids by using a continuum model.

An rf field and magnetic field are able to spatiotemporally trap electrons and ions. Effective pressure of the feed gas under an rf or a magnetic field (or magnetic fields) will become higher as the path of the electron is curved in such a way that the mean free path may be reduced by an increasing possibility of collision with the feed gas. In a magnetic field strong enough to confine charged particles, it is possible to describe transport by the Boltzmann equation, but the fluid description becomes increasingly difficult at lower pressure due to the increase of the transport coefficient and the complex behavior of the charged particles.

Under such conditions the so-called particle model is better. The model may be applied at higher pressure as well. In principle, this model treats individual particles, and it is based on some form of Monte Carlo simulation [10] and treatment of individual collisions in the gas phase and at surfaces. The main advantage of a particle model as opposed to a fluid model is that no assumption for the velocity or energy distribution of charged particles is made implicitly.

7.3.1 Monte Carlo Simulations (MCSs)

The binary collision between particles in a low-temperature plasma is a stochastic process described by the set of collision cross sections (i.e., collision probabilities). The collision events are simulated by Monte Carlo Simulation (MCS) by using the pseudo-random numbers generated in a computer under the database of the set of collision cross sections in the feed gas (see Figure 7.4). MCS traces the trajectory of a number of particles under the external electric $\boldsymbol{E}$ or magnetic field $\boldsymbol{B}$ by considering the collision event as a function of kinetic energy. In the case of an electron at $(\boldsymbol{v}, \boldsymbol{r}, t)$, for example, the trajectory of the electron at $t + \Delta t$ is tracked by Newton's equation

$$\frac{d^2}{dt^2}\boldsymbol{r}(t) = \frac{e}{m}\left[\boldsymbol{E}(\boldsymbol{r}, t) + \boldsymbol{v} \times \boldsymbol{B}(\boldsymbol{r}, t)\right], \tag{7.23}$$

and during the short period Δt, the collision occurs when the total collisional probability $NQ_T(\varepsilon)v\Delta t$ satisfies

$$NQ_T(\varepsilon)v\Delta t \geq \xi_r. \tag{7.24}$$

Here, $\{\xi_r\}$ is the random number distributed uniformly in $[0, 1]$. Of course, the time step Δt must satisfy the condition

$$\Delta t \ll \frac{1}{NQ_T(\varepsilon)v}. \tag{7.25}$$

Problem 7.3.1

The probability that the electron will have a collision in the time interval $(0, t)$ *is given by*

$$p(t) = N\nu_T(\varepsilon(t))exp\left[-\int_0^t N\nu_T(\varepsilon(t))dt\right],$$

where $\nu_T(\varepsilon(t))$ *is the time-dependent total collisional frequency and given by* $Q_T(\varepsilon(t))v(t)$*. Derive the above relation.*

At the same time, the scattering angle (θ, ϕ) after the collision must be determined by the random number. In the case of a collision, the energy and the velocity of the particle change, subject to the momentum and the energy conservation rule between two colliding particles (see Chapter 4). The trace of the electron with $(\boldsymbol{v}, \boldsymbol{r}, t + \Delta t)$ after the collision restarts during the next Δt, and the same procedure is carried out.

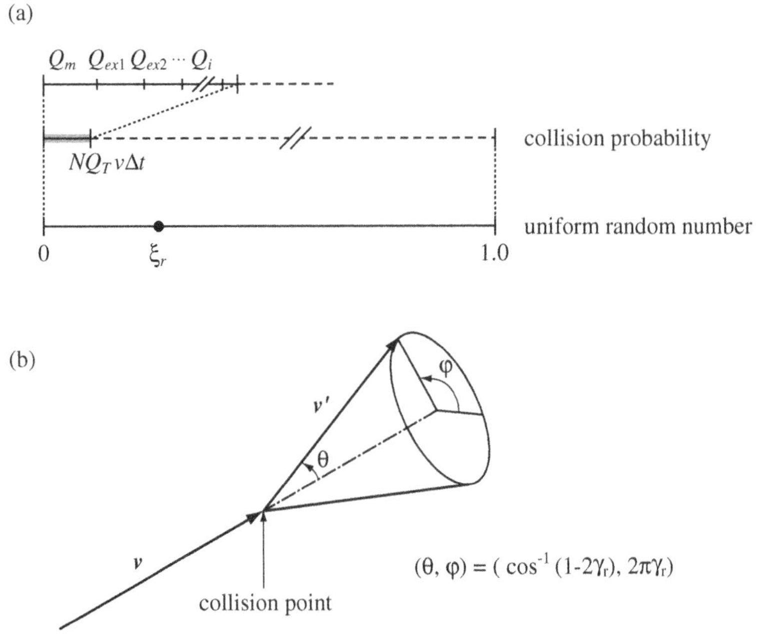

FIGURE 7.4 Determination of the collision and the type by the random number $\{\xi_r\}$ (a) and the scattering angle at collision $\{\gamma_r\}$ (b).

Exercise 7.3.1

An electron has a binary collision with a neutral molecule and is scattered isotropically. Derive the distribution of the scattering angle (θ, ϕ) with respect to the incident direction given by

$$p(\theta) = \frac{1}{2}\sin\theta, \tag{7.26}$$

$$p(\phi) = \frac{1}{2\pi}. \tag{7.27}$$

The function $p(\theta, \phi)$ (the probability density function) distributed uniformly in a unit sphere $(1, \theta, \phi)$ is expressed by

$$p(\theta, \phi) = \frac{\sin\theta d\theta d\phi}{4\pi}.$$

The θ and ϕ in the spherical coordinates are independent of each other, and we can write

$$p(\theta, \phi)d\theta d\phi = p_\theta(\theta)d\theta\, p_\phi(\phi)d\phi,$$

where $p_\theta(\theta)$ and $p_\phi(\phi)$ are normalized to unity:

$$\int_0^\pi p_\theta(\theta)d\theta = 1 \quad \text{and} \quad \int_0^{2\pi} p_\phi(\phi)d\phi = 1. \tag{7.28}$$

Therefore,

$$p_\theta(\theta)d\theta \int_0^{2\pi} p_\phi(\phi)d\phi = \frac{2\pi \sin\theta d\theta}{4\pi} = \frac{1}{2}\sin\theta d\theta, \tag{7.29}$$

$$p_\phi(\phi)d\phi \int_0^{\pi} p_\theta(\theta)d\theta = \frac{2d\phi}{4\pi} = \frac{1}{2\pi}d\phi\,. \tag{7.30}$$

Exercise 7.3.2
Obtain the scattering angle (θ, ϕ) distributed isotropically by using a series of random numbers $\{\gamma_r\}$ uniformly distributed between $[0, 1]$.
From Equation 7.29,

$$\int_0^{\theta} \frac{1}{2}\sin\theta d\theta = \gamma_r \quad \rightarrow \quad \cos\theta = 1 - 2\gamma_r. \tag{7.31}$$

In the same way,

$$\int_0^{\phi} \frac{1}{2\pi}d\phi = \gamma_r \quad \rightarrow \quad \phi = 2\pi\gamma_r. \tag{7.32}$$

An expression of the stochastic judgment is possible by using additional random numbers to more physically precise or complicated collision processes, for example, anisotropic scattering or energy share to two electrons after ionization.

The formulae used to sample transport coefficients are taken from the stochastic theory. We can also determine the energy distribution and velocity distribution of electrons. The transport coefficient of electrons is determined from the position $\boldsymbol{r}$ in the mean position of the electron swarm. That is, the effective ionization rate is given by

$$R_0 = \frac{1}{n_e(\boldsymbol{r},t)}\frac{d}{dt}n_e(\boldsymbol{r},t), \tag{7.33}$$

the drift velocity is obtained as

$$\boldsymbol{v}_d = \frac{d}{dt}\langle \boldsymbol{r}(t)\rangle\,, \tag{7.34}$$

and the diffusion tensor is given by

$$\mathbf{D} = \frac{1}{2}\frac{d}{dt}\left\langle (\boldsymbol{r}(t) - \langle \boldsymbol{r}(t)\rangle)^2 \right\rangle. \tag{7.35}$$

7.3.2 Particle-in-Cell (PIC) and Particle-in-Cell/Monte Carlo Simulation (PIC/MCS) Models

A high-density plasma source in a low-temperature plasma is operated at low pressure, sometimes under the support of an external magnetic field. A strongly ionized plasma consists of electrons and ions interacting through the

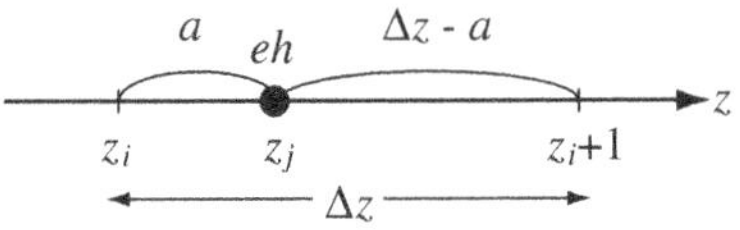

FIGURE 7.5 Superparticle and the charge sharing between the nearest two grids.

electromagnetic fields. The charged particles are subject to the long-range Coulomb interaction among particles. These plasmas are classified as collisionless plasmas. It is quite difficult to trace the motion of a huge number of charged particles in the collisionless plasma. One of the practical methods to overcome this difficulty is the use of the particle-in-cell (PIC) model [8]. The PIC model considers:

1. A finite number of superparticles, consisting of h electrons (ions) with constant (e/m) rather than all the electrons and ions in the reactor; and
2. A particle cloud with a finite spread in place of all the independent particles.

Then, the superparticle with charge eh is traced by using Newton's equation under the external field at time step Δt. When a superparticle is found at z_j, at distance a from the lattice z_j in Figure 7.5, the charge is divided into aeh and $(\Delta z - a)eh$ at z_j and z_{j+1}, respectively. The electric field at the position of the superparticle is given by linear interpolation, after the field is calculated by Poisson's equation at each of the lattices

$$E(z_j) = aE(z_i) + (\Delta z - a)E(z_{i+1}) \; . \tag{7.36}$$

In order to avoid numerical instability in the explicit particle simulation, the time step Δt and mesh space Δz are respectively taken as

$$\Delta t < \omega_{pe}^{-1} \; , \tag{7.37}$$

$$\Delta z < \lambda_D \; , \tag{7.38}$$

where ω_{pe} and λ_D are the plasma frequency and Debye length in a collisionless plasma (see Chapter 3).

When a plasma is maintained by electron impact ionization of a feed gas molecule, we must consider the short-range interaction even in the case where a strongly ionized plasma is mainly subject to the long-range interaction. The binary collision is considered by MCS. Thus the model is classified as a PIC/MCS model. PIC/MCS is widely used in low-pressure plasmas. When the operating pressure of a plasma increases, the number of collision events between the electron and feed gas molecule is huge, and thus the continuum fluid model is much more appropriate for describing the system. The main advantage of the particle model as opposed to the fluid model is that, in the former, no assumption for the velocity or energy distribution of charged particles is made implicitly. In compensation, the cost is high because the PIC model is very time consuming.

7.4 HYBRID MODELS

As described in the section on particle models, one of the difficulties in applying the particle model to a practical system is the cost of the computational time. One of the proposed means of overcoming this difficulty is to combine the continuum model with the particle model.

An example of such combination is the LFA model coupled with the MCS of electrons to identify the nonhydrodynamic behavior of the transport coefficient. Another is the system combining the PIC/MCS model of electrons and the RCT model of massive ions in a low-pressure plasma, especially in a magnetron plasma, where the drift and diffusion of electrons exhibit local and complex characteristics (see Chapter 11).

7.5 CIRCUIT MODEL

Electrical properties, such as the sustaining voltage $V_{sus}(t)$, discharge current $I(t)$, dissipated power $W(t)$, and space potential $V(z)$ are the basis for the understanding of discharge plasmas. These electrical properties have been modeled by using the electrical impedance of the sheath and bulk plasma [11, 12]. A knowledge of the impedance is related to the plasma parameters and is used for the impedance matching of the rf power supply (see Chapter 9). The estimation of the plasma characteristics through the electrical impedance is achieved with an equivalent circuit model of the rf plasma (see Figure 7.6).

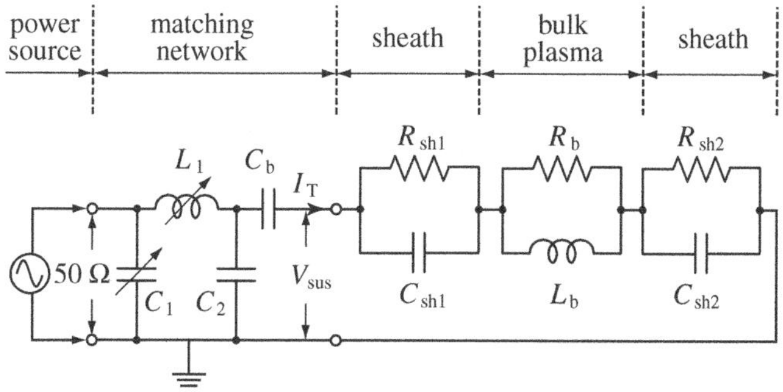

FIGURE 7.6 Typical equivalent circuit of a capacitively coupled rf plasma with an external impedance matching network.

7.5.1 Equivalent Circuit Model in CCP

We first consider a typical CCP sustained between two electrodes. There exist a quasi-neutral bulk plasma and two ion sheaths in front of both electrodes. In a uniform bulk plasma with electrons and positive ions, the current is composed of the conduction current of electrons and ions, $J_e(z,t)$ and $J_p(z,t)$, and the displacement current, $J_{Dis}(z,t)$. Due to the great difference of the mass between the electron and the ion, the ion current will be neglected as compared with the electron in electropositive plasmas. Then,

$$\boldsymbol{J}_T(z,t) \sim \boldsymbol{J}_e(z,t) + \boldsymbol{J}_{Dis}(z,t).$$

In the uniform bulk plasma, a drift current is dominant and by considering the phase delay ϕ of the drift velocity of electrons $\boldsymbol{v}_{de}(t)$ with respect to the field with sinusoidal waveform $\boldsymbol{E}_b \exp(j\omega t)$, the momentum conservation of electrons is

$$\frac{d}{dt}\{m\boldsymbol{v}_{de} \exp j(\omega t - \phi)\} = e\boldsymbol{E}_b \exp(j\omega t) - m\boldsymbol{v}_{de} \exp[j(\omega t - \phi)]R_m,$$

where R_m is the total collision rate of electrons. Then, the electron conduction current density is given as

$$\begin{aligned} \boldsymbol{J}_e(t) &= en_e\boldsymbol{v}_{de}(t) = \frac{n_e e^2}{m(R_m + j\omega)}\boldsymbol{E}_b \exp(j\omega t) \\ &= \frac{n_e e^2}{m\omega}\left(\frac{R_m}{\omega}\frac{1}{1+(\frac{R_m}{\omega})^2} - j\frac{1}{1+(\frac{R_m}{\omega})^2}\right)\boldsymbol{E}_b \exp(j\omega t) . \end{aligned}$$

Then, the total bulk plasma current density is

$$\boldsymbol{J}_T(t) = \frac{n_e e^2}{m\omega}\left(\frac{R_m}{\omega}\frac{1}{1+(\frac{R_m}{\omega})^2} - j\frac{1}{1+(\frac{R_m}{\omega})^2} + j\frac{\varepsilon_0 m}{n_e e^2}\omega^2\right)\boldsymbol{E}_b \exp(j\omega t). \tag{7.39}$$

As a result, the first in-phase oscillation term gives a power dissipation at the resistive impedance (plasma resistance R_b), the second and third out-of-phase oscillations correspond to an inductive and capacitive impedance ($j\omega L_b$ and $1/j\omega C_b$) with no power dissipation (reactance).

$$R_b = \frac{m\omega}{n_e e^2}\frac{\omega}{R_m}\left(1+\left(\frac{R_m}{\omega}\right)^2\right)\frac{d_b}{S_b} = \left(\frac{\omega}{\omega_{pe}}\right)^2\frac{1}{R_m}\left(1+\left(\frac{R_m}{\omega}\right)^2\right)\left(\frac{\varepsilon_0 S_b}{d_b}\right)^{-1}, \tag{7.40}$$

$$L_b = \frac{m\omega_0}{n_e e^2}\left(1+\left(\frac{R_m}{\omega}\right)^2\right)\frac{d_b}{S_b} = \frac{1}{\omega_{pe}^2}\left(1+\left(\frac{R_m}{\omega}\right)^2\right)\left(\frac{\varepsilon_0 S_b}{d_b}\right)^{-1}, \tag{7.41}$$

$$C_b = \frac{\epsilon_0 S_b}{d_b}, \tag{7.42}$$

where $\omega_{pe}^2 (= e^2 n_e / m\varepsilon_0)$ is the electron plasma frequency. S_b and d_b are the cross sectional area and thickness of the bulk plasma. In fact, the third in Equation 7.39, displacement component will be negligible in electropositive plasmas. In addition, in a collision-dominated plasma with $R_m \gg \omega$, the impedance of the bulk plasma is represented by a resistive component R_b.

In the sheath of a typical rf discharge plasma, the circumstances change the relation completely among the current components, J_e, J_p, and J_{Dis}. The sheath region having $n_p \gg n_e$ is formed in front of electrodes or a wafer in a discharge plasma, and the displacement current occupies a large part of $\boldsymbol{J}_T$, although finite currents of positive ions and electrons flow under a DC self-bias voltage in a time-averaged fashion with $\overline{(J_e(t) + J_p(t))} = 0$. As a result, the equivalent circuit of the sheath having a voltage drop V_{sh} is composed of the capacitance $C_{sh} (= \varepsilon_0 S_{sh} / d_{sh})$ and resistance R_{sh}, where S_{sh} and d_{sh} are the cross sectional area and thickness of the sheath. The power is dissipated at the resistance R_{sh} by a collisional effect between the charged particle and feed gas molecule.

$$\boldsymbol{J}_T(t) \sim J_{Dis} + J_p \sim \left(j\omega C_{sh} + \frac{1}{R_{sh}} \right) V_{sh} . \tag{7.43}$$

We show a typical equivalent circuit of the CCP with metallic electrodes connected to an external impedance matching network in Figure 7.6.

When there is a dielectric substrate on a metallic electrode, we have to add a series capacitor between the sheath circuit and the matching network (or grounded-metal electrode). In front of the reactor wall, a wall sheath is formed and is modeled by the capacitance similar to the above model. The equivalent circuit model is usually created by using the passive elements (i.e., resistance, capacitance, and inductance). Sometimes, active elements such as the diode and current source may be included in the circuit to realize the detailed system.

7.5.2 Equivalent Circuit Model in ICP

A plasma externally driven by an rf coil current is classified as an ICP or transformer coupled plasma (see Chapter 10). The equivalent primary and secondary circuits of the transformer are used to model the electrical properties of the total system of an ICP as shown in Figure 7.7. That is, the equivalent circuit is described as a nonideal transformer with a coupling coefficient $k = (M/L_c L_p)^{1/2}$ between the inductance of the current coil (L_c) and of the plasma (L_p) with resistance (R_p) caused by collisions of electrons. The plasma current flows through the inductor with L_p and resistor of R_p.

7.5.3 Transmission-Line Model (TLM)

A very large area processing with a wafer of (sub) meter in size is needed in industrial applications of plasma deposition or etching. With increasing the

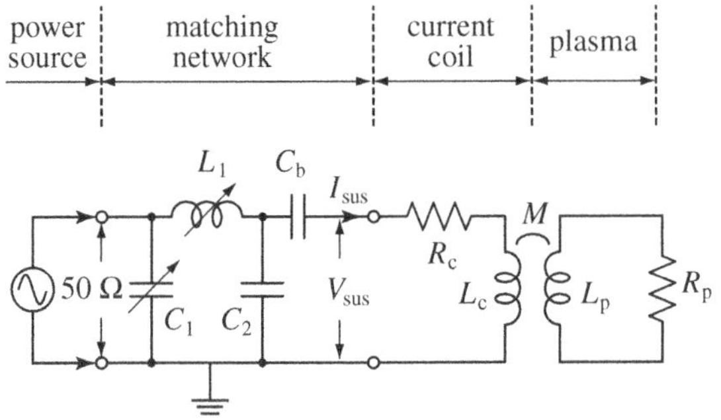

FIGURE 7.7 Typical equivalent circuit of an inductively coupled rf plasma with an external impedance matching network.

frequency of the external rf source, the dimension of the plasma reactor becomes comparable with a quarter of the free-space wavelength of the rf power source. Under these circumstances, it is of first importance to estimate the local distribution of the surface potential on the powered electrode and to perform the plasma uniformity, because the surface potential is influenced by the size of and rf feed position on the electrode. Transmission-line model (TLM) is appropriate to the estimation of the local surface potential as a function of feed position of an rf power on the electrode [13]. An equivalent circuit of a rod to plane electrode system is shown in Figure 7.8 as a transmission line in a small length Δx along the rod. The circuit is named the T-type equivalent circuit, and the series impedance and parallel admittance per unit length are written as $\boldsymbol{Z}$ and $\boldsymbol{Y}$. In Figure 7.8a, we have two relations between $\boldsymbol{V}(x)$ and $\boldsymbol{I}(x)$,

$$\frac{d\boldsymbol{V}(x)}{dx} = -\boldsymbol{Z}\boldsymbol{I}(x) \quad \text{and} \quad \frac{d\boldsymbol{I}(x)}{dx} = -\boldsymbol{Y}\boldsymbol{V}(x).$$

These are arranged as the Helmholtz equation in the one-dimensional rod to plane electrode system,

$$\frac{d^2\boldsymbol{V}(x)}{dx^2} = \boldsymbol{Z}\boldsymbol{Y}\boldsymbol{V}(x). \tag{7.44}$$

Here note that $\boldsymbol{V}(x)$, $\boldsymbol{I}(x)$, $\boldsymbol{Z}$, and $\boldsymbol{Y}$ are complex numbers, expressing the amplitude and phase shift. In the TLM, three types of T-type circuits are usually used (see Figure 7.8). We deal with a plasma sustained between the rod and plane with distance l. The series impedance $\boldsymbol{Z}$ and parallel admittance $\boldsymbol{Y}$ have to be investigated. The widths of the sheath and bulk plasma are d_{sh} and d_b, perpendicular to the electrodes, respectively; that is, $l = 2d_{sh} + d_b$. Then, the impedance $\boldsymbol{Z}$ is written as

$$\begin{aligned} \boldsymbol{Z} &= R + j\omega L\ , \\ &= \rho/(2\pi r t)\ +\ j\omega\frac{\mu_0}{2\pi}\left(\log\left(\frac{1+\sqrt{r_0^2+1}}{r_0}\right) - \sqrt{r_0^2+1} + r_0\right), \end{aligned} \tag{7.45}$$

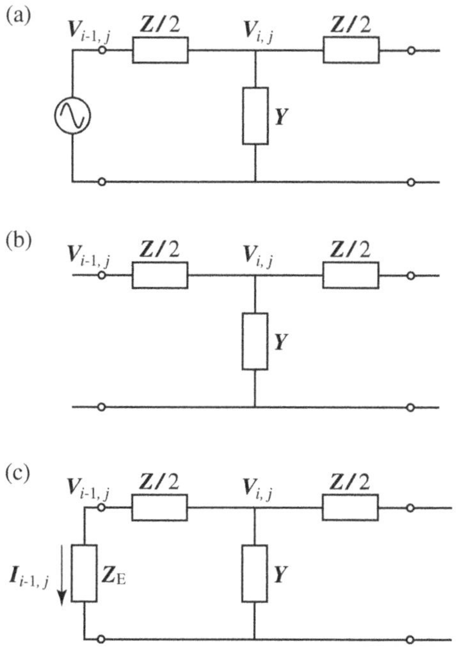

FIGURE 7.8 Transmission line, modeled by T-type equivalent circuit (a) and the other elemental circuits (b) and (c).

where ρ is the resistivity of the rod electrode with radius r_0. The rf current flows in a thin region $r_0 \geq r \geq (r_0 - d_{skin})$ along the rod. d_{skin} is the skin depth. μ_0 is the permeability in vacuum. The plasma admittance including the sheath $\boldsymbol{Y}$ per unit length is given in the form similar to Section 7.5.1,

$$\frac{1}{\boldsymbol{Y}} = \frac{R_b}{1 + \left(\frac{R_b}{\omega L_b}\right)^2} + j\frac{1}{1 + \left(\frac{R_b}{\omega L_b}\right)^2}\frac{R_b}{\omega L_b} - j\frac{2}{\omega C_{sh}}, \tag{7.46}$$

where R_b and L_b are given by Equations 7.40 and 7.41, respectively.

7.6 ELECTROMAGNETIC FIELDS AND MAXWELL'S EQUATIONS

In this section we give a brief summary of some of the laws in electromagnetic theory that are relevant for understanding generation and maintenance of plasmas.

7.6.1 Coulomb's Law, Gauss's Law, and Poisson's Equation

Electric field $\boldsymbol{E}(\boldsymbol{r})$ surrounding a charge e is determined from the electrostatic force $\boldsymbol{F}(\boldsymbol{r})$ on a test particle (of charge q) and it is given by Coulomb's law

$$\frac{\boldsymbol{F}(\boldsymbol{r})}{q} = \boldsymbol{E}(\boldsymbol{r}) = \frac{e}{4\pi\varepsilon_0 r^2}\frac{\boldsymbol{r}}{r}, \qquad (7.47)$$

where ε_0 is the permittivity of vacuum. Equation 7.47 allows us to determine the integral of the magnitude of the field over the surface of a sphere at a distance r

$$|\boldsymbol{E}(r)|4\pi r^2 = \frac{e}{\varepsilon_0},$$

which can be written in an integral form

$$\int \boldsymbol{E}(\boldsymbol{r}) \cdot \boldsymbol{n} dS = \frac{e}{\varepsilon_0}, \qquad (7.48)$$

that is often known as Gauss's law. If we have more than one charge inside the volume and based on the linear additivity of electric field, we write Gauss's law in a more general form, which can be further expanded by using Gauss's theorem

$$\int_S \boldsymbol{E}(\boldsymbol{r}) \cdot \boldsymbol{n} dS = \int_V \mathrm{div}\, \boldsymbol{E}(\boldsymbol{r}) \mathrm{dV} = \frac{1}{\varepsilon_0}\int_V \rho(\boldsymbol{r}) \mathrm{dV}.$$

From this equation we may directly obtain the differential form that is often known as Poisson's equation:

$$\mathrm{div}\, \boldsymbol{E}(\boldsymbol{r}) = \frac{1}{\varepsilon_0}\rho(\boldsymbol{r}) \,. \qquad (7.49)$$

In principle, Gauss's law is valid for all fields that have sources, and it states that the flux of the field over a closed surface is equal to the charges that are inside the volume defined by that surface. Because a magnetic field has no (known) particles that may be its source, we have

$$\mathrm{div}\, \boldsymbol{B}(\boldsymbol{r}) = 0. \qquad (7.50)$$

Poisson's equation is often used in plasma physics to determine the field in a system with space charge.

7.6.2 Faraday's Law

In the year 1831 Faraday established that a time-varying magnetic field induces voltage $u(V)$ or current $I(A)$ in a closed conductor with resistance $R(\Omega)$. See Figure 7.9. The induced current is actually proportional to the time derivative of flux of magnetic field $\Phi(t)$,

$$u = IR = -\frac{d\Phi(t)}{dt}. \qquad (7.51)$$

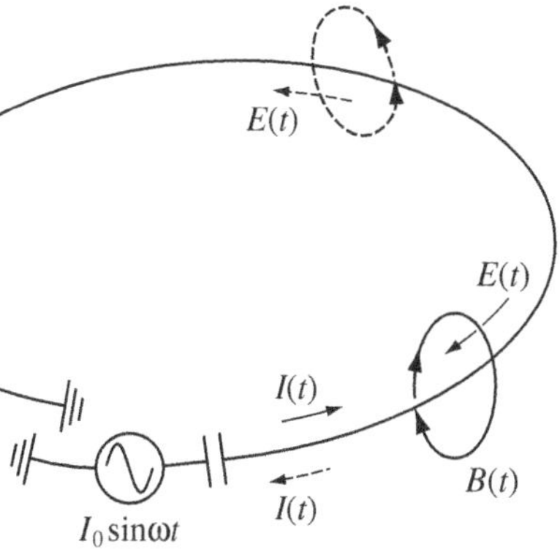

FIGURE 7.9 Schematic representation of the effect of a time-varying current $I_0 \sin \omega t$ that induces external field $E(t)$. However, the external time-varying magnetic field $B(t)$ induces a field in the conductor (dashed lines) and affects the resulting current.

The total induced potential is equal to the line integral of the local induced electric field $\boldsymbol{E}(\boldsymbol{r}, t)$. The flux of magnetic field $\Phi(t)$ is defined in a similar way as the flux of an electric field. Thus we have

$$u = \int \boldsymbol{E}(\boldsymbol{r}, t) \cdot d\boldsymbol{l}$$

$$\Phi = \int_S \boldsymbol{B}(\boldsymbol{r}, t) \cdot \boldsymbol{n} dS. \tag{7.52}$$

If we combine Equations 7.51 and 7.52 we obtain

$$u = \int \boldsymbol{E}(\boldsymbol{r}, t) \cdot d\boldsymbol{l} = -\frac{d}{dt} \int_S \boldsymbol{B}(\boldsymbol{r}, t) \cdot \boldsymbol{n} dS .$$

When we replace the line integral by a surface integral by using Stokes's theorem, we obtain

$$\int_S \left\{ \mathrm{rot} \boldsymbol{E}(\boldsymbol{r}, \mathrm{t}) + \frac{\partial}{\partial \mathrm{t}} \boldsymbol{B}(\boldsymbol{r}, \mathrm{t}) \right\} \cdot \boldsymbol{n} dS = 0,$$

which gives directly

$$\mathrm{rot} \boldsymbol{E}(\boldsymbol{r}, \mathrm{t}) + \frac{\partial}{\partial \mathrm{t}} \boldsymbol{B}(\boldsymbol{r}, \mathrm{t}) = 0, \tag{7.53}$$

known as Faraday's law.

7.6.3 Ampere's Law

If, on the other hand, we wish to calculate the magnetic field induced by a conductor that has a current I, we obtain the line integral of magnetic field according to

$$\int \boldsymbol{B}(\boldsymbol{r}) \cdot d\boldsymbol{l} = \mu_0 I, \tag{7.54}$$

where μ_0 is the magnetic permeability in vacuum. The current is separated into two terms, convection and displacement currents, and the total current that passes through a closed loop defined in an area S is obtained by integrating the current density $\boldsymbol{j}(\boldsymbol{r}, t)$ and the displacement current defined through the time-varying electric field $\boldsymbol{E}(t)$ as

$$I(\boldsymbol{r}) = \int \left(\boldsymbol{j}(\boldsymbol{r}, t) + \varepsilon_0 \frac{\partial \boldsymbol{E}(\boldsymbol{r}, t)}{\partial t} \right) \cdot \boldsymbol{n} dS, \tag{7.55}$$

which when combined with Equation 7.54 leads to

$$\int \boldsymbol{B}(\boldsymbol{r}, t) \cdot d\boldsymbol{l} = \mu_0 \int_S \left\{ \boldsymbol{j}(\boldsymbol{r}, t) + \varepsilon_0 \frac{\partial \boldsymbol{E}(\boldsymbol{r}, t)}{\partial t} \right\} \cdot \boldsymbol{n} dS.$$

If we replace the term on the left-hand side by using Stokes's theorem we obtain

$$\int \boldsymbol{B}(\boldsymbol{r}, t) \cdot d\boldsymbol{l} = \int_S \mathrm{rot} \boldsymbol{B}(\boldsymbol{r}, \mathrm{t}) \cdot \boldsymbol{n} \mathrm{dS}\,.$$

The same equation may be written in a differential form

$$\frac{1}{\mu_0} \mathrm{rot} \boldsymbol{B}(\boldsymbol{r}, \mathrm{t}) = \boldsymbol{j}(\boldsymbol{r}, \mathrm{t}) + \varepsilon_0 \frac{\partial \boldsymbol{E}(\boldsymbol{r}, \mathrm{t})}{\partial \mathrm{t}}, \tag{7.56}$$

which is also known as Ampere's (or the Ampere–Maxwell) law.

7.6.4 Maxwell's Equations

A complete set of differential equations giving the relationship between electric and magnetic fields and their external sources ($\boldsymbol{j}$ and ρ) is given as

$$\begin{aligned} \mathrm{rot} \boldsymbol{E}(\boldsymbol{r}, \mathrm{t}) &= -\frac{\partial \boldsymbol{B}(\boldsymbol{r}, t)}{\partial t} \\ \frac{1}{\mu_0} \mathrm{rot} \boldsymbol{B}(\boldsymbol{r}, \mathrm{t}) &= \boldsymbol{j}(\boldsymbol{r}, t) + \varepsilon_0 \frac{\partial \boldsymbol{E}(\boldsymbol{r}, t)}{\partial t} \\ \mathrm{div} \boldsymbol{E}(\boldsymbol{r}, \mathrm{t}) &= \rho(\boldsymbol{r}) / \varepsilon_0 \\ \mathrm{div} \boldsymbol{B}(\boldsymbol{r}, \mathrm{t}) &= 0\,. \end{aligned} \tag{7.57}$$

These are Maxwell's equations, which provide a basis for understanding the development of electric and magnetic fields.

References

[1] Chua, L.O. and Lin, P.M. 1975. *Computer-Aided Analysis of Electronic Circuit.* Englewood Cliffs, NJ: Prentice-Hall.

[2] Makabe, T. Ed. 2002. *Advances in Low Temperature RF Plasmas: Basis for Process Design.* Amsterdam: Elsevier.

[3] Kim, H.C., Iza, F., Yang, S.S., Radjenovic, M.R., and Lee, J.K. 2005. *J. Phys. D. Topical Review* 38: R283.

[4] Boeuf, J.-P. 1987. *Phys. Rev.* A 36: 2782.

[5] Graves, D.B. 1987. *J. Appl. Phys.* 62: 88.

[6] Okazaki, K., Makabe, T., and Yamaguchi, Y. 1989. *Appl. Phys. Lett.* 54: 1742. Makabe, T., Nakano, N. and Yamaguchi, Y. 1992. *Phys. Rev.* A 45: 2520.

[7] Sommerer T.J., Hitchon, W.N.G., and Lawler, J.E. 1989. *Phys. Rev. Lett.* 63: 2361.

[8] Birdsall, C.K. and Langdon, A.B. 1985. *Plasma Physics via Computer Simulation.* New York: McGraw-Hill.

[9] Kushner, M.J. 1986. *IEEE Trans. Plasma Sci.* PS-14: 188.

[10] Binder, K. and Heermann, B.W. 1988. *Monte Carlo Simulation in Statistical Physics.* Berlin: Springer-Verlag.

[11] Koenig, H.R. and Maissel, L.I. 1970. *IBM J. Res. Develop.* 14: 168.

[12] Piejak, R.B., Godyak, V.A., and Alexandrovich, B.M. 1992. *Plasma Sources Sci. Technol.* 1: 179.

[13] Satake, K., Yamakoshi, H., and Noda, M. 2004. *Plasma Sources Sci. Technol.* 13:436.

CHAPTER 8

Numerical Procedure of Modeling

8.1 TIME CONSTANT OF THE SYSTEM

The low-temperature radio frequency (rf) plasma used for a microelectronic device fabrication includes a number of physical and chemical collision/reaction/transport processes with very different time constants or relaxation times, and the system is expressed by stiff differential equations. This stiffness constitutes a fundamental numerical problem, and we consider methods for overcoming it (see Figure 8.1).

8.1.1 Collision-Oriented Relaxation Time

In a collision-dominated plasma, the electron changes the direction of motion in a field by a collision with a neutral molecule. Rf plasma, sustained by an external rf power source, changes the direction of the bulk field once per period. Each of the bulk electrons can follow the local field with a finite time delay through a two-body collision, and the time constant of the momentum change is the collisional momentum relaxation time $\langle \tau_m \rangle$, which is derived as

$$\frac{1}{\langle \tau_m \rangle} = R_m + \Sigma R_j, \tag{8.1}$$

where R_m and R_j are the rates of the elastic momentum transfer and the inelastic scattering with threshold energy ϵ_j, respectively. The energy relaxation time of electrons by two-body collisions in a time-varying field are obtained from the conservation of energy in Chapter 7:

$$\frac{1}{\langle \tau_\epsilon \rangle} = \frac{2m}{M} R_m + \frac{\Sigma \epsilon_j R_j}{\langle \epsilon_m \rangle}, \tag{8.2}$$

where m and M are the mass of the electron and the molecule, respectively, and $\langle \epsilon_m \rangle$ is the averaged energy of electrons.

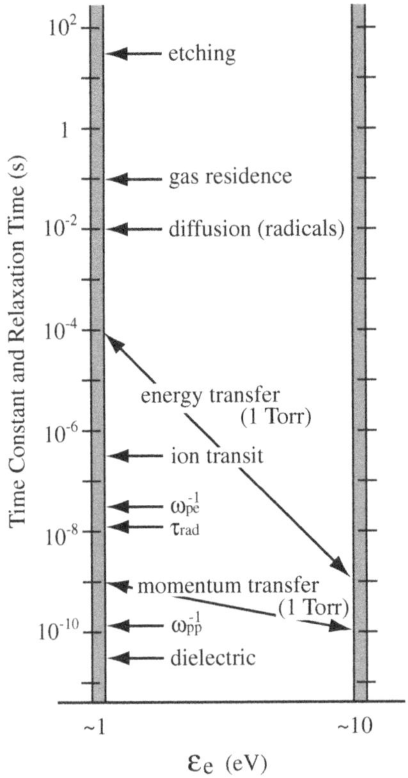

FIGURE 8.1 Time constant and relaxation time of each of the processes in a low-temperature plasma.

8.1.2 Plasma Species-Oriented Time Constant

Characteristics of the species in a plasma in a periodic steady state depend on the transport. For neutral species, the characteristic diffusion time,

$$\tau_D = \frac{\Lambda^2}{D}, \tag{8.3}$$

is necessary to obtain a steady-state profile of neutral molecules. Here, D and Λ are the diffusion coefficient and characteristic diffusion length, respectively. In the case of a system including a reaction to another molecule, the effective diffusion time τ_{Dr} is shortened from τ_D to

$$\tau_{Dr} = \frac{\Lambda^2}{D} \frac{1}{1 + \frac{k_r N_r \Lambda^2}{D}}, \tag{8.4}$$

where k_r is the reaction rate constant, and N_r the reactant number density. For charged particles, the characteristic drift time is given by

$$\tau_d = \frac{L}{v_d}, \tag{8.5}$$

where v_d is the drift velocity and L is the characteristic drift length.

The degree of activation (i.e., ionization, dissociation to radical) of feed gases introduced to a plasma reactor depends on the residence time. The residence time is defined as

$$\tau_{res} = \frac{p\, V_{vol}}{Q_{flow}}, \tag{8.6}$$

where p, V_{vol}, and Q_{flow} are the pressure, effective plasma volume, and total mass flow rate, respectively.

Exercise 8.1.1

A plasma reactor with volume 10^3 cm^3 is operated at pressure 1 Torr and flow rate 50 sccm (standard cubic cm per minute). Estimate the residence time of the feed gas.

$$\tau_{res} = \frac{\frac{1 \text{ Torr}}{760 \text{ Torr}} \times 10^3 \text{cm}^3}{\frac{50 \text{ cm}^3}{60 \text{ s}}} \sim 1.6 \text{ s}.$$

Problem 8.1.1

Derive Equation 8.4 from the continuity equation of the system,

$$\frac{\partial}{\partial t} N_j(\mathbf{r}, t) = -k_r N_k N_j + D \nabla^2 N_j. \tag{8.7}$$

In Equation 8.4, $k_r N_r \Lambda^2 / D$ is called the Damkohler number $\tilde{D}$. Show that the system is controlled by diffusion in the case of $\tilde{D} \ll 1$ and by reaction in the case of $\tilde{D} \gg 1$.

8.1.3 Plasma-Oriented Time Constant/Dielectric Relaxation Time

In general, when we simulate a plasma structure with a number density n by using a time-development method, the time step Δt is proportional to $1/n$. The dielectric relaxation, however, gives the relationship between Δt and $1/n$. Here we consider a system without production or loss of charged particles and with an almost uniform number density of electrons and ions. The continuity equation for total charge is expressed by

$$\frac{\partial \rho_T(t)}{\partial t} + div(e\mathbf{\Gamma}_T) = 0\ , \tag{8.8}$$

where $\rho_T(t) = e\{n_p(t) - n_e(t)\}$. Under these circumstances, $\mathbf{\Gamma}_T$ is a drift flux only because the uniformity of the charge distribution and the drift flux may be related to the electric field by apparent mobility μ:

$$\mathbf{\Gamma}_T = n_p \boldsymbol{v}_{dp} - n_e \boldsymbol{v}_{de} = (\mu_p n_p + \mu_e n_e)\boldsymbol{E}.$$

When we consider Poisson's equation, $\mathrm{div}\boldsymbol{E} = \rho_{\mathrm{T}}(\mathrm{t})/\varepsilon_0$, Equation 8.8 may be converted to

$$\frac{\partial \rho_T(t)}{\partial t} \sim -e(\mu_p n_p + \mu_e n_e)\mathrm{div}\boldsymbol{E} = -\frac{e(\mu_p n_p + \mu_e n_e)}{\varepsilon_0}\rho_T(t) \ , \tag{8.9}$$

which forms the relaxation equation for total charge, leading to the solution

$$\rho_T(t) = \rho_{T_0}\exp\left(-\frac{t}{\tau_d}\right),$$

and therefore we may define the dielectric relaxation time from Equation 8.9 as

$$\tau_d = \frac{\varepsilon_0}{e(\mu_p n_p + \mu_e n_e)}. \tag{8.10}$$

τ_d is also known as the Maxwell relaxation time. Solutions to the system are obtained only if the time step Δt is shorter than the relaxation time (during relaxation time τ_d the charge may decay by a factor of $1/e$):

$$\Delta t < \tau_d \left(= \frac{\varepsilon_0}{e(\mu_p n_p + \mu_e n_e)} \right), \tag{8.11}$$

which gives an additional limitation on the time steps in a numerical calculation. The dependence of the dielectric relaxation time limit as a function of charged particle density is given in Figure 8.2.

Exercise 8.1.2

Estimate the dielectric relaxation time for plasmas with density (a) $10^9\,\mathrm{cm}^{-3}$ and (b) $10^{12}\,\mathrm{cm}^{-3}$.

$$\begin{aligned}\tau_d &= \frac{\varepsilon_0}{e(\mu_p n_p + \mu_e n_e)} \sim \frac{\varepsilon_0}{e\mu_e n_e(1 + \frac{\mu_p n_p}{\mu_e n_e})} \\ &\sim \frac{8.854 \times 10^{-12}}{1.6 \times 10^{-19}} \times \frac{1}{\mu_e[m^2 s^{-1} V^{-1}] n_e[m^{-3}]}(s),\end{aligned}$$

where for $\mu_e \sim 70\,\mathrm{m}^2/\mathrm{sV}$, the calculated values are $(a)\tau_d \sim 5.5 \times 10^7 \times 1.4 \times 10^{-2} \times 10^{-15} \sim 10^{-10}$ s and $(b)\tau_d \sim 10^{-13}$ s.

The dielectric relaxation time becomes very small with increasing plasma density. For example, for $n \sim 10^{12}\,\mathrm{cm}^{-3}$, the relaxation time is $\tau_d \sim 10^{-13}\,\mathrm{s}$, which creates a difficult situation in the time-development modeling method due to the excessive CPU time required.

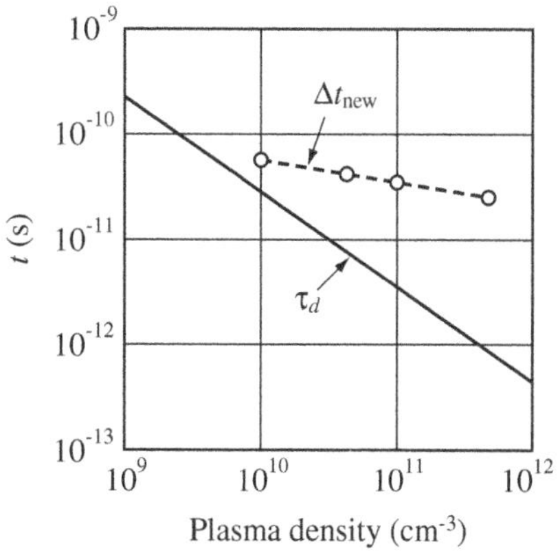

FIGURE 8.2 Dielectric relaxation time constant τ_d and the time steps used in the semi-empirical method as a function of plasma density.

8.2 NUMERICAL TECHNIQUES TO SOLVE THE TIME-DEPENDENT DRIFT-DIFFUSION EQUATION

The numerical techniques involved in solving continuity equations and understanding their accuracy and convergence are discussed below.

8.2.1 Time-Evolution Method

The governing equation systems of low-temperature plasmas are both time and space dependent, as described in Chapter 7. Using a time-development solution to solve the governing equations, we can simulate a plasma with a periodic steady-state structure and periodic steady-state characteristics, in addition to simulating the initiation of the discharge. The time-evolution method consists of the following solutions from time t_m to time t_{m+1}. The present-day computer has become a powerful tool that allows modeling of rf plasmas by following both time and space profiles in a reasonable amount of CPU time. However, it is beyond the capability of present-day computers to obtain a time solution and follow the time modulation in microwave plasmas at 2.45 GHz. Under these circumstances, the computer does have sufficient memory and speed to allow development of two- and three-dimensional codes in space for plasmas excited by external sources from direct current (DC) to very high frequency (VHF).

The low-temperature rf plasma used for a microelectronic device fabrication includes a number of physical and chemical collision/reaction/transport processes with very different time constants or relaxation times, and the system is expressed by stiff differential equations. This stiffness constitutes a fundamental numerical problem, and we consider methods for overcoming it below.

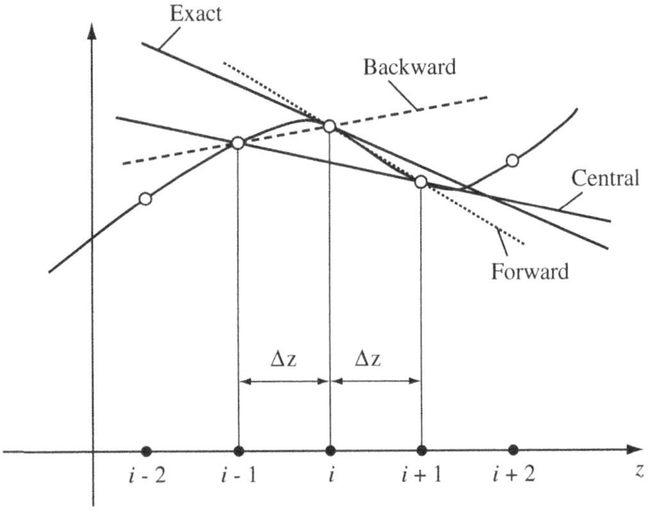

FIGURE 8.3 Example of a finite difference scheme.

8.2.1.1 Finite Difference

The modeling of a low-temperature plasma basically requires a numerical solution to a time-dependent conservation equation, specifically, the partial differential drift-diffusion equation. We briefly review the discretization method for obtaining the finite difference between the first and second derivatives of the conservation equation. The forward (FD), central (CD), and backward (BD) difference schemes in Figure 8.3 are expressed in the case of uniform grid size ($\Delta z = z_{i+1} - z_i = \text{const.}$) as

$$\frac{\partial}{\partial z} n(z,t) \sim \frac{n_{i+1} - n_i}{\Delta z} - \frac{\Delta z}{2} \left(\frac{\partial^2 n}{\partial z^2} \right)_i \text{(FD);} \tag{8.12}$$

$$\frac{\partial}{\partial z} n(z,t) \sim \frac{n_{i+1} - n_{i-1}}{2\Delta z} - \frac{(\Delta z)^2}{6} \left(\frac{\partial^3 n}{\partial z^3} \right)_i \text{(CD);} \tag{8.13}$$

$$\frac{\partial}{\partial z} n(z,t) \sim \frac{n_i - n_{i-1}}{\Delta z} + \frac{\Delta z}{2} \left(\frac{\partial^2 n}{\partial z^2} \right)_i \text{(BD);} \tag{8.14}$$

$$\frac{\partial^2}{\partial z^2} n(z,t) \sim \frac{n_{i+1} - 2n_i + n_{i-1}}{(\Delta z)^2} - \frac{(\Delta z)^2}{12} \left(\frac{\partial^4 n}{\partial z^4} \right)_i \text{(CD).} \tag{8.15}$$

Exercise 8.2.1

A low-temperature rf plasma is described by a system of stiff differential equations. It is known that Gear's algorithms of orders 1 to 6 are stiffly stable [2]. Discuss the algorithms.

For example, for a differential equation $\partial n(z,t)/\partial t = d(n(z,t),t)$, Gear's algorithms for the first three orders are

$$\begin{aligned} n_{i+1} &= n_i + \Delta t[d(n_{i+1}, t_{i+1})] \quad \text{(first order)} \\ &= \frac{4}{3}n_i - \frac{1}{3}n_{i-1} + \Delta t\left[\frac{2}{3}d(n_{i+1}, t_{i+1})\right] \quad \text{(second order)} \\ &= \frac{18}{11}n_i - \frac{9}{11}n_{i-1} + \frac{2}{11}n_{i-2} + \Delta t\left[\frac{6}{11}d(n_{i+1}, t_{i+1})\right] \quad \text{(third order).} \end{aligned}$$

Note that the kth-order Gear's algorithm requires k starting values, $n_i, n_{i-1}, \ldots, n_{i-(k-1)}$.

8.2.1.2 Digitalization and Stabilization

The drift-diffusion equation without production of new charged particles is given in the one-dimensional space,

$$\frac{\partial n}{\partial t} = -v_d\frac{\partial n}{\partial z} + D\frac{\partial^2 n}{\partial z^2}. \tag{8.16}$$

Here, we use the central difference for uniform spatial mesh (index i) and calculate the spatial density $n_i^{m+1} = n(t_{m+1}, z_i)$ at time $t_{m+1} = t_m + \Delta t$ as

$$n_i^{m+1} = n_i^m + \left[-v_d\frac{n_{i+1}^m - n_{i-1}^m}{2\Delta z} + D\frac{n_{i+1}^m + n_{i-1}^m - 2n_i^m}{(\Delta z)^2}\right]\Delta t, \tag{8.17}$$

where Δz is the spatial step. The solution is rewritten as [1],

$$n_i^{m+1} = (1-2d)n_i^m + \left(d - \frac{c}{2}\right)n_{i+1}^m + \left(d + \frac{c}{2}\right)n_{i-1}^m, \tag{8.18}$$

where

$$d = D\,\frac{\Delta t}{(\Delta z)^2} \quad (> 0). \tag{8.19}$$

d is the ratio between the time step Δt and the characteristic diffusion time $(\Delta z)^2/D$ (i.e., the time it takes to move by Δz due to diffusion).

$$c = v_d\,\frac{\Delta t}{\Delta z} \quad (> 0) \tag{8.20}$$

is the ratio between the time step and the characteristic drift time $(\Delta z/v_d)$, and c is also known as the Courant number. Finally, we rewrite Equation 8.18 as

$$n_i^{m+1} = A\, n_i^m, \tag{8.21}$$

where A is a tridiagonal matrix. The distribution at t_{m+1}, n^{m+1} may be obtained by applying matrix A $(m+1)$ times starting from the initial assumed distribution n^0 and bearing in mind that the values of the matrix elements change both with time and with changing conditions in the spatial profile of the plasma. The difference between the densities obtained in subsequent iterations is:

$$\varepsilon = \|n_i^m - n_i^{m-1}\| = \sqrt{\sum_i \left\{n_i^m - n_i^{m-1}\right\}^2}, \tag{8.22}$$

and the difference ε may be associated with the error of calculations. If we consider eigenvalue λ of the matrix A, the solution for the number density may be written as

$$n_i^m = \lambda^m \exp(j\alpha i), \tag{8.23}$$

where $j^2 = -1$ and α is an arbitrary constant in wavenumber units. If λ is less than 1 then ε will disappear. Combining Equations 8.18 and 8.23 we obtain the solution for the eigenvalue λ:

$$\lambda = 1 + 2d(\cos\alpha - 1) - jc\,\sin\alpha, \tag{8.24}$$

where the number density n_i^m is stable under $|\lambda|^2 < 1$. There are three possible situations (see Figure 8.4).

i. Without diffusion ($d = 0$), λ is always greater than 1 for arbitrary values of c, and the solution is stable;

ii. Without drift ($c = 0$), λ has a maximum for $\cos\alpha = -1$, and the solution is stable for $d < 1/2$;

iii. With diffusion and drift, the system is stable for $d < 1/2$ and $c < 2d$, is:

$$d < 1/2, \quad c < 2d \quad (\text{i.e., } c < 1). \tag{8.25}$$

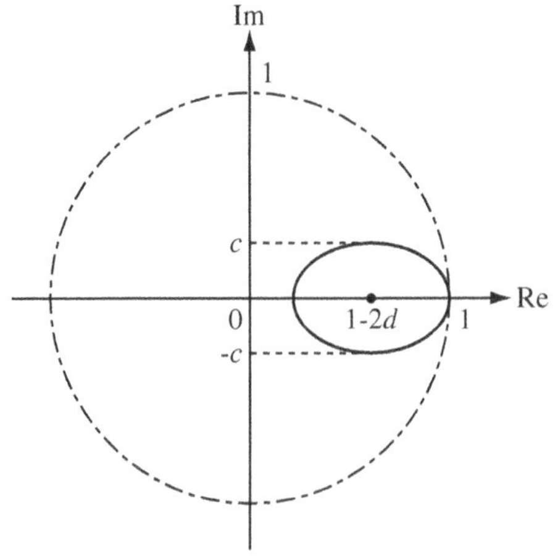

FIGURE 8.4 Trajectory of the eigenvalue Equation 8.25.

Problem 8.2.1

Derive the relations in Equation 8.25 by reference to the trajectory of Equation 8.24 in the complex plane in Figure 8.4.

By combining Equations 8.19 and 8.25 we obtain

$$\Delta t < \frac{(\Delta z)^2}{2D}. \tag{8.26}$$

This equation indicates that if we reduce the size of the spatial mash Δz (or increase the spatial resolution) by a factor of 2, the time step should be reduced by a factor of 1/4, and therefore the overall CPU time should increase by a factor of 8. Also, by combining Equations 8.20 and 8.25 we have

$$\Delta t < \frac{\Delta z}{v_d}. \tag{8.27}$$

To determine the proper temporal resolution for solving the drift-diffusion equations, we set the following condition:

$$\Delta t < \min\left(\frac{\Delta z}{v_d}, \frac{(\Delta z)^2}{2D}\right). \tag{8.28}$$

Sometimes we may redefine the condition 8.25, $c < 2d$, by using the Peclet number P_e, which is defined as

$$P_e = \frac{c}{d} = \frac{v_d \Delta z}{D} < 2\,. \tag{8.29}$$

It can be shown that no oscillation of the solution (i.e., stabilization) occurs when the local Peclet number is $P_c < 2$.

8.2.1.3 Time Discretization and Accuracy

For the analysis of time development, the drift-diffusion equation may be rewritten as

$$\frac{du(t)}{dt} = \beta u(t), \tag{8.30}$$

and Equation 8.30 is solved explicitly by Euler's method. We can then write

$$\frac{u_i^{m+1} - u_i^m}{\Delta t} = \beta u_i^m.$$

This means that at time t_{m+1} the solution u^{m+1} may be represented by the previous time t_m so that

$$u_i^{m+1} = (1 + \beta\Delta t)u_i^m, \tag{8.31}$$

and if we start at $t = 0$ from the initial distribution $u(0)$ after m steps or after time $t = m\Delta t (= T)$, we obtain

$$u_i^m = (1 + \beta\Delta t)^m u_i(0). \tag{8.32}$$

Convergence of the procedure (stability) is achieved when

$$|1 + \beta \Delta t| \leq 1, \tag{8.33}$$

and therefore

$$-\beta \Delta t \leq 2. \tag{8.34}$$

From Equation 8.32 and in the limit $\Delta t \to 0$ one obtains a continuous solution

$$(1 + \beta \Delta t)^m = (1 + \beta \Delta t)^{T/\Delta t} = \left\{(1 + \beta \Delta t)^{1/\beta \Delta t}\right\}^{\beta T} \to e^{\beta T},$$

in which we use the definition of exponential function. As a result, we obtain

$$\frac{(1 + \beta \Delta t)^{T/\Delta t}}{e^{\beta T}} \sim \exp\left(-\frac{\beta^2}{2} T \Delta t\right) \sim 1 - \frac{\beta^2}{2} T \Delta t,$$

which may be used to define the relative error $|\delta|$,

$$|\delta| = \frac{1}{2}(\beta T)(\beta \Delta t). \tag{8.35}$$

When we choose very small Δt for large $T = m\Delta t$, the relative error will approach 1. The relation 8.35 may be rewritten as a condition of accuracy:

$$\Delta t < \frac{2\delta_0}{\beta^2 T}. \tag{8.36}$$

8.2.2 Scharfetter–Gummel Method

When we simulate a discharge plasma we have to consider both the bulk plasma and the ion sheath region close to the electrodes, wafer, or walls. In the region close to the electrode, the variations of fluxes of charged particles may be very pronounced. In the bulk plasma the flow of particles is dominated by diffusion, whereas in the sheaths it is dominated by drift.

When gas pressure is relatively high, the flux due to diffusion becomes small. In the case of a low-pressure plasma sustained between two parallel plates, the drift flux will be dominant in the positive ion sheath. In that case, a region satisfying $D\partial n/\partial z \ll v_d n$ appears in front of the electrodes, wafer, or side wall. Then, the Peclet number P_e rapidly increases and stability of the solution of the drift-diffusion equation is not achieved (see Equation 8.29). Under these circumstances the drift-diffusion equation becomes inappropriate, so one has to use the drift equation

$$\frac{\partial n}{\partial t} = -v_d \frac{\partial n}{\partial z}. \tag{8.37}$$

The drift equation should be solved from the upwind to the downwind direction. There are a few methods that may be used to achieve stability

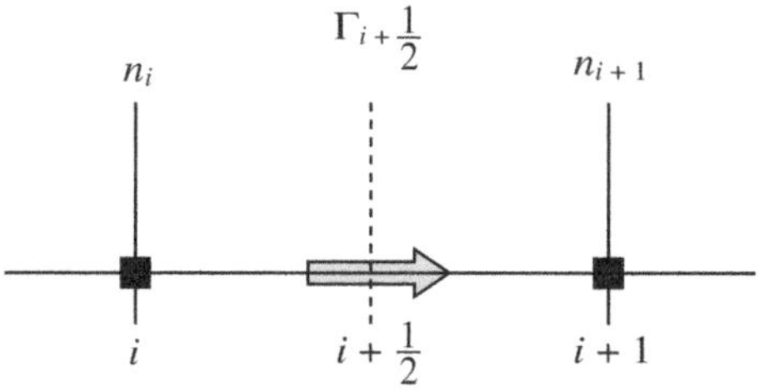

FIGURE 8.5 Staggered mesh method. Scalar quantities are defined on integer i mesh points and vector quantities on mid $i + 1/2$ points.

of the drift-diffusion Equation 8.16 when condition 8.29 is not met. One is known as the Patankar difference method, but the practical application that is discussed here is the so-called Scharfetter–Gummel (SG) method [3]. In this procedure we use the staggered mesh method for spatial distribution; in this method, all scalar quantities are defined on integer mesh points and vector quantities are defined on midpoints, as described in Figure 8.5.

The flux at time t_m is a combination of drift and diffusion fluxes:

$$\Gamma^m = n^m v_d^m - D^m \frac{\partial n^m}{\partial z}, \tag{8.38}$$

where the values of Γ^m and v_d^m are represented at the midpoint $(i + 1/2)$ between the mesh point i and $i + 1$. The first-order differential Equation 8.38 is analytically solved in the range of $z(0 < z < \Delta z)$ as

$$n^m(z) = C \exp\left[\frac{v^m_{d\ i+1/2}}{D^m_{i+1/2}} \Delta z\right] + \frac{\Gamma^m_{i+1/2}}{{v_d}^m_{i+1/2}}, \tag{8.39}$$

and we take into account the boundary condition $n^m(0) = n_i^m$ at i. We derive the value of the constant as

$$C = n_i^m - \frac{\Gamma^m_{i+1/2}}{v^m_{d\ i+1/2}}. \tag{8.40}$$

As a result, $n^m(\Delta z) = n^m_{i+1}$ is expressed as

$$n^m_{i+1} = \left(n_i^m - \frac{\Gamma^m_{i+1/2}}{v^m_{d\ i+1/2}}\right) \exp\left[\frac{v^m_{d\ i+1/2}}{D^m_{i+1/2}} \Delta z\right] + \frac{\Gamma^m_{i+1/2}}{v^m_{d\ i+1/2}}, \tag{8.41}$$

and this equation may be used to solve for $\Gamma^m_{i+1/2}$ at the midpoint $i + 1/2$:

$$\Gamma^m_{i+1/2} = \frac{n_i^m \exp\left[\frac{v^m_{d\ i+1/2}}{D^m_{i+1/2}} \Delta z\right] - n^m_{i+1}}{\exp\left[\frac{v^m_{d\ i+1/2}}{D^m_{i+1/2}} \Delta z\right] - 1} v^m_{d i+1/2}. \tag{8.42}$$

Equation 8.42 is the so-called SG expression for drift-diffusion flux. By applying the SG expression to the r.h.s. of Equation 8.16, the spatial derivative of the flux expressed at the midpoints $i-1/2$ and $i+1/2$ is estimated by the quantities for $i-1$, i, and $i+1$:

$$-\frac{\Gamma_{i+1/2}^{m}-\Gamma_{i-1/2}^{m}}{\Delta z}=\frac{n_{i+1}^{m}-\left[\exp\left(\frac{v_{di}^{m}}{D_{i}^{m}}\Delta z\right)+1\right]n_{i}^{m}+\exp\left(\frac{v_{di}^{m}}{D_{i}^{m}}\Delta z\right)n_{i-1}^{m}}{\left[\exp\left(\frac{v_{di}^{m}}{D_{i}^{m}}\Delta z\right)-1\right]\frac{\Delta z}{v_{di}^{m}}}. \tag{8.43}$$

Problem 8.2.2

Prove that Equation 8.43 reduces to: (i) for $P_e\to\infty$, then the equation is expressed by:

$$-\frac{v_{di}^{m}}{\Delta z}\left(n_{i}^{m}-n_{i-1}^{m}\right) \tag{8.44}$$

and we have the backward difference equation for the drift flux, and (ii) for $P_e\to 0$, we obtain the diffusion flux

$$\frac{D_{i}^{m}}{(\Delta z)^2}\left(n_{i+1}^{m}-2n_{i}^{m}+n_{i-1}^{m}\right), \tag{8.45}$$

where v_d and D have no change in the small range between $i-1$ and $i+1$.

Equation 8.43 is thus able to represent the whole range of conditions occurring in the sheath and in the bulk where both drift-dominated and diffusion-dominated fluxes occur.

Exercise 8.2.2

Show that the backward difference Equation 8.44 is equivalent to a drift-diffusion equation with an artificial coefficient $D=v_d\Delta z/2$.

If we rearrange Equation 8.44 as

$$-\frac{v_{di}^{m}}{\Delta z}\left(n_{i}^{m}-n_{i-1}^{m}\right)=-v_{di}^{m}\frac{n_{i+1}^{m}-n_{i-1}^{m}}{2\Delta z}+\frac{v_{di}^{m}\Delta z}{2}\frac{n_{i+1}^{m}-2n_{i}^{m}+n_{i-1}^{m}}{(\Delta z)^2}, \tag{8.46}$$

we obtain two terms, respectively describing the derivatives of the drift flux and the effective diffusion flux with the artificial diffusion coefficient, $v_d\Delta z/2$, in the central difference scheme. The truncation error with order $O(\Delta z)^2$ in Equation 8.46 has the effect of an artificial diffusion.

Problem 8.2.3

When the drift is along the positive z-axis, show that Equation 8.44 should be expressed as

$$-\frac{|v_d|}{\Delta z}\left\{n_{i}^{m}-n_{i-1}^{m}\right\}.$$

On the other hand, when the drift is along the negative z-axis, show that it is given as

$$-\frac{|v_d|}{\Delta z}\left\{n_i^m - n_{i+1}^m\right\}.$$

This method, in which the artificial diffusion in the central diffusion scheme is reduced by checking the flow direction, is called the upwind scheme.

8.2.3 Cubic Interpolated Pseudoparticle Method

The cubic interpolated pseudoparticle (CIP) method is applied to the drift problem typical of the propagation of a soliton, which is a single wave with a constant spatial profile in a plasma [4]. In Exercise 8.2.2, we have shown that the numerical solution of the drift equation will bring artificial diffusion in the procedure. Let us return to a consideration of the drift equation with drift velocity v_d:

$$\frac{\partial n(z,t)}{\partial t} = -v_d \frac{\partial n(z,t)}{\partial z}. \tag{8.47}$$

Equation 8.47 has an analytical solution $n(z,t) = n(z - v_d t, t)$, which denotes the density propagation to the negative direction with the same spatial form (see Figure 8.6). The CIP method allows us to consider the analytical solution. That is, the density at $(m+1)\Delta t$ under a constant v_d is expressed by

$$n_i^{m+1}(z_i, t+\Delta t) \simeq n_i^m(z_i - v_d \Delta t, t). \tag{8.48}$$

$n(z)$ at $z_i \leq z \leq z_{i+1}$ is given by the interpolation as

$$\tilde{n}(z) = n_i^m(z_i) + \frac{dn_i^m(z)}{dz}(z - z_i) + h_i(z - z_i)^3 + k_i(z - z_i)^2, \tag{8.49}$$

where the constants h_i and k_i are obtained by $n(z_{i+1})$ and $dn(z_{i+1})/dt$ in Equation 8.49. Then,

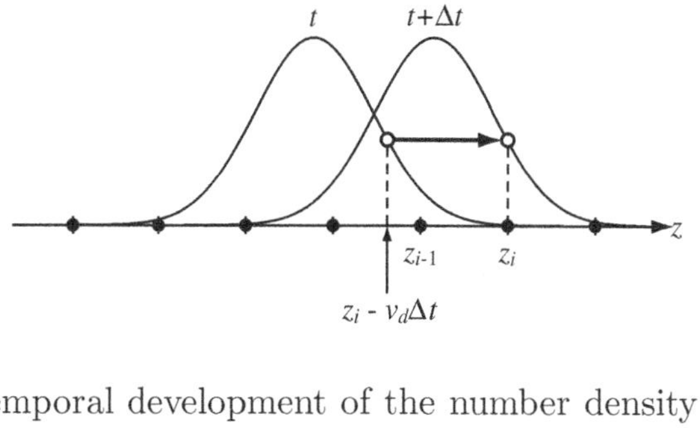

FIGURE 8.6 Temporal development of the number density subject only to the drift equation.

$$n_i^{m+1} = n_i^m + \left(\left(-h_i^m v_{di}^m \Delta t + k_i^m \right) v_{di}^m \Delta t - \frac{dn_i^m}{dz} \right) v_{di}^m \Delta t \tag{8.50}$$

$$\frac{dn_i^{m+1}}{dz} = \frac{dn_i^m}{dz} + \left(3h_i^m v_{di}^m \Delta t - 2k_i^m \right) v_{di}^m \Delta t, \tag{8.51}$$

where by using the sign, that is, isgn $= 1$ at $v_d < 0$ and isgn $= -1$ at $v_d > 0$, we have

$$h_i^m = \frac{\frac{d}{dz}\left(n_i^m + n_{i+\mathrm{isgn}}^m\right)}{(\Delta z)^2} - \frac{2\left(n_{i+\mathrm{isgn}}^m - n_i^m\right)}{(\Delta z)^3} \times \mathrm{isgn}$$

$$k_i^m = -\frac{\frac{d}{dz}\left(2n_i^m + n_{i+\mathrm{isgn}}^m\right)}{\Delta z} \times \mathrm{isgn} + \frac{3\left(n_{i+\mathrm{isgn}}^m - n_i^m\right)}{(\Delta z)^2}.$$

The diffusion part of the drift-diffusion equation is numerically calculated in an implicit form by using Equation 8.50

$$\frac{n_i^{m+1} - n_i^*}{\Delta t} = D_{i+1/2}^m \frac{n_{i+1}^{m+1} - n_i^{m+1}}{(\Delta z)^2} - D_{i-1/2}^m \frac{n_i^{m+1} - n_{i-1}^{m+1}}{(\Delta z)^2} + S_i^m. \tag{8.52}$$

Then, by collecting the terms in Equation 8.52 with respect to the position grids, $i-1, i,$ and $i+1$, we obtain

$$-\frac{D_{i-1/2}^m \Delta t}{\Delta z^2} n_{i-1}^{m+1} + \left(1 + \frac{D_{i+1/2}^m \Delta t}{\Delta z^2} + \frac{D_{i-1/2}^m \Delta t}{\Delta z^2}\right) n_i^{m+1} - \frac{D_{i+1/2}^m \Delta t}{\Delta z^2} n_{i+1}^{m+1} = n_i^* + S_i^m \Delta t. \tag{8.53}$$

After we write the Equation 8.53 for all the spatial meshes, the simultaneous equations are expressed in the matrix form

$$\begin{pmatrix} b_1 & c_1 & & & & \mathbf{0} \\ a_2 & b_2 & c_2 & & & \\ & \cdots & \cdots & & & \\ & & \cdots & \cdots & & \\ & & a_{I-2} & b_{I-2} & c_{I-2} & \\ \mathbf{0} & & & b_{I-1} & c_{I-1} & \end{pmatrix} \begin{pmatrix} n_1^{m+1} \\ n_2^{m+1} \\ \vdots \\ \vdots \\ n_{I-2}^{m+1} \\ n_{I-1}^{m+1} \end{pmatrix} = \begin{pmatrix} d_1 \\ d_2 \\ \vdots \\ \vdots \\ d_{I-2} \\ d_{I-1} \end{pmatrix}. \tag{8.54}$$

The above matrix is solved by the appropriate numerical method, and we obtain the solution n_i^{m+1} of the drift-diffusion equation in Equation 8.16. $\partial n_i^{m+1}/\partial z$ is given by

$$\frac{\partial n_i^{m+1}}{\partial z} = \frac{\partial n_i^*}{\partial z} + \frac{\left(n_{i+1}^{m+1} - n_{i-1}^{m+1} - n_{i+1}^* + n_{i-1}^{m+1}\right)}{2\Delta z}. \tag{8.55}$$

The CIP calculation in the next time step proceeds by using the set $(n_i^{m+1}, \partial n_i^{m+1}/\partial z)$.

8.2.4 Semi-Implicit Method for Solving Poisson's Equation

In the explicit method, the selected value of the time step Δt must be much smaller than the dielectric relaxation time, and it is difficult to apply the time-development method of the governing equation with increasing plasma density. We here present a method for optimizing this calculation [1]. By performing a Taylor expansion of the r.h.s. of Poisson's equation, the total charge density ρ_T is defined as

$$-\varepsilon_0 \nabla^2 V(z, t + \Delta t) = \rho_T(z, t + \Delta t) \sim \rho_T(z, t) + \frac{\partial \rho_T(z, t)}{\partial t} \Delta t. \tag{8.56}$$

Here, we use $\rho_T(z, t) = -\varepsilon_0 \nabla^2 V(z, t)$ and $\partial \rho_T / \partial t = -\mathrm{div}(\mathrm{e}\mathbf{\Gamma}_\mathrm{T}(t))$, and thus we obtain

$$-\varepsilon_0 \nabla^2 V(z, t + \Delta t) = -\varepsilon_0 \nabla^2 V(z, t) - e\Delta t\, \mathrm{div}\mathbf{\Gamma}_\mathrm{T}(z, t) \tag{8.57}$$

and, consequently,

$$\frac{\nabla^2 \{V(z, t + \Delta t) - V(z, t)\}}{\Delta t} = \frac{e}{\varepsilon_0} \mathrm{div}\mathbf{\Gamma}_\mathrm{T}(z, t). \tag{8.58}$$

This means that the potential $V(z, t)$ at time $t + \Delta t$ is expressed by the value at the time t (semi-implicit method) and from the potential we determine the electric field. The technique described here allows more efficient calculation of the potentials and fields, as can be seen in Figure 8.2.

8.3 BOUNDARY CONDITIONS

Nonequilibrium plasmas are maintained in reactors that have electrodes and walls. The properties of the plasma may be strongly affected by the nature of the interactions with these surfaces. It is thus very important to properly specify the boundary conditions. The key physical and chemical processes in plasma etching, deposition, and many other plasma technologies occur at surfaces and may even involve material transport. Therefore, in this section we discuss methods for defining the boundary conditions for plasma models.

8.3.1 Ideal Boundary – Without Surface Interactions

8.3.1.1 Dirichlet Condition

In some cases, the value of the function in the governing equation is given in the Dirichlet-type boundary. For example, the boundary condition at $\boldsymbol{r} = \boldsymbol{r}_0$ is given for the electron number density as

$$n_e(\boldsymbol{r}, t)|_{\boldsymbol{r}_0} = \; const. \tag{8.59}$$

Note that the reflection (as well the absorption) at the surface affects the velocity distribution function close to the surface, $g(\boldsymbol{v}, \boldsymbol{r}, t)$. At the perfect

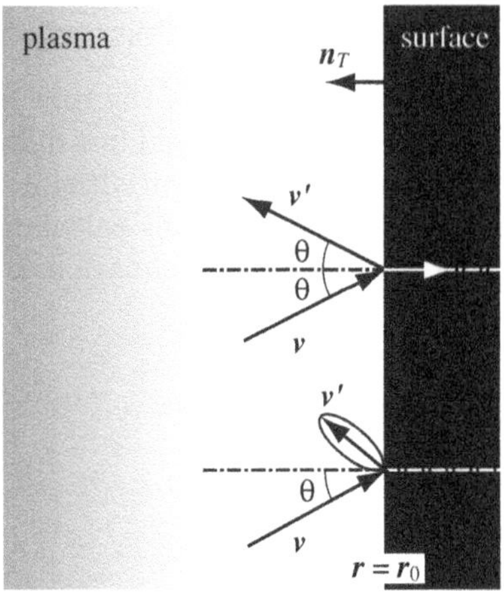

FIGURE 8.7 Reflection of particles from the surface: reflection of a flux of particles, velocities before **v** and after reflection **v**′ against an ideally flat and against a realistic irregular surface.

absorbing boundary, the velocity distribution function satisfies

$$g(\boldsymbol{v}, \boldsymbol{r}, t; \quad \boldsymbol{v} \cdot \boldsymbol{n}_T > 0)|_{\boldsymbol{r}_0} \;=\; 0. \tag{8.60}$$

In the case of the ideal perfect reflection surface, the velocity after reflection $\boldsymbol{v}'$ only changes direction perpendicular to the surface:

$$\begin{aligned} g(\boldsymbol{v}', \boldsymbol{r}, t)|_{\boldsymbol{r}_0} &= g(\boldsymbol{v}, \boldsymbol{r}, t)|_{\boldsymbol{r}_0}, \\ \text{where} \qquad \boldsymbol{v}' &= \boldsymbol{v} + 2(\boldsymbol{v} \cdot \boldsymbol{n}_T)\boldsymbol{n}_T. \end{aligned} \tag{8.61}$$

The general surface including random reflection may be expressed in terms of the reflection coefficient $\alpha_{ref}(\boldsymbol{v})$ as

$$g(\boldsymbol{v}', \boldsymbol{r}, t)|_{\boldsymbol{r}_0} = \alpha_{ref}(\boldsymbol{v}) \; g(\boldsymbol{v}, \boldsymbol{r}, t)|_{\boldsymbol{r}_0}, \tag{8.62}$$

where $\alpha_{ref} = 0$ for the perfect absorbing boundary, and $\alpha_{ref} = 1$ for specular reflection in Equation 8.61.

In Figure 8.7 we schematically show the effect of reflection on the flux of particles and on the velocity. In addition, this schematic shows that, due to surface irregularities, the direction of the particles after scattering may not have the same angles as the incoming particles.

8.3.1.2 Neumann Condition

The boundary condition for the first derivative is known as the Neumann condition. For example, for the electron density, this condition is

$$\left. \frac{\partial n_e(\boldsymbol{r}, t)}{\partial \boldsymbol{r}} \right|_{\boldsymbol{r}_0} = const. \tag{8.63}$$

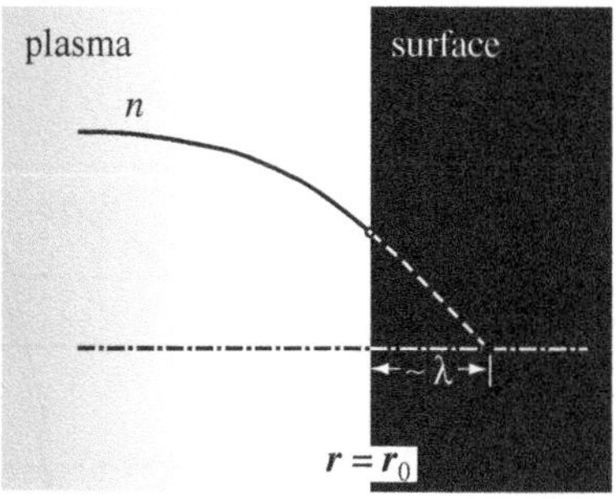

FIGURE 8.8 Boundary condition of Chantry and Phelps and the particle density close to the surface.

Chantry introduced the effect of the reflection of the species incident on the boundary,

$$\left.\frac{\partial n(\boldsymbol{r},t)}{\partial \boldsymbol{r}}\right|_{\boldsymbol{r}_0} = \frac{n(\boldsymbol{r},t)}{\beta} \qquad \beta = \lambda \cos\theta \frac{1+\alpha_{ref}}{1-\alpha_{ref}}, \tag{8.64}$$

where λ is the mean free path, and α_{ref} is the reflection coefficient of the species. β is defined as a distance between $\boldsymbol{r}_0$ and the point where the tangential line of n intercepts the $\boldsymbol{r}$-axis (see Figure 8.8). The sticking coefficient s_t on the boundary surface is frequently used instead of the reflecton α_{ref} in the field of thin film deposition. Then,

$$s_t = 1 - \alpha_{ref} \tag{8.65}$$

A value $2/3$ or $1/\sqrt{3}$ is taken for $\cos\theta$ [6, 7].

8.3.1.3 Periodicity Condition

In an rf plasma externally sustained at a voltage source $V_0 \sin\omega t$, the plasma characteristics $\Phi(\boldsymbol{v},\boldsymbol{r},t)$ in a periodic steady state satisfy periodic boundary conditions in time:

$$\Phi(\boldsymbol{v},\boldsymbol{r},t) = \Phi\left(\boldsymbol{v},\boldsymbol{r},t+\frac{i\pi}{\omega}\right) \qquad i = 1 \text{ or } 2. \tag{8.66}$$

The series of structures on a wafer will come into contact with a plasma having a spatial periodicity defined as

$$\Phi(\boldsymbol{v},\boldsymbol{r},t)|_{\boldsymbol{r}_0} = \Phi(\boldsymbol{v},\boldsymbol{r}+i\boldsymbol{k},t)|_{\boldsymbol{r}=0} \quad i = 1, 2, \ldots, \tag{8.67}$$

where $\boldsymbol{k}$ is the spatial wavenumber of the surface structure.

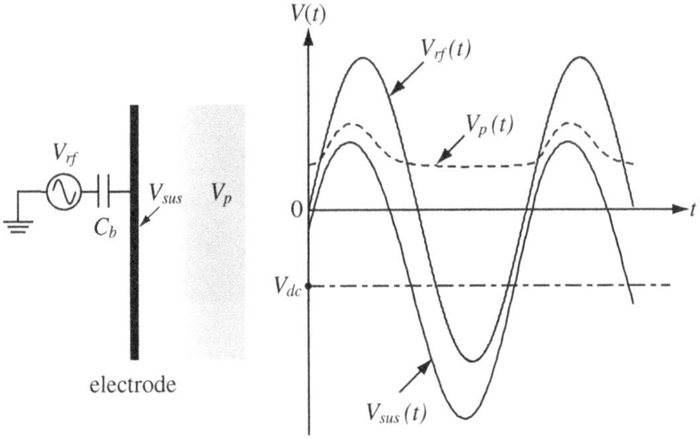

FIGURE 8.9 Potentials at points around an electrode with voltage source V_{rf}.

8.3.2 Electrode Surface

There exists practically some type of electrode, wafer, or substrate to be processed in plasma processing for material fabrications.

8.3.2.1 *Metallic Electrode*

On a metallic electrode the incident-charged particles are immediately recombined. In most cases, the metallic electrode is biased through the blocking capacitor C_b to the external source having $V_{rf}(t) = V_0 \sin \omega t$ (see Figure 8.9). Then, the metallic surface potential is different from the external source,

$$V_{sus}(t) = V_0 \sin \omega t - V_{self}(t), \tag{8.68}$$

where the self-bias voltage $V_{self}(t)$ is given by the incident fluxes of electrons, and positive and negative ions, Γ_e, Γ_p, and Γ_n, from the plasma to the metallic surface S, and the displacement current density J_{dis},

$$V_{self}(t) = Q(t)|_{sur}/C_b, \tag{8.69}$$

where

$$Q(t)|_{sur} = \int_{-\infty}^{t} \int_S e\{\Gamma_p(\boldsymbol{r},t) - \Gamma_e(\boldsymbol{r},t) - \Gamma_n(\boldsymbol{r},t) + \frac{1}{e}J_{dis}(\boldsymbol{r},t)\}\, dSdt. \tag{8.70}$$

Usually, $V_{self}(t)$ is temporally modulated and DC self-bias voltage V_{dc} is experimentally defined as

$$V_{dc} = \frac{1}{T}\int_0^T V_{sus}(t)dt = \frac{1}{T}\int_0^T -V_{self}(t)dt. \tag{8.71}$$

8.3.2.2 Dielectric Electrode

On the other hand, when a dielectric electrode or wafer is exposed to a plasma, the incident-charged particles are locally accumulated on the surface,

$$\sigma(\boldsymbol{r},t)|_{sur} = \int_0^t e\{\Gamma_p(\boldsymbol{r},t) - \Gamma_e(\boldsymbol{r},t) - \Gamma_n(\boldsymbol{r},t)\}dt. \qquad (8.72)$$

When an insulated wafer set on a metallic electrode is biased externally, then the surface potential is considered as a series of capacitors, C_b and C_{wafer}, in circuit theory. The insulator thickness of the wafer is usually very thin, and the voltage drop-off of the wafer will be very low as compared with that in the blocking capacitor C_b.

Exercise 8.3.1

There is a trench structure with a finite surface conductivity σ on an SiO_2 wafer. The trench is exposed by fluxes of electrons and positive ions from a plasma. When the surface etching and deposition are negligible, derive the equation of the local number density of the surface charge.

By using the local surface fluxes incident from the plasma, $\Gamma_e(\boldsymbol{r},t)$ and $\Gamma_p(\boldsymbol{r},t)$, a simple continuity equation of electrons and positive ions at the surface can be given as

$$\frac{\partial n_e}{\partial t} + \mathrm{div}\Gamma_e(\boldsymbol{r},t) = -\mathrm{div}[\boldsymbol{j}_{e_{surf}}(\boldsymbol{r},t)/e] - R_r n_e n_p, \qquad (8.73)$$

$$\frac{\partial n_p}{\partial t} + \mathrm{div}\Gamma_p(\boldsymbol{r},t) = -R_r n_e n_p, \qquad (8.74)$$

where $\boldsymbol{j}_{e_{surf}}(\boldsymbol{r},t) = \sigma(\boldsymbol{r},t)\boldsymbol{E}_{suf}(\boldsymbol{r},t)$ is the surface electron current density, and R_r is the surface recombination coefficient. The Equations 8.73 and 8.74 are simultaneously solved with Poisson's equation. A self-consistent solution is obtained with respect to the potential of the metallic electrode behind the wafer. The Laplace equation must, of course, be used inside the dielectric SiO_2.

8.3.3 Boundary Conditions with Charge Exchange

Normally, charged particles are lost or reflected at the surface. Sometimes they may also be generated at surfaces. Secondary production of electrons by the impact of positive ions (p) with the surface proceeds through two processes (see Chapter 4): Auger potential ejection and kinetic mechanisms. In a similar fashion, fluxes of fast neutrals (Γ_n), photons (Γ_{ph}), and metastables (Γ_m) induce electrons exclusively through either a kinetic (n) or a potential mechanism (ph, m). Under normal conditions, we have fluxes of all of these particles arriving at the electrode in a plasma, and then the produced electrons

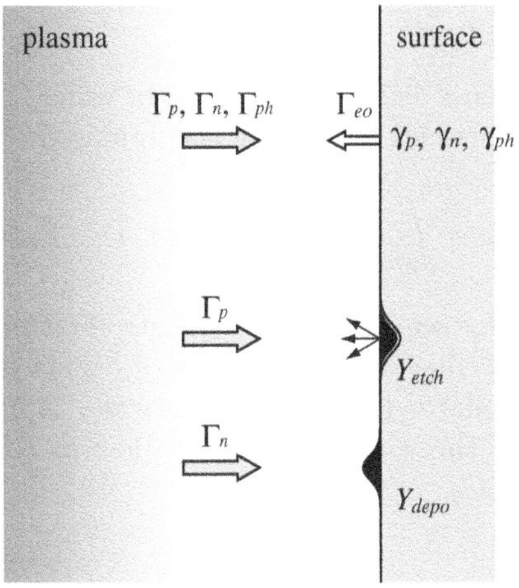

FIGURE 8.10 Boundary conditions with charge exchange and mass transfer.

are ejected and contribute in part to further kinetics in a plasma. The resulting flux of electrons is

$$\Gamma_e(\boldsymbol{r},t)|_{\boldsymbol{r}_0} = \{\gamma_p\Gamma_p(\boldsymbol{r},t) + \gamma_n\Gamma_n(\boldsymbol{r},t) + \gamma_m\Gamma_m(\boldsymbol{r},t) + \gamma_{ph}\Gamma_{ph}(\boldsymbol{r},t)\}|_{\boldsymbol{r}_0}, \quad (8.75)$$

where γ_p, γ_n, γ_{ph}, and γ_m are the secondary electron emission yields for positive ions, fast neutrals, photons, and metastables, respectively. These processes are shown in the upper panel of Figure 8.10.

8.3.4 Boundary Conditions with Mass Transport

The most important goal in the application of plasmas to integrated circuit production is to perform plasma etching. In the case of plasma etching, sputtering, or deposition, material is transported between the plasma and surface. This material usually originates from a surface but may also be generated in the gas phase of the plasma. The transport of material is induced or facilitated by fluxes of physically or chemically active ions and neutrals that interact with the surface, which leads to the modification of the surface profile as well as superficial changes in the substrate material.

8.3.4.1 Plasma Etching

When there exist high-energy ions with a mass of M_p, velocity distribution $g_p(\boldsymbol{v},\boldsymbol{r},t)$, number density $n_p(\boldsymbol{r},t)$, and chemically active molecules (i.e., radicals) with M_r, $g_r(\boldsymbol{v},\boldsymbol{r},t)$ and $n_r(\boldsymbol{r},t)$ incident on a wafer, we may calculate the flux of ejected materials from the surface by dry etching as

$$N_sV_s|_{\boldsymbol{r}_0} = n_p(\boldsymbol{r}_0)\int Y^p_{etch}(\varepsilon_p)\left(\frac{2\varepsilon_p}{M_p}\right)^{1/2} g_p(\boldsymbol{v},\boldsymbol{r}_0,t)d\boldsymbol{v} + n_r(\boldsymbol{r}_0)\int Y^c_{etch}(\varepsilon_r)\left(\frac{2\varepsilon_r}{M_r}\right)^{1/2} g_r(\boldsymbol{v},\boldsymbol{r}_0,t)d\boldsymbol{v}, \quad (8.76)$$

where $Y^p_{etch}(\epsilon_p)$ and $Y^c_{etch}(\epsilon_r)$ are the etching yield of ions (p) and radicals (r), and N_s and V_s the number density and velocity of the molecule ejected from the wafer. The r.h.s of Equation 8.76 represents the sum of the contributions from the physical etching and chemical etching. The velocity distribution functions are normalized as $\int g_p(\boldsymbol{v},\boldsymbol{r},t)d\boldsymbol{v} = 1$ and $\int g_r(\boldsymbol{v},\boldsymbol{r},t)d\boldsymbol{v} = 1$.

The rate of change of the material surface with atomic density ρ by etching, that is, the etch rate, is

$$R_{etch} = \frac{n_p(\boldsymbol{r}_0)}{\rho}\int Y^p_{etch}(\varepsilon_p)\left(\frac{2\varepsilon_p}{M_p}\right)^{1/2} g_p(\boldsymbol{v},\boldsymbol{r}_0,t)d\boldsymbol{v} + \frac{n_r(\boldsymbol{r}_0)}{\rho}\int Y^c_{etch}(\varepsilon_r)\left(\frac{2\varepsilon_r}{M_r}\right)^{1/2} g_r(\boldsymbol{v},\boldsymbol{r}_0,t)d\boldsymbol{v}. \quad (8.77)$$

In general, $g_p(\boldsymbol{v},\boldsymbol{r},t)$ is a beamlike, high anisotropic distribution, whereas $g_r(\boldsymbol{v},\boldsymbol{r},t)$ is an isotropic thermal distribution (Maxwellian) in front of the wafer or electrode.

8.3.4.2 *Plasma Deposition*

Dissociated neutral radical species generated in a plasma may be deposited on the surface. This is known as plasma deposition. Often the precursors of deposition are first deposited on a surface that may be activated by ion bombardment. When the incident flux of neutral radicals is denoted by $\Gamma_r(z)$, and the sticking coefficient as S_t, the rate of deposition to the surface is

$$R_{depo} = \frac{S_t}{\rho}\Gamma_r(\boldsymbol{r})\Big|_{\boldsymbol{r}_0}, \quad (8.78)$$

where ρ is the atomic density of the deposition film.

8.3.4.3 *Plasma Sputtering*

Plasma sputtering is a purely physical process of surface etching, and the sputtering rate is defined by an equation similar to Equation 8.77:

$$R_{sputter} = \frac{n_p(\boldsymbol{r}_0)}{\rho}\int Y^p_{sputt}(\varepsilon_p)\left(\frac{2\varepsilon_p}{M_p}\right)^{1/2} g_p(\boldsymbol{v},\boldsymbol{r}_0,t)d\boldsymbol{v}. \quad (8.79)$$

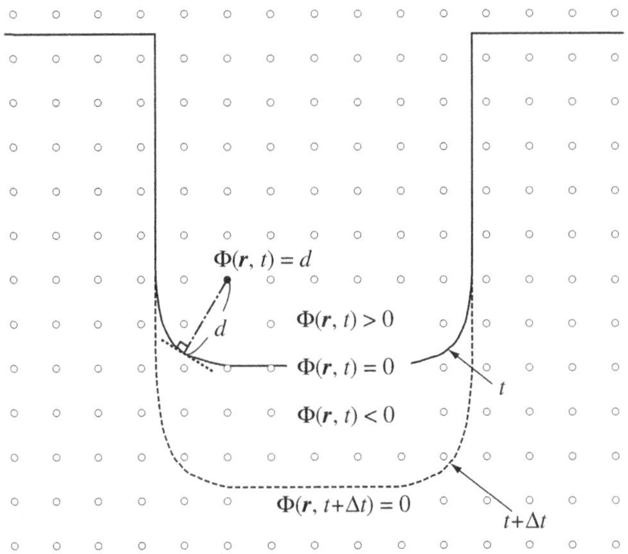

FIGURE 8.11 The moving boundary in material processing.

8.3.5 Moving Boundary under Processing

A surface affected by etching, sputtering, or deposition in a plasma will undergo a change in geometrical and physical profiles, and then the surface boundary is classified as a moving boundary. A spatiotemporally varying surface boundary as shown in Figure 8.11 can be generally described by the surface evolution equation known as the Hamilton–Jacobi equation [8]:

$$\frac{\partial}{\partial t}\Phi(\boldsymbol{r},t) = R_{react}(\boldsymbol{r},t)\left|\frac{\partial \Phi(\boldsymbol{r},t)}{\partial \boldsymbol{r}}\right|, \qquad (8.80)$$

where $\Phi(\boldsymbol{r},t)$ is the surface function defined as

$$\begin{aligned} \Phi(\boldsymbol{r},t) \quad &> \quad 0 \quad \text{in gas phase;} \\ &= \quad 0 \quad \text{on a boundary surface; and} \\ &< \quad 0 \quad \text{inside a material.} \end{aligned} \qquad (8.81)$$

Note that the local surface reaction rate as the local surface velocity $R_{react}(\boldsymbol{r},t)$ is a negative value in etching and sputtering (see Equations 8.77 and 8.79) and a positive value in deposition (see Equation 8.78).

An approach to the surface evolution based on the numerical solution of the Hamilton–Jacobi Equation 8.80 is named the Level Set method. The Level Set method was developed in [8] and is robust in two- and three-dimensional evolution problems.

References

[1] Ferziger, J.H. and Penc, M. 1996. *Computational Methods for Fluid Dynamics.* Berlin: Springer Verlag.

[2] Chua, O.L. and Lin, P.M. 1975. *Computer-Aided Analysis of Electronic Circuit.* Englewood Cliffs, NJ: Prentice-Hall.

[3] Sharfetter, D.L. and Gummel, H.K. 1969. *IEEE Trans. on Electron Devices* ED-16:64.

[4] Takewaki, H., Nishiguchi, A., and Yabe, T. 1985. *J. Comput. Phys.* 61:261–268. Nakamura, T. and Yabe, T. 1999. *Comput. Phys. Commun.* 120:125–154.

[5] Ventzek, P.L.G., Hoekstra, R.T., and Kushner, M.J. 1994. *J. Vac. Sci. Technol.* B12:461.

[6] Chantry, P.J. 1987. *J. Appl. Phys.* 62:1141.

[7] Phelps, A.V. 1990. *J. Res. Natl. Inst. Stand. Technol.* 95:407.

[8] Osher, S.J. and Sethian, J.A. 1988. *J. Comput. Phys.* 79:129. Sethian, J.A. and Strain, J. 1992. *J. Comput. Phys.* 98:231.

CHAPTER 9

Capacitively Coupled Plasma

9.1 RADIO-FREQUENCY CAPACITIVE COUPLING

Plasma is generally sustained capacitively or inductively by a radio-frequency (rf) power supply. Part of the rf power input to the plasma is reflected back to the power supply as a reactive power. The maximum power dissipation in a plasma is supplied by an external rf source when the discharge plasma impedance $\boldsymbol{Z}$ is equal to the impedance at the external power source $\boldsymbol{z}$ (the maximum-power transfer theorem). Plasma impedance is determined self-consistently both by external plasma parameters and by the internal fundamental property of feed gas molecules. An impedance matching network is, therefore, required between the rf electrode and the power supply. A matching network with an equivalent capacitor, C_b (blocking capacitor), is typical. The discharge plasma connected by way of a capacitor to the rf power source is named capacitively coupled plasma (CCP). The value of C_b should be much greater than the sheath capacitance of the reactor.

9.2 MECHANISM OF PLASMA MAINTENANCE

In this chapter, we consider a parallel plates reactor made by metallic electrodes as shown in Figure 9.1. A powered electrode is connected to an rf voltage source, $V_{rf}(t)$, defined as

$$V_{rf}(t) = V_0 \sin \omega t, \tag{9.1}$$

through a blocking capacitor C_b. Here, V_0 and ω are the amplitude and angular frequency, $\omega(= 2\pi f)$, of the rf power supply, respectively. The other electrode as well as the side metallic wall is grounded to the earth.

When an rf voltage $V_{rf}(t)$ is applied to the parallel plate reactor in a vacuum, the current $I_T(t)$ between electrodes leads the applied rf voltage waveform $V_{rf}(t)$ by $\pi/2$. The current is known as the displacement current

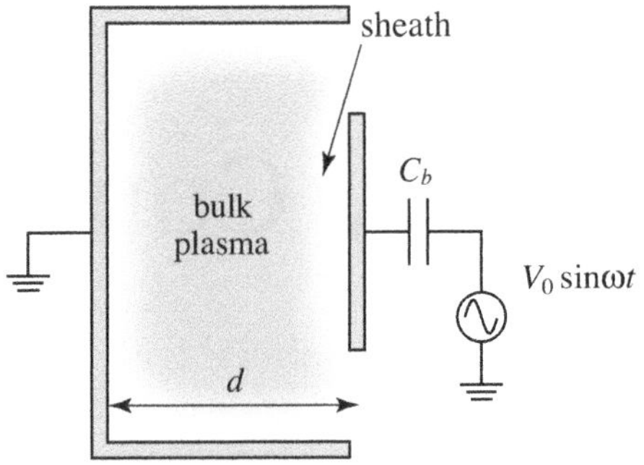

FIGURE 9.1 A typical CCP reactor.

or charging/discharging current between the two electrodes, and the external source has no power dissipation to the reactor; that is, $\overline{\boldsymbol{V}_{rf}(t) \cdot \boldsymbol{I}_T(t)} = 0$.

When an rf voltage appropriate to sustain a discharge plasma is supplied between the two electrodes filled with feed gas molecules at pressure p, a finite discharge current flows through the closed circuit shown in Figure 9.1. Of course, the total current $I_T(t)$ is continuous as a function of axial position, whereas the magnitude of each of the elements of $I_T(t)$ is different in each position (see Chapter 7). When a low-temperature plasma is sustained between electrodes, the sustaining voltage at the powered electrode $V_{sus}(t)$ is given by

$$V_{sus}(t) = V_{rf}(t) - \frac{1}{C_b}\int_{-\infty}^{t} I_T(t)dt, \tag{9.2}$$

where the second term on the right-hand side is the voltage drop between C_b to ensure a net direct current (DC) of zero in a periodic steady state. As a result, the sustaining voltage has a finite negative bias voltage expressed by the second term on the right-hand side of Equation 9.2.

Capacitively coupled rf plasma is generally classified into two regions by the phase-shift between the total current $I_T(t)$ and the sustaining voltage $V_{sus}(t)$. Low-frequency CCP is defined simply when the ions in the plasma satisfy the relation

$$V_d^{eff}\,\frac{1}{2f} > d, \tag{9.3}$$

where V_d^{eff} is the effective drift velocity of positive ions between parallel plates with distance d. That is, at low-frequency plasma, the ions can follow the local field change in time, and the current-sustaining voltage characteristics are resistive without phase difference. Note that the sheath behaves resistively at low-frequency CCP.

On the other hand, at high-frequency plasma, it is difficult for massive ions to follow the instantaneous local field,

$$V_d^{eff}\,\frac{1}{2f} \ll d. \tag{9.4}$$

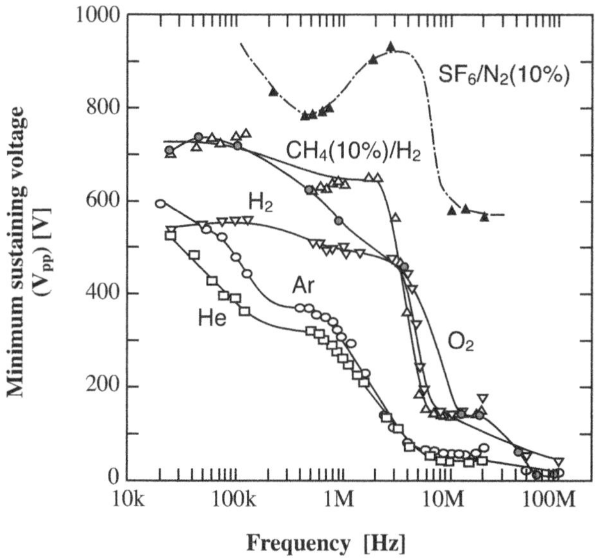

FIGURE 9.2 Minimum sustaining voltage (experimental) in a quasi-symmetric parallel plate CCP as a function of applied frequency at 1 Torr and $d = 2$ cm.

At high-frequency CCP, the total current $I_T(t)$ has a finite phase lead with respect to the sustaining voltage waveform $V_{sus}(t)$, and the sheath behaves capacitively. Usually there exists a boundary between the low- and high-frequency plasmas at several MHz of the external frequency (see Figure 9.2), and the sustaining mechanism of the discharge plasma is completely different between the two regions.

Exercise 9.2.1

Discuss the maximum-power transfer theorem in the CCP system sustained by a sinusoidal voltage in Equation 9.1 in Figure 9.1.

The circuit current $\boldsymbol{I}_T(t)$ in the series circuit with plasma impedance $\boldsymbol{Z}$ and external variable impedance $\boldsymbol{z}$ is given by $\boldsymbol{V}_{rf}(t)/(\boldsymbol{z}+\boldsymbol{Z})$, and the average power dissipated in the discharge plasma, P_{av}, is expressed by

$$P_{av} = \frac{1}{T}\int_0^T \boldsymbol{V}_{sus}(t)\cdot \boldsymbol{I}_T(t)dt = \frac{1}{2}V_0^2\frac{\mathrm{Re}(\boldsymbol{Z})}{|\boldsymbol{z}+\boldsymbol{Z}|^2}, \tag{9.5}$$

where T is the period of the external rf voltage source. The maximum power dissipation to the discharge plasma is obtained at the condition of $\boldsymbol{Z} = \overline{\boldsymbol{z}}$ (impedance matching). That is, the impedance of the discharge plasma $\boldsymbol{Z}$ must be matched to the conjugate of the power supply, $\overline{\boldsymbol{z}}$ (maximum-power transfer theorem).

9.2.1 Low-Frequency Plasma

As the amplitude of the applied voltage waveform is increased, the space between parallel plates filled by neutral gas will change from the Laplace field to Poisson's field by way of the transition region from the Townsend discharge to glow discharge having a plasma phase (see [1,2]). Low-frequency plasma is sustained by the electron multiplication of the secondary electrons ejected mainly by the impact of positive ions on the powered electrode by Auger potential ejection (see Section 4.11.2.4). The maximum flux and kinetic energy of ions incident on the powered electrode from the plasma occur at the phase of the lowest surface potential, that is, at $3\pi/2$ in Equation 9.2. In order to keep a maximum current condition under a given rf voltage at the powered electrode, the space between the two electrodes is bifurcated into two regions, an ion sheaths region and a bulk plasma region. In front of the powered electrode, an ion sheath is constructed so as to allow a sufficient amount of electron ejection from the electrode and a sufficient acceleration of the secondary electrons, which generate ionization multiplications to maintain the low-frequency plasma. Therefore, a high field with a strong potential gradient is essential to the ion sheath (active sheath). A small ion sheath appears in front of the grounded electrode in the conventional parallel plates reactor. Note that, in an ideal symmetric electrode system, a pair of active ion sheaths is constructed in front of both electrodes during one rf period (see Figure 9.3). A bulk plasma region with quasi-neutrality ($n_e \sim n_p$) having a low field is formed between the two ion sheaths. The whole system is completely controlled by Poisson's field.

The total current in the sheath, which consists mainly of a conduction component of positive ions and electrons, coincides with the sustaining voltage waveform V_{sus}(t) without phase difference. Even in the case where the

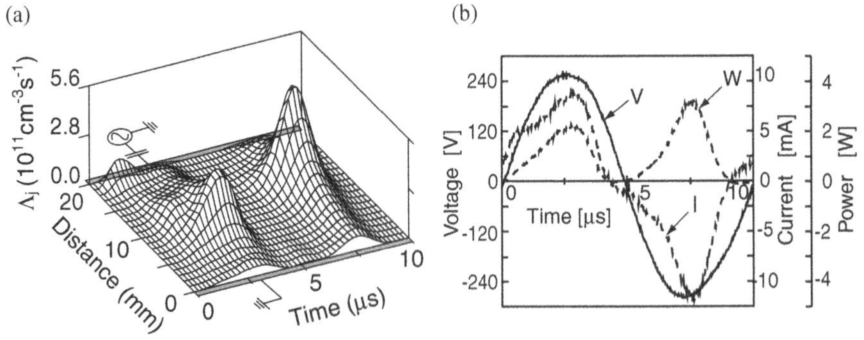

FIGURE 9.3 Typical example of the net excitation rate (a) and voltage-current characteristics (b) during one period in a low-frequency CCP at 100 kHz at 1 Torr in H_2.

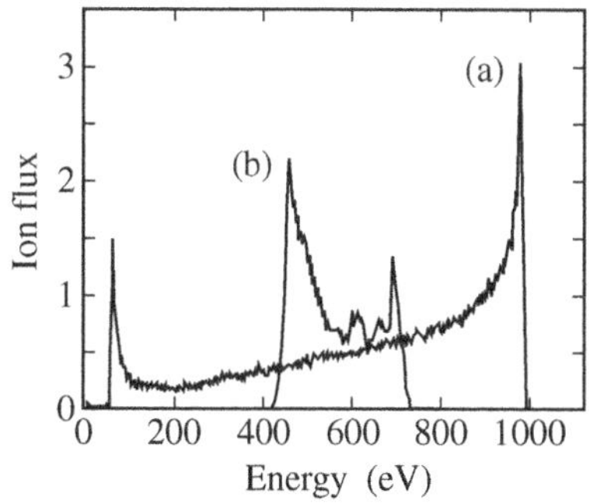

FIGURE 9.4 Time-averaged ion energy distribution incident on the powered electrode at a low-pressure rf plasma. The (a) low- and (b) high-frequency cases are shown.

sustaining voltage is a sinusoidal waveform, the total current is nonsinusoidal in time, and the plasma potential behaves nonsinusoidally. Under the condition of a low-frequency CCP in which ions are influenced by the instantaneous sheath field, the energy of ions incident on the electrode through the sheath is characterized by a saddle-shaped profile, that is, bimodal distribution (see Figure 9.4). The energy dispersion between the maximum and minimum energy peaks decreases as the frequency of the rf source increases to the high-frequency region.

The high degree of loss of ions from the bulk plasma to both electrodes or the side wall during a half-period of the low-frequency plasma exhibits some peculiar characteristics:

i. Formation of low-density plasma with a strong and thick sheath;

ii. Production of high-energy ions incident on the powered electrode; and

iii. A considerably high value of the sustaining voltage.

Problem 9.2.1

A parallel plates reactor is connected to a low-frequency voltage source by a matching network consisting of inductance L_m. Discuss the DC self-bias voltage at the powered electrode.

The effective surface area of the powered electrode connected to an external rf source with C_b is usually different from that of the opposite electrode in a CCP. It is typical when the opposite metallic electrode is grounded to the earth as well as the grounded chamber wall. Under these circumstances, the asymmetry of the discharge plasma is enhanced as there exists a large difference in the effective area between the two electrodes.

Exercise 9.2.2

Consider the pair of asymmetric parallel plate electrodes with the surface areas A_P and A_G shown in Figure 9.1. Estimate the ratio of each of the drops in sheath potential in front of both electrodes, V_{shP} and V_{shG}, in the case of a space-charge-limited regime at low frequency.

In a space-charge-limited regime, the ion current density incident on the sheath with width of d is given by the Child–Langmuir expression,

$$J_p = \frac{K_o V_{sh}^{3/2}}{M_p^{1/2} d^2}, \tag{9.6}$$

where K_o is constant, and M_p is the mass of the ion. Provided that the ion current is dominant in the sheath, the ion currents incident on both electrodes during one period are equal to each other; that is, $J_{pP}A_P = J_{pG}A_G$. It is assumed that the time-averaged potential drops in the sheaths with capacitances C_P and C_G are related to the total charge Q flowing to each of the sheaths as

$$Q = C_P V_{shP} = C_G V_{shG}, \tag{9.7}$$

where the two sheath capacitances are given by $C_P = \epsilon_0 A_P/d_P$ and $C_G = \epsilon_0 A_G/d_G$, respectively. The ratio of the sheath voltage is derived from the above two equations as a function of the electrode area:

$$\frac{V_{shP}}{V_{shG}} = \left(\frac{A_G}{A_P}\right)^2. \tag{9.8}$$

This relation is valid for low-frequency plasma or DC glow discharge under the condition that the main sheath current is supplied by the ion conduction.

9.2.2 High-Frequency Plasma

As the frequency of the voltage supply increases, the ions will have a finite phase delay with respect to a time-varying instantaneous field. This means that the ion flux incident on the powered electrode and the number of the secondary electrons will decrease as the rf frequency satisfies the relation (9.4). At the same time, due to the difference of the phase between $I_T(t)$ and $V_{sur}(t)$, the number of electrons and ions released to the electrodes will decrease. In turn, electrons released from the bulk plasma toward the powered electrode will be effectively reflected in the positive ion sheath as wave-riding electrons. The reflected electrons will be accelerated toward the bulk plasma through the sheath, and ionization multiplication will occur at the boundary between the sheath and bulk plasma (see Figure 9.5). This is a sustaining mechanism in a high-frequency plasma. As the external frequency increases, electrons in addition to ions will gradually enter into their trap in space. Due to the presence of electron trapping inside the reactor, an apparent field of ambipolar diffusion will drop in front of the powered electrode in a time-averaged fashion. This will introduce a reduction of the minimum sustaining voltages as shown in

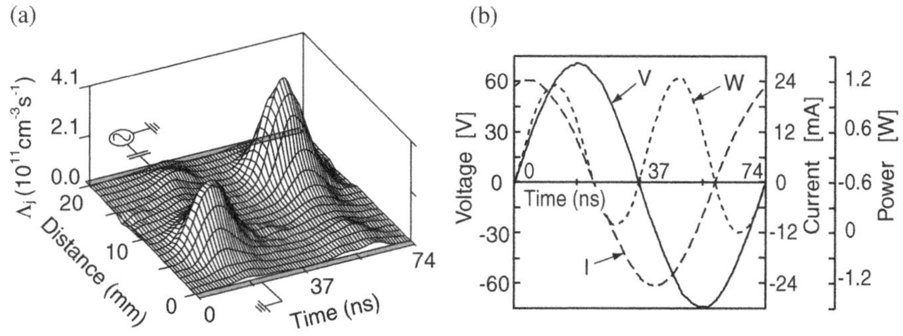

FIGURE 9.5 A typical example of the net excitation rate (a) and voltage-current characteristics (b) during one period of a high-frequency CCP at 13.56 MHz in H_2.

Figure 9.2. The boundary between the low- and high-frequency plasma under conventional external conditions will appear at approximately several MHz.

When a sinusoidal high-frequency voltage is supplied at the powered electrode, the total current in the sheath, consisting of the predominant displacement current and the conduction of electrons and ions, has a sinusoidal waveform. The capacitive sheath results in a sinusoidal variation of the plasma potential. The high-frequency plasma has several proper characteristics:

i. Formation of high-density plasma with a weak thin sheath;

ii. A relatively low sustaining voltage;

iii. Production of low-energy ions incident on an electrode; and

iv. A sinusoidal current waveform $I_T(t)$ leading the sustaining voltage waveform $V_{sus}(t)$ is formed, and the discharge is defined as capacitive.

The electron charge flowing into the blocking capacitance C_b through a small powered electrode during the positive potential should be equal to the positive ions during the rest of one period in a periodic steady state. The great difference between the mass of the electron and that of the ion will cause an excess negative charge in the capacitor during one period. Therefore, a negative bias voltage V_{dc} to the small electrode is needed in order to keep the zero net DC current in a periodic steady state through the C_b. That is, the surface of the powered electrode is negatively biased as

$$V_{sus}(t) = V_{rf}(t) - V_{dc}(t). \tag{9.9}$$

Here, $-V_{dc}(t)$ is the negative self-bias voltage with time variation. The time-averaged DC self-bias voltage,

$$V_{dc} = \overline{V_{dc}(t)} = \frac{1}{2\pi}\int_0^{2\pi} V_{sus}(t)\, d(\omega t), \tag{9.10}$$

is practically observed in the experiment.

TABLE 9.1 Classification of Sheaths Formed in a Collisional Plasma

Item	Active Sheath	Passive Sheath
Governing law	Time-averaged ambipolar diffusion	Time-averaged ambipolar diffusion
Ideal function	Plasma confinement	Plasma confinement w/o ionization
	Plasma maintenance with ionization	Ion acceleration to surface
Control of		
		Ion flux Γ_{ion}
	$g_{\text{e}}(\boldsymbol{v})$ of electrons	$g_{\text{ion}}(\boldsymbol{v})$ of p-ions
	Space potential	Species at specific cond.
		Sheath length & potential
		DC-self-bias voltage
Position formed		
	in front of powered-electrode	in front of biased-wafer
	in front of DC cathode	in front of reactor wall
		around electric probe

At high-frequency CCP, under which the ion transit time across the sheath is sufficiently longer than the half-period of the external source, the ion incident on the electrode reflects the time-averaged sheath characteristics (see Table 9.1). As a result, a single-peaked energy distribution is formed with a maximum at about the time-averaged sheath voltage (see Figure 9.4). The sheath formed in discharge plasmas is classified in Table 9.1.

Problem 9.2.2

At low pressure a collisionless sheath with thickness d_{sh}, bulk electrons with average velocity $< \boldsymbol{v} >$ diffusing to the plasma sheath boundary will interact electrically with the moving boundary having a velocity $\boldsymbol{V}_{sh}(t)$. This is not a binary collision between the electron and the neutral molecule but a wavelike interaction resulting from the long-range Coulomb interaction. Provided that the interaction is perfectly elastic under $d_{sh}/v > 2\pi/\omega$, the reflected electrons will change their velocity from

$$- < \boldsymbol{v} > \quad \rightarrow \quad < \boldsymbol{v} > + \boldsymbol{V}_{sh}, \tag{9.11}$$

close at instantaneous cathode phase. Therefore, the phenomenon is a collisionless heating and is referred to as a stochastic heating of electrons, Fermi heating, or wave riding. Derive the relation 9.11 from the momentum balance of the electrons.

9.2.3 Electronegative Plasma

A conventional low-temperature rf plasma consists of electrons, positive ions, and feed gas molecules as the components. In a reactive plasma for etching, electronegative gases are widely used due to a strong chemical reactivity of the feed gas or the dissociated molecules (radicals) on a surface. There are hidden and interesting characteristics of an rf plasma for dry etching. That is, these reactive gases have the ability to produce negative ions by dissociative or nondissociative electron attachment with a threshold of several eV or thermal energy (see Section 4.8.2). With increasing pressure, the percentage of negative ions in the plasma increases due to the increase of the degree of spatial trapping of massive negative ions and of the collisional chance of electron attachment (see Table 9.2). The quasi-neutrality in the bulk plasma is realized as

$$n_p \sim (n_e + n_n). \tag{9.12}$$

The electronegativity H_{en} is defined by the terms of the number density of negative and positive ions, n_n and n_p, as

$$H_{en} = \frac{n_n}{n_p}. \tag{9.13}$$

Fully negative ion plasma is performed under $H_{en} = 1$, though the plasma is not maintained without electrons. In an electronegative plasma with densities n_e, n_p, n_n, and N, the macroscopic measure of the plasma is characterized by both the electronegativity H_{en} and the degree of ionization H_{di} given by

$$H_{di} = \frac{n_p}{N}. \tag{9.14}$$

Let us consider again the CCP system consisting of light electrons and massive positive ions as the components of the charged particle. The system

TABLE 9.2 Negative Ions in Processing Plasmas

Feed Gas	Negative Ion	Material Processing
SF_6	SF_j^- $(6 \geq j \geq 1)$, F^-	Si, W etching Reactor cleaning
Cl_2	Cl^-	Si, Al etching
BCl_3	Cl^-	Si, Al etching
CF_4	F^-, CF_j^-	SiO_2 etching
C_4F_8	F^-, CF_j^-	SiO_2 etching
CF_3I	I^-, CF_j^-	SiO_2 etching
O_2	O^-, O_2^-	Oxidization, photoresist ashing Surface trimming
SiH_4	SiH^-	a-Si:H deposition
H_2/N_2	H^-	Reduction, low-k(organic) etching
HCl	Cl^-	Si etching

has typical plasma characteristics with a bulk plasma under quasi-neutrality ($n_e \sim n_p$) and with positive ion sheaths (i.e., electropositive plasma). That is, one is the presence of a positive space potential (plasma potential) V_s in the bulk plasma. The other is the strong sheath potential V_{sh}, which is essential to maintaining the plasma by the electron impact ionization. Both are caused by the great difference in speed between light electrons and massive positive ions. The general characteristics in the electropositive plasma will gradually change to a different phase in the presence of negative ions. Generally, in a high-frequency electronegative plasma consisting primarily of positive and negative ions, with fewer electrons, a thin, higher sheath field is realized as well as the presence of a strengthened bulk field. This is mainly caused by massive negative ions having a drift velocity of two orders of magnitude less than electrons. As a result, there exist three different regions of the plasma production during the half-period in electronegative plasmas at higher pressure, as shown in Figure 9.6a. One is the production by reflected electrons in front of the instantaneous cathode (I), another is by the bulk electrons in a high field (II), and the other is by electrons accelerated through a double layer close to the instantaneous anode (III). Also the phase difference between the total current $I_T(t)$ and the sustaining voltage $V_{sur}(t)$ is shortened with increasing electronegativity H_{en}, as compared with electropositive plasma at all frequencies. In other words, the characteristics of an rf plasma change from capacitive to resistive in a strong electronegative plasma (see Figure 9.6).

A typical electronegative plasma is formed in O_2, SF_6, Cl_2, CF_4, and so on with a finite collision cross section of electron attachment at high pressure. Electron capture by molecules excited to electronic or vibrational states is known to have a larger cross section as compared with that by the ground-state molecule. Negative ion production by way of electronically or vibrationally excited states will make a considerable contribution to the electronegativity in plasma, particularly in a high-density plasma. $O_2(a^1\Delta)$ in oxygen plasma is an example of the excited states that strongly contribute to the electron

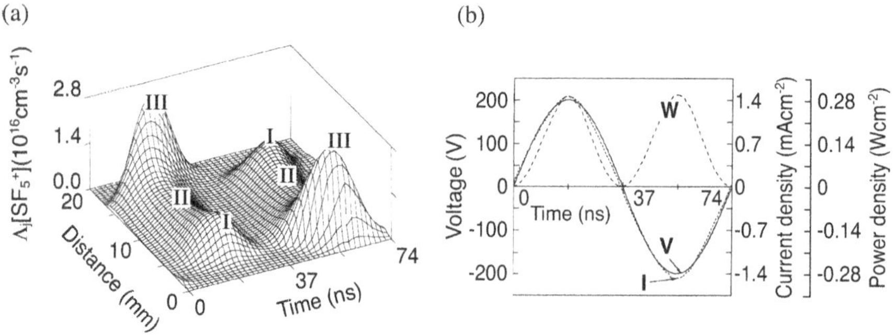

FIGURE 9.6 Typical example of net excitation rate in electronegative high-frequency CCP in $SF_6/N_2(10\%)$ at 1 Torr.

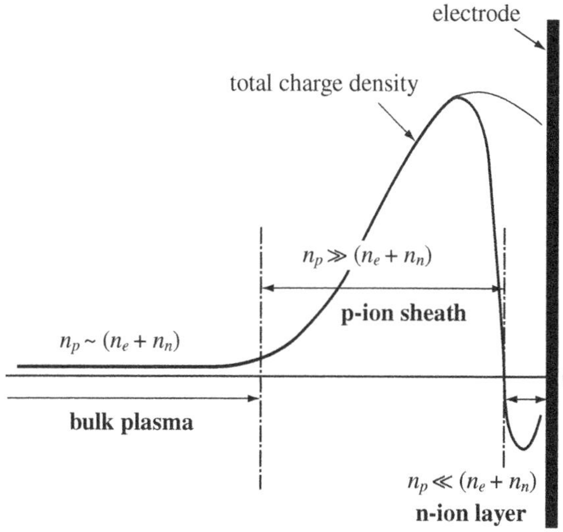

FIGURE 9.7 Double layer formation in front of the instantaneous anode in an rf electronegative plasma.

attachment, because the peak of the attachment cross section is several times higher than the the ground state $O_2(^3\Sigma_g^-)$, and the threshold energy stands at 2.76 eV (see Section 4.8.2).

Exercise 9.2.3

In an rf electronegative plasma at a range of Torr, a negative charge-layer is formed, facing the electrode in the instantaneous anode phase (see Figure 9.7). Discuss the negative layer formation (it is usually called double layer formation).

Minority electrons in an rf plasma with $H_{en} \sim 1$ vibrate between the two electrodes. At the instantaneous anode phase, the total charge in front of the anode, $\rho = n_p + n_n + n_e$, is negative due to the flow of electrons from the bulk plasma to the sheath, though the sheath is positive except for the anodic phase.

Problem 9.2.3

In an electronegative plasma with electronegativity $H_{en} \sim 1$, negative ion density at low power is the rate-limiting step between the production by an electron attachment to the feed gas molecule and the loss by a recombination with positive ion. Then, derive the simple relation between n_n and n_e as

$$n_n \simeq \left(\frac{R_a n_e}{R_r} N\right)^{1/2}, \tag{9.15}$$

where R_a and R_r are the electron attachment rate and recombination rate, respectively.

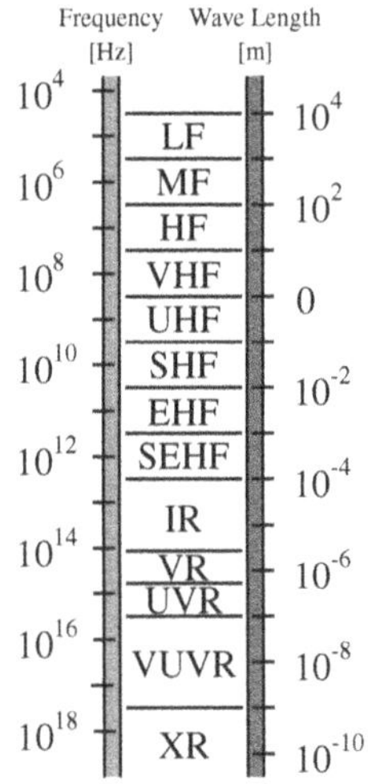

FIGURE 9.8 Frequency and its wavelength of electromagnetic wave.

9.2.4 Very High-Frequency Plasma

A low-temperature plasma source has developed the range of driving frequency from low frequency (LF) ($f <\sim$ MHz) to ultra-high frequency UHF (3×10^8 Hz $\leq f \leq 3 \times 10^9$ Hz), through high frequency (HF) (3×10^6 Hz $\leq f \leq 3 \times 10^7$ Hz) and very high frequency VHF (3×10^7Hz $\leq f \leq 3 \times 10^8$ Hz) (see Figure 9.8). The most pronounced characteristic expected for a parallel plate CCP is radial uniformity of the plasma structure derived from the uniform surface potential of the electrode connected to an rf power source.

The VHF source has the advantage of maintaining a high-density plasma under a low self-bias voltage at the powered electrode and is used primarily to obtain a good quality film at high deposition rate and a high-efficiency etching. One of the issues in a large area and high-density plasma sustained using the VHF technique is the radial nonuniformity of the surface voltage due to a standing wave effect, when the dimension of the powered electrode exceeds or is comparable with $\lambda/4$ or $\lambda/8$ of the free space wavelength λ of the VHF. The Si wafer in ultra-large scale integration (ULSI) has grown to a large size from 100 mm in diameter in 1977 to 300 mm in 2000. The glass substrate of 1 m^2 is standard in a-Si:H deposition for thin film solar cells and in plasma display panels and active matrix liquid crystal displays. In the UHF source, the plasma is sustained by the aid of an adequate antenna or waveguide due to a wavelength that is comparable with or shorter than the reactor geometry.

Figure 9.9 shows the experimental amplitude of the surface potential on a metallic rod electrode coupled at the edge to a VHF (100 MHz) [3]. The rod electrode is faced to a large area substrate, and VHF plasma is maintained between them. Here, it is shown that a surface standing wave propagates with a reduced short wavelength. As shown in the example, radial uniformity of

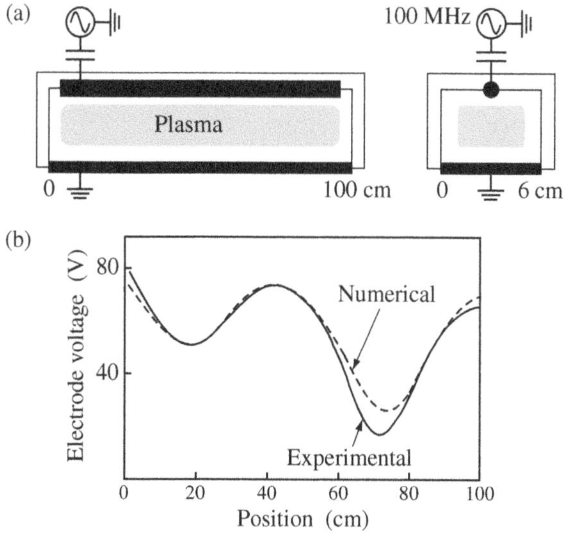

FIGURE 9.9 Surface potential of a powered electrode exposed to a VHF plasma at 100 MHz at 13 Pa in Ar. Power is supplied at $z = 0$, and $z = 100$ cm is the open end.

the plasma structure in a large-size reactor is disturbed by the presence of harmonics of the external source frequency in the range of VHF. Adequate multicoupling of the VHF source on the electrode is one of the practical methods to maintain the uniformity.

The external electrical power supplied by rf sources is first deposited to the electron in the plasma in the form of transport/conduction. The external power is transferred to emission of various kinds of lights by way of molecules excited by the electron impact. The heavy particle collisions between the ion and feed gas molecule and between the excited molecule and the feed gas transfer their kinetic or potential energy to the feed gas translational energy. These lead to the heating of the feed gas molecule, gas heating. The total dissipated power is divided into the element, electron, positive ion, and negative ion. In the bulk plasma, almost all power is deposited to electrons in electropositive gases under a low bulk field. However, in the case of electronegative gases, the circumstances change. With increasing the electronegativity $H_{en} = n_n/n_p$, both the negative ion density and the bulk field increase, because a high field is needed to carry the majority of negative ions in the bulk plasma. Then, the negative ions in the bulk plasma share the power among the electron and positive ions. In the sheath, the power is deposited into positive ions in front of both electrodes. The asymmetry of the power deposition in each of the sheaths arises from the presence of the DC self-bias voltage at the powered electrode as compared with the grounded electrode. Various applications in a low-temperature plasma technology utilize the plasma element in a form of maximizing or optimizing the function of the element. Then, the control of the

TABLE 9.3 Time-Averaged Power Deposition

Feed Gas	Active Sheath	Bulk Plasma	Passive Sheath
Electro positive	P-ions	Electrons	P-ions
Electro negative			
(Weakly)	P-ions	Electrons	P-ions
(Strongly)	P-ions	Electrons N-ions, P-ions	P-ions

power deposition in space and in time is of first importance to the technology. See Table 9.3.

Problem 9.2.4
Scaling lows of internal plasma parameters (plasma density, sheath width, etc.) are prepared with respect to the external parameters (frequency, amplitude, pressure, etc.) in a simplified condition. Discuss the relationship between plasma density and driving frequency in a high-density plasma at a sufficiently low pressure.

Exercise 9.2.4
There exists an external electromagnetic wave with frequency ω, which is lower than the plasma (electron) frequency $\omega < \omega_{pe}$; then, it is difficult for the external electromagnetic wave to penetrate the plasma. Discuss that this does not mean none of plasma production/maintenance in the collisional plasma. By applying the rf wave with frequency ω and amplitude V_0 through an antenna to a neutral gas molecule, the interaction with the gas molecule will cause the electron multiplication when we increase the V_0. A periodic steady state plasma will be formed through an active sheath in front of the antenna. Then, inside of the formed bulk plasma a cut-off density $n_c(= \omega_{pe}^2 m\epsilon e^{-2})$ will exist.

9.2.5 Two-Frequency Plasma

A two-frequency CCP reactor has two types of arrangement of rf power sources. One is the configuration of a different rf source at each of two electrodes (see Figure 9.10). The other is the system of an electrode with two rf sources and a grounded electrode. For the choice of the rf frequency, it is of the first importance to clarify the function in the plasma. For example, in a low-temperature plasma reactor for plasma etching of SiO_2, a high-density plasma production and high-energy ions incident on a wafer electrode are two functions that are required simultaneously. External power sources sometimes interact with each other in a similar frequency range. Both functions are competitive in a two-frequency (2f) CCP reactor for sustaining a high-density plasma at VHF with low-voltage amplitude and for biasing the wafer at LF with high amplitude in each of the electrodes at low pressure. One example of the functional separation is shown in Figure 9.11 in terms of net ionization rate as a function of bias frequency (i.e. 1 MHz and 10 MHz).

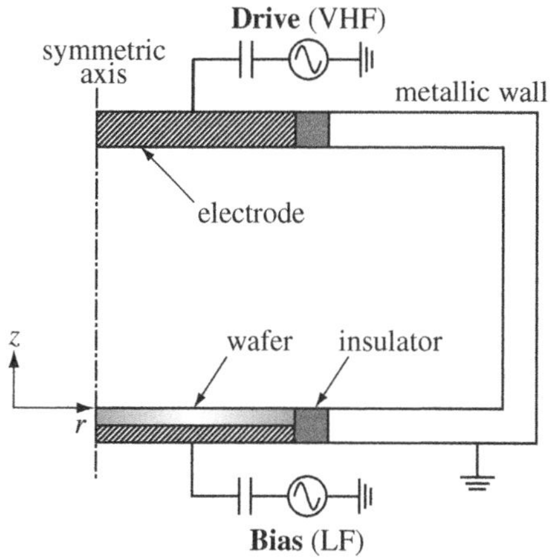

FIGURE 9.10 Schematics of a 2f-CCP reactor for SiO_2 etching.

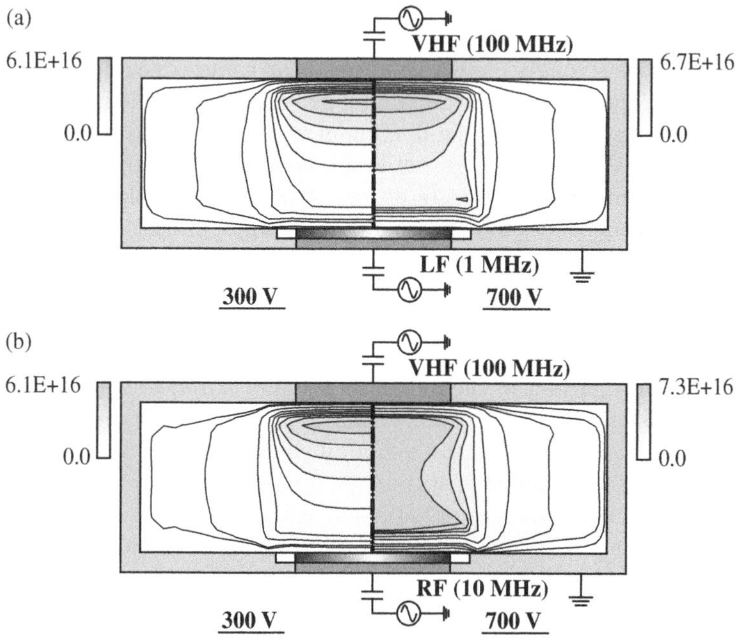

FIGURE 9.11 Example of the functional separation in a 2f-CCP in CF_4(5%)/Ar at 50 mTorr in terms of net ionization rate.

The values of the negative self-bias voltage V_{dc} at each of the electrodes are considerably different, with the value being very high at LF and very low at VHF. That is,

i. A high V_{dc} at the bias electrode is capable of generating the beamlike ions with an anisotropic velocity distribution at the surface;

ii. High-energy secondary electrons produced by the high-energy ion bombardment are not confined by a very low V_{dc} at the opposite VHF electrode; and

iii. The contribution of the fast secondary electrons to an excess degree of dissociation is significantly reduced in the 2f-CCP at low pressure.

In particular, high V_{dc} is realized at the biased electrode without disturbance of the plasma profile in the VHF/LF-CCP system. A narrow gap 2f-CCP is operated by three potentials, V_{sur} at the powered and biased electrodes, and a wall potential usually grounded to the earth. The plasma potential in the system is kept considerably high. One example of the sheath potential drop in front of the wafer and the surrounding wall is shown in Figure 9.12 as a function of radial position. The sheath potential drop is defined as the difference between the plasma potential and the surface potential at the surface. The potential drop is strongly distorted between the wafer edge and the grounded metallic wall. The phase of the maximum potential drop depends on the radial position. On the SiO_2 wafer the maximum value appears at $\omega t = 3\pi/2$ when the wafer surface has the lowest potential, and on the grounded metallic side wall the maximum value appears at $\omega t = \pi/2$ when the plasma potential is highest. The magnitude and the energy of charged particles incident on the surface are influenced by the temporal change of the radial field caused by the edge characteristics of the potential drop.

The velocity distribution of ions incident on the wafer gives critical information for the reactive ion etching of chemically active plasma. There are a number of investigations of the influence of the bias frequency on the incident ion energy distribution. It is known that at high-frequency regions the ions can not relax their energy to the temporal change of the wafer potential; the energy distribution exhibits a stationary profile. However, at low-frequency, ions follow the instantaneous potential change at the wafer, and a strong time modulation arises on the energy distribution with a maximum and minimum at low pressure (see Figures 9.4 and 9.14). Note that the radial characteristics of the velocity distribution of the incident ions are the key to realizing a uniform etch process on a wafer. The vicinity of the wafer edge is strongly influenced by the potential difference between the wafer and the outer grounded wall, whereas the central part of the wafer keeps the intrinsic characteristics of the ion velocity distribution as a function of sheath potential drop.

In the other system of 2f-CCP, both of the power sources are supplied to the same electrode and the second electrode is grounded. This system is less functional for the material etching device.

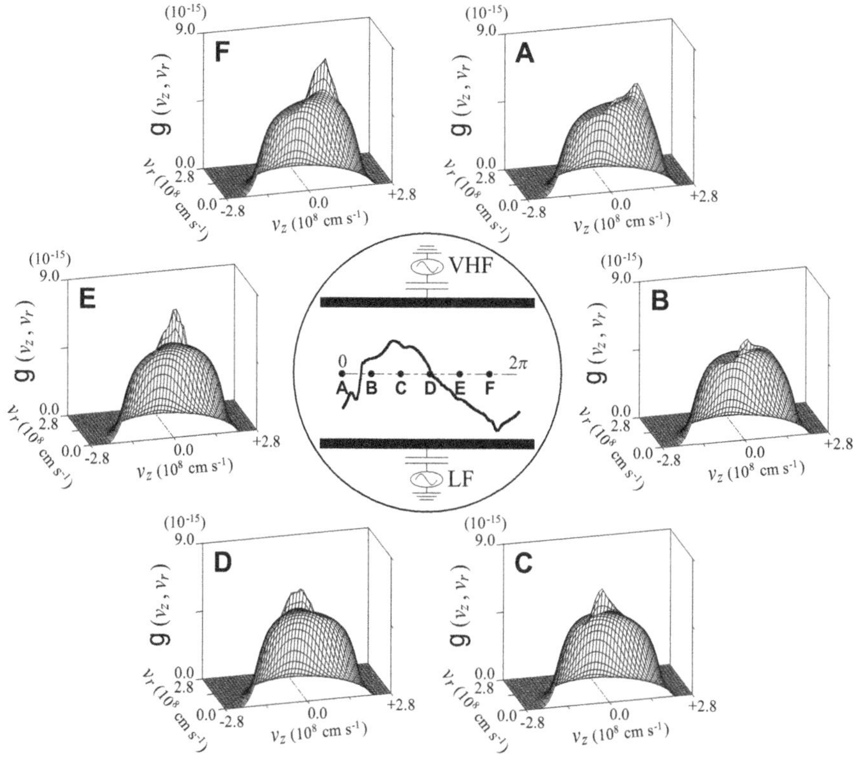

FIGURE 9.12 Temporal behavior of $g^0(\boldsymbol{v}, \omega t)$ of electrons in the bulk plasma during one period of the VHF at $\omega t_{LF} = \pi/2$ at LF of 700 V. External conditions are in Figure 9.11(a). *From Makabe, T, and Yagisawa, T. Plasma Sources Science and Technology, "Low-pressure nonequilibrium plasma for a top-down nanoprocess", Volume 20, Number 2, 2011. With permission.*

9.2.6 Pulsed Two-Frequency Plasma

A modern plasma reactor has to be equipped with a multifunction for processing. One is a selective injection of particles accumulated to a sufficient density in a bulk plasma. In a CCP, even in a 2f-CCP at low pressure, it is difficult to accelerate negative charges, electrons, or negative ions to a wafer surface, because the sheath field in front of the wafer is always formed to decelerate the particles in a cw operation. Pulsed operation of plasma production by a VHF source makes it possible to inject negative charged particles in an acceleration mode to the biased wafer in a 2f-CCP (see Figure 9.15). It is found that in Figure 9.15b secondary electrons produced by a high-energy ionimpact on

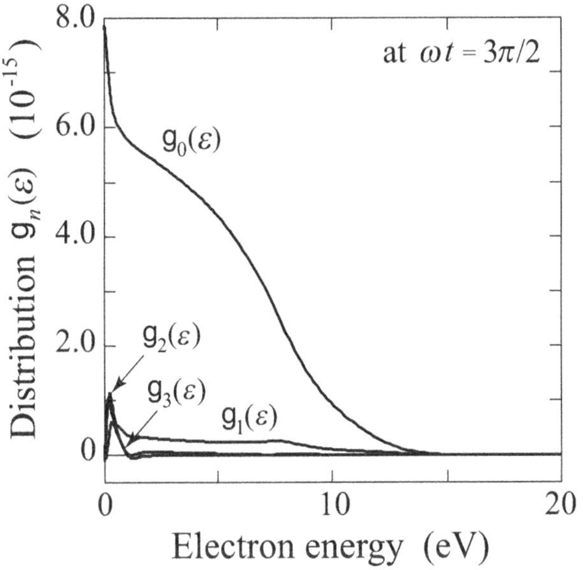

FIGURE 9.13 Each component $g_n(\epsilon)$ in the Legendre expansion of $g^0(\boldsymbol{v}; \omega t_{\text{VHF}} = 3\pi/2)$ of electrons in Figure 9.12. *From Makabe, T, and Yagisawa, T. Plasma Sources Science and Technology, "Low-pressure nonequilibrium plasma for a top-down nanoprocess", Volume 20, Number 2, 2011. With permission.*

TABLE 9.4 Effect of Bias Voltage Form, $V(t) = V_{\text{LF}} \sin \omega_{\text{LF}} t$ on $g_{\text{ion}}(\boldsymbol{v})$ in a 2f-CCP

External Plasma Parameter	**Structure of $g_{\text{ion}}(\boldsymbol{v})$**
Frequency	at increase in ω_{LF}; ϵ_{max} shifts to lower
	Width decreases
	Peak structure is simplified
	Incident angle becomes wider
Amplitude	at increase in V_{LF}; ϵ_{max} shifts to higher
	Width increases
	Incident angle becomes narrower
Gas pressure	at increase in p; ϵ_{max} shifts to lower
	Peak structure diffuses
	Both peaks at ϵ_{max} and ϵ_{min} decrease
	Incident angle becomes wider
Ion mass	at smaller M;
	ϵ_{max} and ϵ_{min} shift to higher
	Width decreases
	Multi-peaks appear widely
	FWHM of incident angle decreases

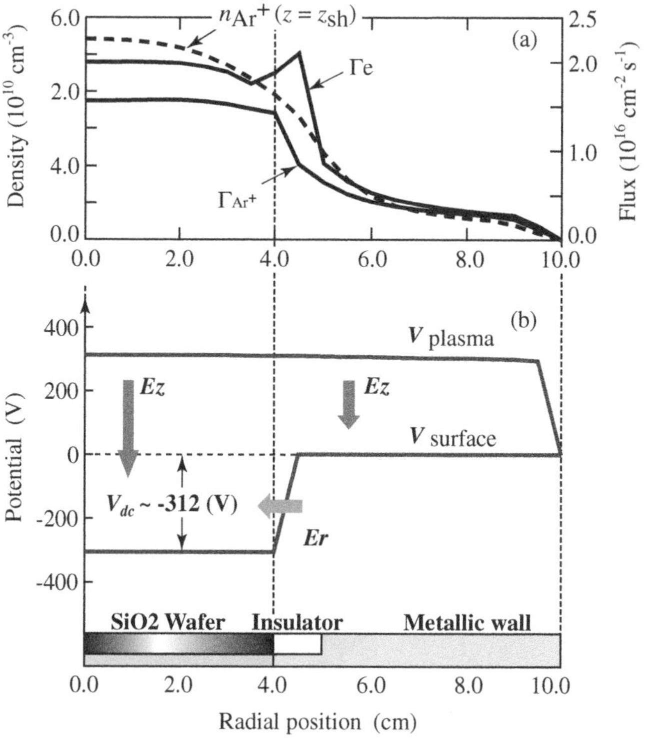

FIGURE 9.14 Time-averaged plasma density and flux (a) and potential (b) in a 2f-CCP maintained under conditions of Figure 9.11a.

the wafer have little influence on the plasma production in a high-density 2f-CCP. An injection of negative charged particles with high energy are much enhanced in a system where both pulsed operations of VHF/LF sources are synchronized in a 2f-CCP [4].

Problem 9.2.5

A plasma is sustained in circumstances including an impure gas. Then, the effect of an impurity on a plasma structure will be negligible under conditions,

$$N_{imp}Q_{m_{imp}}(\epsilon)\epsilon^{1/2} \ll N_oQ_{m_o}(\epsilon)\epsilon^{1/2},$$

$$\left(\frac{2m}{M_{imp}}N_{imp}Q_{m_{imp}}(\epsilon) + \sum \epsilon_{j_{imp}}N_{imp}Q_{j_{imp}}(\epsilon)\right)$$
$$\ll \left(\frac{2m}{M_o}N_oQ_{m_o}(\epsilon) + \sum \epsilon_{j_o}N_oQ_{j_o}(\epsilon)\right),$$

$$\left(n_eN_{imp}Q_{i_{imp}}\epsilon^{1/2}, N_oN_{imp}Q_{PI}\epsilon_r^{1/2}, n_eN_{imp}Q_{a_{imp}}\epsilon^{1/2}\right) \ll n_eN_oQ_{i_o}\epsilon^{1/2},$$

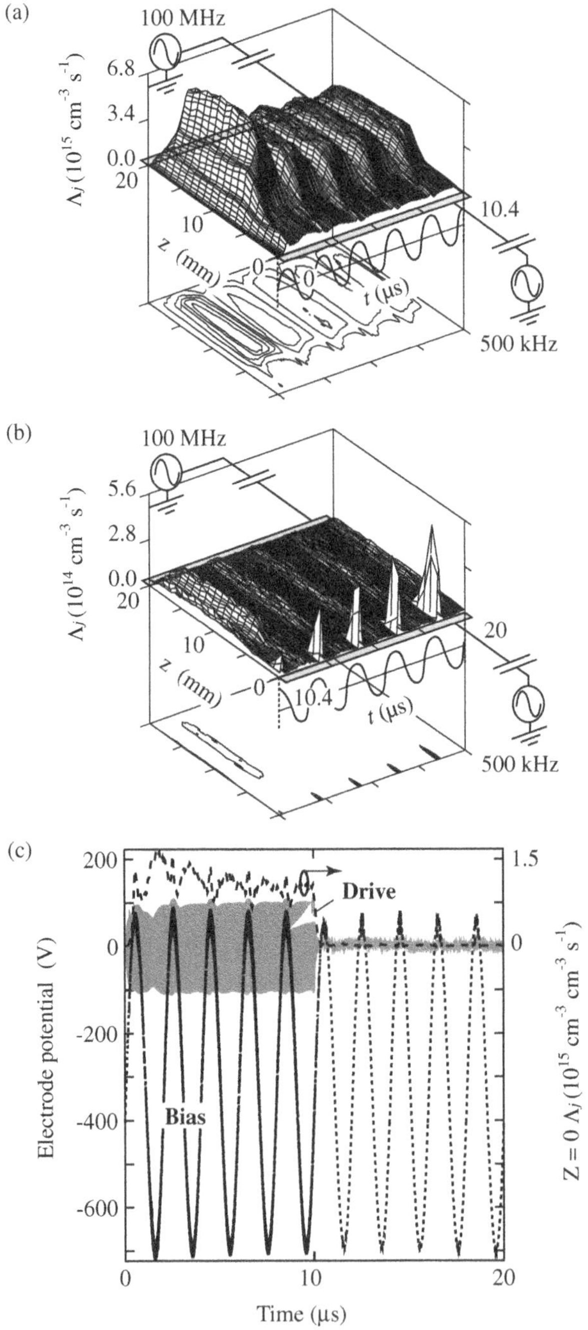

FIGURE 9.15 Pulsed-plasma source operation in 2f-CCP with VHF (100 MHz) and LF (1 MHz) at 25 mTorr in CF_4/Ar: (a) net excitation rate during on phase and (b) during off phase and (c) driving voltage on both electrodes.

TABLE 9.5 Observation of $g_{\text{ion}}(\boldsymbol{v})$ in a Low Temperature Plasma

Method	Analysis	Comments
Quadropole Mass Spectrometry (QMS) with energy analyzer	Mass & Energy	at electrode or wall
Time-of-Flight (TOF) technique with energy analyzer	Mass & Energy	at electrode or wall
Laser Induced Fluorescence (LIF)	State & Energy	in a plasma (contactless)
Electrostatic Probes	Energy & Density	in a plasma (contact)

where the subscript "o" and "imp" show the number density of the feed gas and impurity, respectively. The set of Inequalities is the conditions for the drift velocity, energy, and number density of electrons, respectively. Find these relations.

Problem 9.2.6

A small metallic wafer is inserted into a diffused uniform DC-plasma without production and loss. The potential of the wafer surface is kept at a dc value, $V_w(V)$. The electron in the passive wall-sheath is in a quasi-equilibrium under the condition that the collision is neglected between the feed gas and the electron. Then, in the sheath the velocity distribution of the electron with mass m and charge e will be subject to the collisionless Boltzmann equation when the incident electron from the plasma is not taken into account. That is, the governing equation of electrons in the one dimesional form is given by

$$v\frac{\partial g(v,x)}{\partial x} + \frac{e}{m}\frac{V(x)}{\partial x}\frac{g(v,x)}{\partial v} = 0,$$

where $g(v,x)$ is the isotropic velocity distribution as a function of velocity v and position x. $V(x)$ is the potential at x from the wafer surface. Find the analytical form of $g(v,x)$.

References

[1] von Engel, A. 1983. *Electric Plasmas Their Nature & Uses*, London: Taylor & Francis.

[2] Hirshi, M.N. and Oskam, H.J. Eds., 1978. *Gaseous Electronics*, Vol. 1. New York: Academic Press.

[3] Satake, K., Yamakoshi, H., and Noda, M. 2004. *Plasma Sources Sci. Technol.* 13:436–445.

[4] Ohmori, T., Goto, T., and Makabe, T. 2004, 2003. *J. Phys. D.* 37:2223, and *Appl. Phys. Lett.* 83:4637.

CHAPTER 10

Inductively Coupled Plasma

10.1 RADIO-FREQUENCY INDUCTIVE COUPLING

Radio-Frequency (rf) plasma is sustained by the rf electromagnetic field radiated from an rf current coil. This is named the inductively coupled plasma (ICP) or transformer coupled plasma (TCP) based on the difference of the current coil arrangement, as shown in Figure 10.1 [1]. Usually ICP is initiated locally by the capacitive coupling between both terminals of the current coil with increasing the voltage amplitude between terminals of the coil (E-mode). A locally sustained low-density plasma diffuses into gas, and an abrupt change occurs from a capacitive coupling between the terminals (E-mode) to an inductive coupling between the plasma and coil (H-mode) when the power (or coil current) is further increased. The phase transition is known as an E–H transition in ICP with a hysteresis loop between a rise and fall in the power. ICP and TCP in Figure 10.1 are examples of the electrodeless plasma. The pressure range of ICP in plasma etching is between a few mTorr to some tens of mTorr, whereas for thin film deposition the feed gas pressure is usually distributed over 100 mTorr and 10 Torr.

10.2 MECHANISM OF PLASMA MAINTENANCE

10.2.1 E-mode and H-mode

The transition between the E-mode and the H-mode is one of the most marked features of the ICP. One example of the E–H transition is shown in Figure 10.2 in terms of the net excitation rate of the short-lived $Ar(2p_1)$ in the coil plane of a cylindrical glass reactor, sustained by a single-turn current coil at 13.56 MHz in Ar, in Figure 10.1a as a function of amplitude of the coil current [2]. A sudden transition of the emission is observed at a point (I) with an increase in intensity of two orders of magnitude, and from the point (II) a gradual increase continues with increasing coil current. A hysteresis is observed

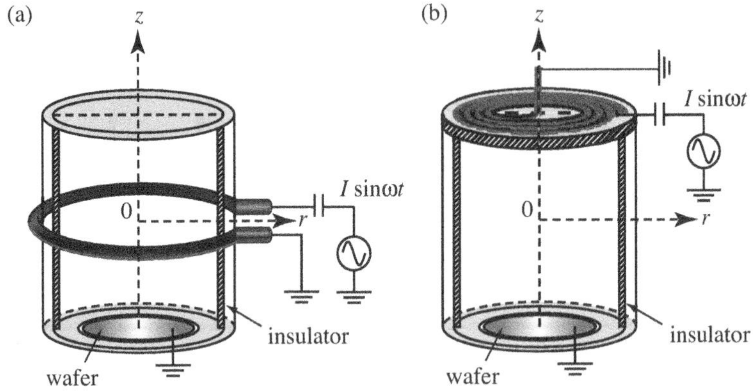

FIGURE 10.1 A typical example of ICP (a) and TCP (b) reactors.

at high pressures of 100 mTorr and 300 mTorr upon the return to smaller currents. The ICP in the H-mode has a threshold for maintenance.

In the E-mode the power deposition in the discharge is low and electron density is low, whereas the mean electron energy is high with a nonequilibrium distribution. In the H-mode high power is dissipated with a low-power reflection. High density of plasma with a lower mean energy of electrons is sustained in the H mode. The differences of the electron energy diminish at low pressures due to the nonlocal transport of electrons.

Figure 10.3 shows the computerized tomography (CT) images of the three-dimensional number density of the excited Ar($2p_1$) in pure Ar at 300 mTorr

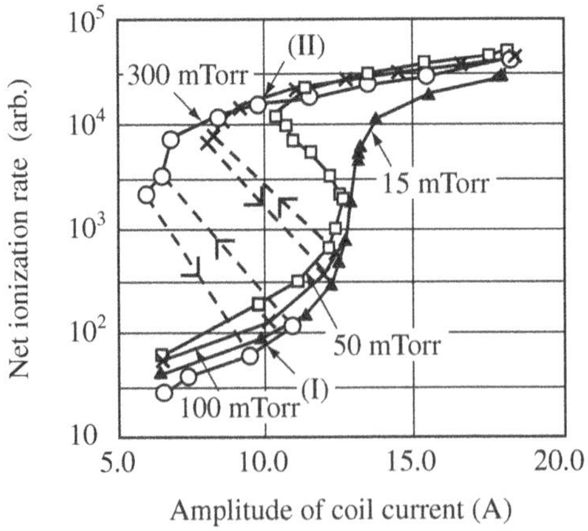

FIGURE 10.2 Net excitation rate of Ar($2p_1$) as a function of coil-current amplitude for pressures of 15 mTorr (Δ), 50 mTorr ($\square$), 100 mTorr ($\times$), and 300 mTorr ($\circ$) in ICP at 13.56 MHz in Ar in the reactor shown in Figure 10.1a.

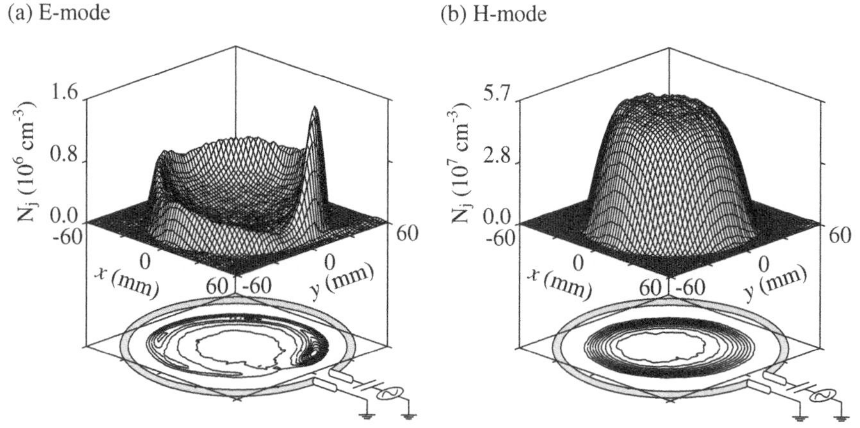

FIGURE 10.3 Emission CT images of short-lived excited atom Ar($2p_1$) in Ar. External conditions are 300 mTorr, 50 sccm, and amplitude 10 A of coil current at 13.56 MHz. (a) E mode (7 W) and (b) H mode (30 W) correspond to (I) and (II) in Figure 10.2.

and 13.56 MHz in E-mode (a) and H-mode (b) [2]. Both images are scanned close to the coil plane. These figures demonstrate that in the E-mode the coupling of energy into the plasma is capacitive and a strong ion sheath is formed. The voltage on the coil is divided between the capacitances of the sheath and the glass wall, so the transfer of power is azimuthally nonuniform. As a result, a local peak appears near the powered terminal. At lower pressure, the nonlocal behavior of electrons distributes the excitation over the entire radius and the azimuthal anisotropy is lost.

10.2.2 Mechanism of Plasma Maintenance

In this section we discuss the sustaining mechanism of an ICP in the H-mode. We consider a cylindrical glass tube wound tightly by a ribbon-shaped coil driven at 13.56 MHz. We then simplify the system to a one-dimensional system as a function of radial position r and time t. The external coil current, $I_0 \sin \omega t$, generates an rf magnetic field $B_z(r,t)$ under Ampere's law,

$$B_z(r,t) = a(r)\mu_0 I_0 \sin \omega t. \tag{10.1}$$

Here, $a(r)$ is the radial component, μ_0 is the permeability, and ω is the angular frequency of the rf current source. At the same time, the azimuthal electric field

$$E_\theta(r,t) = -\frac{\int Sra(r)dr}{r}\omega\mu_\theta I_\phi T_\theta \cos \omega t \tag{10.2}$$

is induced by Faraday's law. Also the ambipolar diffusion field $E_r(r,t)$ is formed close to the wall and results in a passive wall sheath in a stationary

ICP. ICP in the H-mode is mainly sustained by the collisional ionization under an electron heating in the θ direction having power absorption

$$P_\theta(t) = \boldsymbol{J}_{e\theta}(t) \cdot \boldsymbol{E} \sim E_\theta^2 \sim n_e \omega^2 I_0^2 (1 + \cos 2\omega t) \tag{10.3}$$

in a very low electromagnetic field, E_θ ~10 Vcm^{-1} and $B_z \sim$ 10 G. Under a high-pressure condition the plasma production is limited in a space close to the glass wall, and at low pressure of several mTorr, nonlocal electrons affect the maintenance of ICP through the ionization collision allover reactor. Supplementary ionization due to the $\boldsymbol{E} \times \boldsymbol{B}$ drift motion of electrons is expected except in the case of low pressure.

Problem 10.2.1
Prove that the ionization event under $\boldsymbol{E} \times \boldsymbol{B}$ fields leads by $\pi/4$ with respect to that in the E_θ-drift motion of electrons.

Exercise 10.2.1
Discuss the measurement method of the power deposition in an ICP sustained by an external coil current.
When the coil is supplied from an rf source with variable power and an appropriate matching box, the power deposited into the ICP is determined by monitoring the coil current $I(t)$ and subtracting $R_{eff} I(t)^2$ from the power $W(t)$ measured by the power meter, that is, $W(t) - R_{eff} I(t)^2$. Here, the effective impedance R_{eff} is determined from the dissipated power $W'(t)$ in a vacuum with no plasma in the same reactor at the same external coil current $I(t)$. It is assumed that the phase difference between $W'(t)$ and $I(t)$ is negligible.

10.2.3 Effect of Metastables

ICP produces a diffusive high-density plasma. When the electron density increases, we may expect the metastable or dissociated radical species to increase as the electron density does. This increase can be estimated by using a numerical modeling based on the fundamental collision and reaction processes of the related species. Then, the stepwise ionization and quenching of the metastable by electrons are the two basic and competitive mechanisms of the maintenance of ICP.

We consider an ICP in Ar driven by a one-turn rf-coil current. The fundamental processes of the metastable Ar(^{3}P$_2$,^{3}P$_0$) in Ar are summarized in Table 10.1. In a periodic steady state, the continuity equation of the metastable density N^* is

$$\begin{aligned} D^* \nabla^2 N^*(r,z) &+ k_{di} n_e(r,z) N_g + k_{si} n_e N^* + k_{mp} N^* N^* \\ &- k_{eq} n_e N^* - k_{2bq} N^* N_g - k_{3bq} N^* N_g N_g = 0, \end{aligned} \tag{10.4}$$

where each of the reaction rate coefficients corresponds to the value in Table 10.1. We compare the rates of ionization caused by some of the processes

TABLE 10.1 Optical Diagnostics for a Plasma Structure

Method	Target Species	Plasma Structure
Optical Emission Spectroscopy (OES) 2(3)D-Emission CT	Short-lived excited species N_{j} (Ar, He, N_2, O_2, O, etc.)	$N_{\mathrm{j}}(\boldsymbol{r}, t)$
Optical Absorption spectroscopy (OAS) Incoherent OAS	Long-lived species N (H, C, N, O, F, Si, C_2, etc.)	N
1(2)D-Laser Absorption CT		
using visible-line	Long-lived metastables (Ar*)	$N^*(\boldsymbol{r})$
using infrared-line	Radicals (SiH_3,CF_3,CF_2,CH_3)	$N_{\mathrm{rad}}(\boldsymbol{r})$
using VUV-line	Atoms (F,C)	N
Cavity ring down spectroscopy	Radicals (SiH_3,SiH,Si,CH_3)	$N_{\mathrm{rad}}(\boldsymbol{r})$
Laser induced fluorescence(LIF)	Metastable ions (Ar^+,Cl_2^+)	$g_{\mathrm{ion}}(\boldsymbol{v}, t)$, $T_{\mathrm{ion}}(\boldsymbol{r})$, $N^{*^+}(\boldsymbol{r}, t)$
	Metastable atoms (Ar*)	$N^*(\boldsymbol{r})$, $T_{\mathrm{g}}(\boldsymbol{r})$
	Radicals (CF_2,CF,H)	$N_{\mathrm{rad}}(\boldsymbol{r})$
	Electric field	$\boldsymbol{E}(\boldsymbol{r})$
Laser Photodetachment (LPD)	Negative ions (H^-, F^-, $C_4F_8^-$)	$N_{\mathrm{Nion}}(\boldsymbol{r})$
Laser Thomson Scatering (LTS)	Electrons	$n_{\mathrm{e}}(\boldsymbol{r})$, $T_{\mathrm{e}}(\boldsymbol{r})$
Fourier Transform Infrared Spectroscopy	Chemical bond of a surface	

TABLE 10.2 Collision and Reaction Processes of Metastable $Ar^*(^3P_2, ^3P_0)$ in Ar

Process	Collision/Reaction	Rate Coefficient(cm^3s^{-1})
Excitation by electron	$e + Ar(^1S_0) \rightarrow e + Ar^*$	B. Eq.
Electron quenching	$e + Ar^* \rightarrow e + Ar(\text{upper state})$	B. Eq. (2×10^{-7})
Superelastic e-quenching	$e + Ar^* \rightarrow e + Ar(^1S_0)$	B. Eq.
Stepwise ionization	$e + Ar^* \rightarrow e + e + Ar^+$	B. Eq.
Elastic collision	$Ar^* + Ar(^1S_0) \rightarrow Ar^* + Ar(^1S_0)$	
Metastable pooling	$Ar^* + Ar^* \rightarrow e + Ar^+ + Ar$	6.2×10^{-10}
Two-body quenching	$Ar^* + Ar(^1S_0) \rightarrow 2Ar(^1S_0)$	2.1×10^{-15}
Impurity quenching	$Ar^* + Cl_2 \rightarrow Ar(^1S_0) + Cl_2$	7.1×10^{-10} (at 300 K)
	$Ar^* + CF_4 \rightarrow Ar(^1S_0) + CF_4$	4×10^{-11} (at 300 K)
	$Ar^* + SF_6 \rightarrow Ar(^1S_0) + SF_6$	1.6×10^{-10} (at 300 K)
	$Ar^* + N_2 \rightarrow Ar(^1S_0) + N_2$	6.0×10^{-11} (at 300 K)
Three-body quenching	$Ar^* + 2Ar(^1S_0) \rightarrow 3Ar(^1S_0)$	$1.1 \times 10^{-31}(cm^6\ s^{-1})$
Surface de-excitation	$Ar^* + \text{Wall} \rightarrow Ar$	
Diffusion coefficient	ND_{Ar*}	$1.5 \times 10^{18}(cm^{-1}\ s^{-1})$

in Table 10.1 to estimate the effect of metastables on the maintenance. The influence of the direct and stepwise ionization and the metastable pooling on the plasma structure are simply estimated by comparison of each of the rates (see Figure 10.4). The ratio of the net rate between the stepwise and direct ionization in Ar is

$$\frac{\Lambda_{si}}{\Lambda_{di}} = \frac{k_{si} n_e N^*}{k_{di} n_e N_g} = \frac{k_{si}}{k_{di}} \frac{N^*}{N_g}, \tag{10.5}$$

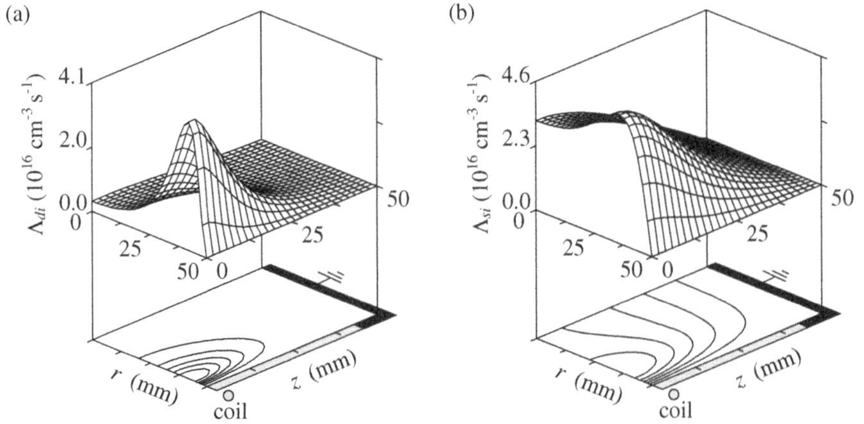

FIGURE 10.4 Two-dimensional net ionization rate in ICP at 13.56 MHz and 50 A at 50 mTorr in Ar.

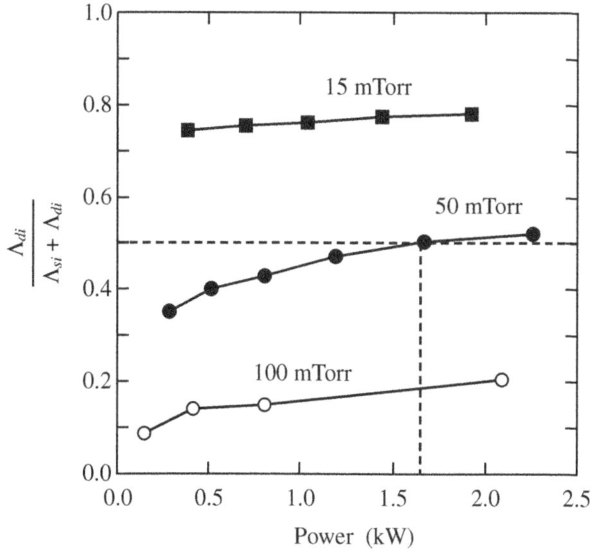

FIGURE 10.5 Contribution of direct ionization to the maintenance of ICP at 13.56 MHz in Ar as a function of dissipated power.

and for the rate of the metastable pooling to the direct ionization,

$$\frac{\Lambda_{mp}}{\Lambda_{di}} = \frac{k_{mp}N^*N^*}{k_{di}n_eN_g} = \frac{k_{mp}}{k_{di}}\left(\frac{N^*}{N_g}\right)^2\left(\frac{N_g}{n_e}\right). \tag{10.6}$$

A notable contribution of the metastable to the plasma maintenance becomes important at low plasma density at high pressure. The influence of the atomic metastable on the plasma is rapidly reduced through the molecular quenching by impurity molecules (see, for example, Table 10.1). Figure 10.5 shows a comparison of the net ionization rate Λ_i among the production terms in the ICP at 13.56 MHz in Ar as a function of power. Notice that the direct ionization is dominant at low pressure (15 mTorr), and the stepwise ionization through Ar* is predominant at high pressure (100 mTorr). At 50 mTorr, with increasing power the dominant process of ionization changes from stepwise to direct ionization. The contribution of the metastable pooling to the ionization is negligible as compared with the conditions mentioned above.

The two-dimensional–density distribution of Ar* in the ICP in Figure 10.1a, predicted by RCT-model, is shown in Figure 10.6 for comparison with the experimental profile by laser absorption spectroscopy. The spatial profile of the metastable Ar* in the ICP reflects the production and the destruction mechanisms. That is, the shallow minimum at the center is mostly caused by electron quenching, and the broad peak in front of the wall is caused by production by the collision between the electron and the feed gas

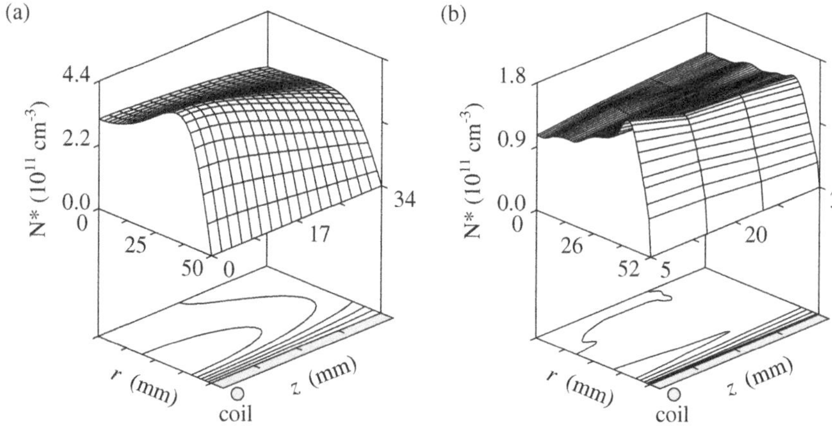

FIGURE 10.6 Time-averaged number density of $Ar^*(r, z)$: (a) numerical value at 386 W (0.49 Wcm^{-3}) and (b) experimental value at 400 W (0.25 Wcm^{-3}). The other external conditions are 15 mTorr in Ar at 13.56 MHz.

$Ar(^1S_0)$. The database of the fundamental process and the absolute value give the first accuracy. The predicted number density of Ar^* coincides with the experimental value within a factor of two. This difference in detail may be attributed to the difference of the dissipated power and the possibility of impurity molecules in the reactor during the experiment. Metastable molecules are produced both by the direct excitation to the metastable state and by the cascade transition from the upper excited states. The number density of the metastables is typically 10^{11} cm^{-3} under the balance between the production and loss by electron quenching and diffusion to the wall. In a high-density plasma with $n_e \sim 10^{11}$ cm^{-3}, the effect of ionization of the metastables by a low-energy electron impact becomes less important under a strong electron quenching. On the other hand, under a low-power and low-density plasma condition, metastable species make a great contribution to the plasma maintenance through a low-energy electron impact ionization (stepwise ionization) or metastable-metastable collision (metastable pooling).

Exercise 10.2.2

An Ar atom has two metastable states, $Ar(^3P_2, ^3P_0)$. Investigate the density ratio of the two metastables in an ICP in pure Ar.

The steady-state value after production depends on the electron quenching. The value is 4×10^{-7} cm^3 s^{-1} for $Ar(^3P_2)$–$Ar(^3P_0)$ and 1.5×10^{-7} cm^3 s^{-1} for $Ar(^3P_0)$–$Ar(^1P_1)$, respectively. As a result, the density ratio $N[Ar(^3P_2)]/N[Ar(^3P_0)]$ is measured at 100 W as 5.7 at 50 mTorr and 7.3 at 5 mTorr, respectively.

TABLE 10.3 Change of the Sustaining Mechanism of a Low Temperature Plasma

Plasma system	Early stage	Stable stage
ICP		
General	E-mode(capacitive)	H-mode(inductive)
High density plasma	Molecular feed gas	Atomic feed gas
(ex.) O_2	$e + O_2 \rightarrow 2e + O_2^+$	$e + O \rightarrow 2e + O^+$
CCP		
High pressure gas	Direct ionization	Step-wise & pooling ionization
(ex.) Micro plasma in Ar	$e + Ar(^1S_0) \rightarrow 2e + Ar^+$	$e + Ar(^3P_2) \rightarrow e + Ar^+ + Ar$ $Ar^* + Ar^* \rightarrow e + Ar^+ + Ar$

10.2.4 Function of ICP

ICPs and TCPs in H-mode are able to keep a high-density plasma in diffusive mode, though the plasma is nonuniform in the production area (see Figure 10.7). This means that a rather long distance is necessary between the production region and wafer in order to realize a radial uniformity of the plasma in front of the substrate. As a result, a high degree of dissociation of the feed gas consisting of polyatomic molecules is induced in the large volume between the plasma production and the wafer. These characteristics in ICPs lead to the marked contrast to the narrow-gap capacitively coupled plasma (CCP).

Problem 10.2.2

The plasma potential in an ICP is usually 10 V to 20 V and the ambipolar diffusion of electrons to the reactor wall controls the plasma density in the case where there is no extinction of electrons in the bulk plasma. Estimate the time constant of electron loss at 50 mTorr in Ar.

Exercise 10.2.3

Estimate the typical Larmor frequency ω_L in an ICP and compare the value with the rf frequency ω and the total collision rate R_t.

The typical ICP in Figure 10.1 is driven at 100 mTorr at 13.56 MHz at a current amplitude of 30 A. Then, the Lamor frequency (cyclotron frequency) ω_L of the electron at 1 cm inside the glass wall is

$$\omega_L = \frac{eB}{m} = \frac{1.60 \cdot 10^{-19} \times 1.26 \cdot 10^{-6} \times 30/(2\pi 10^{-2})}{9.11 \cdot 10^{-31}} = 1.06 \times 10^8 \text{ s}^{-1}.$$

On the other hand, the external frequency ω is 8.5×10^7 s^{-1}, and R_T is usually 10^8 s^{-1}. The relation among these three frequencies is $\omega_L \sim \omega \sim R_T$.

Problem 10.2.3

There exists a uniform plasma with density n_e in Ar at density N_o. A small amount of O_2 flows into Ar plasma at the rate Γ under the condition that n_e

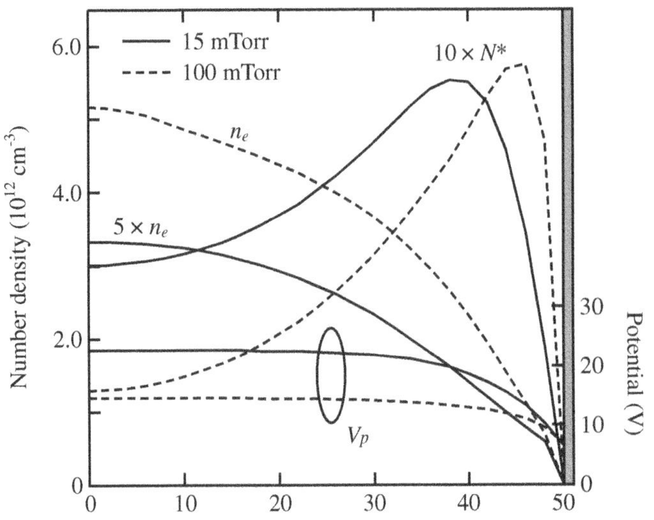

FIGURE 10.7 Typical plasma density and potential distribution at coil plane in ICP sustained at 13.56 MHz in Ar: (a) 15 mTorr and 50 A, (b) 100 mTorr and 30 A.

and N_o are not changed. Find the simple relation of the number density N^*_{Ar} of the Ar-metastable (Ar^*) in the steady state,

$$N^*_{Ar} = \frac{k_{jm} N_o n_e}{k_{qm} n_e + k_{qO_2} \times \frac{\Gamma T_r}{1 - k_d T_r n_e}},$$

where k_{jm} and k_{qm} are the rate constants of the production and loss of Ar^* by the electron impact, respectively. k_{qO_2} is the quenching rate constant of Ar^* by O_2. k_d is the dissociation rate of O_2 by electrons. τ_r is the residence time of O_2.

10.3 PHASE TRANSITION BETWEEN E-MODE AND H-MODE IN AN ICP

It is known that an ICP sustained by an rf current coil has a mode transition and hysteresis characteristics of the plasma inner parameter as a function of external plasma parameter. The E-to-H transition is characterized by two time-constants of electrons: establishment of an axisymmetric distribution by electron diffusion and the accumulation of symmetric high-density electrons in order to sustain the inductive discharge under a weak electromagnetic field. The change in the spatiotemporal structure of the E-to-H transition of an ICP driven at 13.56 MHz by an external single-turn current coil at 300 mTorr in Ar is shown in Figure 10.8, which is optically investigated by using the

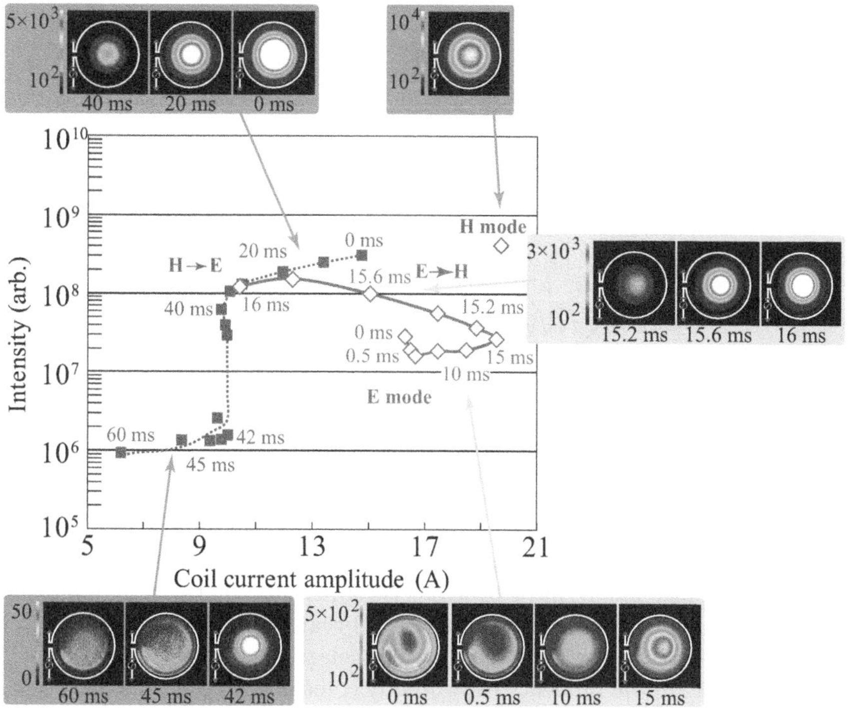

FIGURE 10.8 Hysteresis in an ICP, demonstrated by 2D net excitation rate of Ar(2p$_1$). ICP is driven at 13.56 MHz at 300 mTorr, 50 sccm in Ar. *From Y. Hayashi et al, "Actions of low- and high-energy electrons on the phase transition between E- and H-modes in an inductively coupled plasma in Ar", New J. Phys. 13, July, 2011. With permission.*

short-lived excited atoms Ar(2p$_1$) as a function of time. Figure 10.8 shows the temporal change from a capacitive to an inductive mode in two-dimensional real space. Also the H-to-E transition from the stable H-mode is demonstrated in Figure 10.8. The H-to-E transition is strongly influenced by the presence of a long-lived excited atom (i.e., metastables).

10.4 WAVE PROPAGATION IN PLASMAS

10.4.1 Plasma and Skin Depth

We have learned that there exist two types of bulk plasma in gases: collisional and colllisionless plasma. Here, the designation of collisional or collisionless is derived from the presence or absence of two-body collisions between the

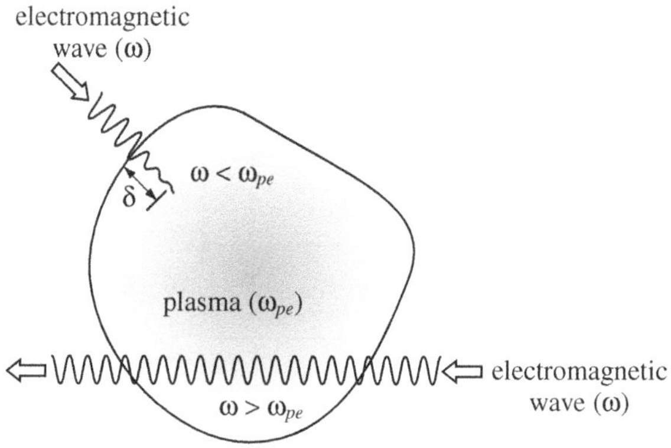

FIGURE 10.9 Propagation, cut-off, and decay of an electromagnetic wave in a uniform plasma.

electron and the neutral molecule in the plasma. In the collision-dominated plasma, the collisional event of electrons is represented by the rate of momentum transfer for the elastic scattering R_m, because the magnitude of R_m is usually much greater than the rate for the inelastic collisions.

We investigate the influence of two-body collisions R_m, plasma density n_e, and frequency ω_{pe} on the propagation of an external electromagnetic wave $\boldsymbol{E}_0 e^{j\omega t}$ incident on a uniform plasma (see Figure 10.9). When we consider the phase delay ϕ of the electron drift velocity $\boldsymbol{v}_{de} \exp j(\omega t - \phi)$ in a plasma, the momentum conservation of electrons is given by

$$\frac{d}{dt}\left(m\boldsymbol{v}_{de}e^{j(\omega t-\phi)}\right) = e\boldsymbol{E}_0 e^{j\omega t} - m\boldsymbol{v}_{de}e^{j(\omega t-\phi)}R_m. \tag{10.7}$$

By using the drift velocity derived from Equation 10.7, the electron current density in the plasma is given as

$$\begin{aligned}\boldsymbol{J}_e(t) &= en_e\boldsymbol{v}_{de}(t) = \frac{n_e e^2}{m(R_m + j\omega)}\boldsymbol{E}_0 \exp(j\omega t) \\ &= \frac{n_e e^2}{m}\left(\frac{R_m}{\omega}\frac{1}{1+\left(\frac{R_m}{\omega}\right)^2} - j\frac{1}{1+\left(\frac{R_m}{\omega}\right)^2}\right)\frac{\boldsymbol{E}_0\exp(j\omega t)}{\omega}. \end{aligned} \tag{10.8}$$

Here, the conduction of massive ions and the displacement current are neglected in an rf wave. The wave propagation is described by the Maxwell equations (see Section 7.6)

$$\mathrm{rot}\boldsymbol{B}(t) = \mu\boldsymbol{J}_e(t) + \frac{1}{c^2}\frac{\partial}{\partial t}\boldsymbol{E}(t), \tag{10.9}$$

$$\mathrm{rot}\boldsymbol{E}(t) = -\frac{\partial}{\partial t}\boldsymbol{B}(t). \tag{10.10}$$

If we take the rotation of Equation 10.10 and substitute it into Equation 10.9, we derive the wave equation

$$\nabla^2 \boldsymbol{E}(t) + \frac{\omega^2}{c^2}\boldsymbol{E}(t) = j\omega\mu \boldsymbol{J}_e(t). \tag{10.11}$$

When the wave propagates in the z-direction in the plasma, the wave equation is divided into the z- and r-directions:

$$\frac{d^2 E_z(t)}{dz^2} + \frac{\omega^2}{c^2}E_z(t) = j\omega\mu J_{ez}(t), \tag{10.12}$$

$$\frac{\omega^2}{c^2}E_r(t) = j\omega\mu J_{er}(t). \tag{10.13}$$

We eliminate the conduction current density $J_e(t)$ in Equation 10.12 by using Equation 10.8, and then the axial component of the wave $E_z(t)$ satisfies

$$\frac{d^2}{dz^2}E_z(t) = -\frac{\omega^2}{c^2}\left(1 - \frac{\omega_{pe}^2}{\omega^2 + R_m^2} - j\frac{\omega_{pe}^2}{\omega^2 + R_m^2}\frac{R_m}{\omega}\right)E_z(t), \tag{10.14}$$

where ω_{pe} is the plasma electron frequency, $(e^2 n_e/m\epsilon_0)^{1/2}$. The solution has the form

$$E_z(t) = E_0 \exp[j(\omega t - kz)]. \tag{10.15}$$

Here, the complex wavenumber k is expressed as

$$k = (k_r - jk_i) = \frac{\omega}{c}\left(1 - \frac{\omega_{pe}^2}{\omega^2 + R_m^2} - j\frac{\omega_{pe}^2}{\omega^2 + R_m^2}\frac{R_m}{\omega}\right)^{1/2}. \tag{10.16}$$

As a result, we rewrite the propagating wave as

$$E_z(z,t) = E_0 e^{-k_i z} e^{j(\omega t - k_r z)},$$

where k_i corresponds to the attenuation constant, and the wave is extinguished during the propagation of a length L in a plasma in the case of $k_i L \gg 1$. The penetration length of the wave into the homogeneous plasma is estimated practically by the distance that the wave decays to e^{-1} of the amplitude of the incident point, referred to as the skin depth (see Figure 10.9). That is, the skin depth δ at $R_m \neq 0$ is given as

$$\delta = \frac{1}{k_i} = \frac{\sqrt{2}c}{\omega}\left(\left[\left(1 - \frac{\omega_{pe}^2}{\omega^2 + R_m^2}\right)^2 + \left(\frac{\omega_{pe}^2}{\omega^2 + R_m^2}\frac{R_m}{\omega}\right)^2\right]^{1/2} - \left(1 - \frac{\omega_{pe}^2}{\omega^2 + R_m^2}\right)\right)^{-1/2}, \tag{10.17}$$

where δ is shown in Figure 10.10 as a function of ω_{pe}/ω and R_m/ω. It is noted that the skin depth is a concept describing the propagation of the electromagnetic wave in a plasma, and is not appropriate for a plasma accompanied by an appreciable amount of collisional ionization of neutral gases such as in the plasma sources. See Table 10.4.

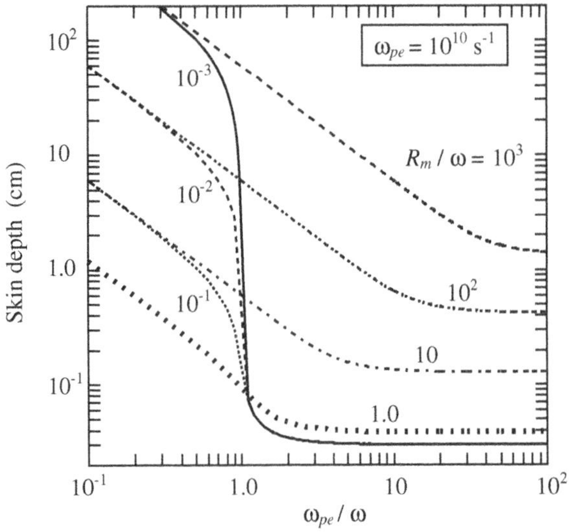

FIGURE 10.10 Skin depth of electromagnetic waves incident on a plasma.

Problem 10.4.1

As noted, the expression of the skin depth, Equation 10.17, in a collisional plasma is exactly valuable for an electromagnetic wave incapable of a plasma production. When an external wave brings about a plasma production, we have to consider the energy conservation of electrons in addition to the above equations in order to investigate the wave propagation into a plasma. Discuss the qualitative characteristics of the wave propagation with or without a plasma production.

TABLE 10.4 Typical Skin Depth as a Function of ω, ω_{pe}, and R_m

Plasma Region	ω and ω_{pe}	Skin Depth
Collisionless ($R_m \ll \omega$)	$\omega_{pe} \gg \omega$	$\sim \frac{c}{\omega_{pe}}$
$R_m \sim \omega$	$\omega_{pe} \gg \omega$	$\sim \frac{2}{(\sqrt{2}+1)^{1/2}} \frac{c}{\omega_{pe}}$
$R_m \sim \omega$	$\omega_{pe} \sim \omega$	$\sim \frac{2}{(\sqrt{2}-1)^{1/2}} \frac{c}{\omega_{pe}}$
Collisional ($R_m \gg \omega$)		$\sim \frac{\sqrt{2}c}{\omega_{pe}} \left(\frac{R_m}{\omega}\right)^{1/2}$

10.4.2 ICP and the Skin Depth

Electrons in an ICP are heated by the azimuthal field induced by an external current coil (see Figure 10.1a). The azimuthal electric field E_θ satisfies the wave equation

$$\frac{\partial}{\partial r}\left\{\frac{1}{r}\frac{\partial}{\partial r}(rE_\theta)\right\} + \frac{\partial^2}{\partial z^2}E_\theta + \frac{\omega^2}{c^2}E_\theta = \frac{\omega_{pe}^2}{c^2}\frac{\omega(\omega + jR_m)}{\omega^2 + R_m^2}, \tag{10.18}$$

and E_θ is given by

$$E_\theta(r, z, t) = E_0 J_1\left(\frac{3.83r}{r_0}\right)\exp[j(\omega t - kz)], \tag{10.19}$$

where J_1 is the first term of the Bessel function of the first kind and r_0 is the radius of the cylindrical reactor. k is the complex wavenumber and is obtained from

$$-\left(\frac{3.83}{r_0}\right)^2 - k^2 + \frac{\omega^2}{c^2} = \frac{\omega_{pe}^2}{c^2}\frac{\omega(\omega + jR_m)}{\omega^2 + R_m^2}.$$

The skin depth is obtained by the reciprocal of the imaginary part of k,

$$\delta = \frac{c}{\omega_{pe}}\left(\frac{2\left[1 + (R_m^2/\omega^2)\right]}{b\left\{1 + \left[1 + (R_m^2/\omega^2 b^2)\right]^{1/2}\right\}}\right)^{1/2}, \tag{10.20}$$

where b is

$$b = \left(1 + \left(\frac{c}{\omega_{pe}}\right)^2\left[\left(\frac{3.83}{r_0}\right)^2 - \left(\frac{\omega}{c}\right)^2\right]\right)\left[1 + (R_m^2/\omega^2)\right] + 1.$$

Exercise 10.4.1

There is a collisionless plasma that has a linear density distribution $n(z) = n_0 z$. Discuss the profile of the wave propagation as a function of z.

In a collisionless plasma ($R_m \ll \omega$) the dispersion relation is given by

$$\xi^2\left(\equiv \frac{c^2k^2}{\omega^2}\right) = 1 - \frac{\omega_{pe}^2}{\omega^2}. \tag{10.21}$$

As shown in Figure 10.11, the propagation properties of electromagnetic waves are classified into three regions with different modes as a function of $n_e(z)$. An incident wave propagates from left to right. As the plasma frequency ω_{pe} is proportional to $n_e(z)^{1/2}$, the wavelength of the electromagnetic wave $2\pi/k$ becomes longer, and at $\omega_{pe} = \omega$ the wavelength is infinity. That is, the wave is almost stationary in space and oscillates with ω in time. In the region of $\omega_{pe} \gg \omega$, k is complex, and the wave is attenuated with the attenuation constant Im$\{k\}$. The frequency at $\omega = \omega_{pe}$ gives the cut-off frequency. The position at $\omega_{pe} = \omega$ acts as the reflective wall, and the wave is reflected to the left. Here, ξ is known as the refractive index.

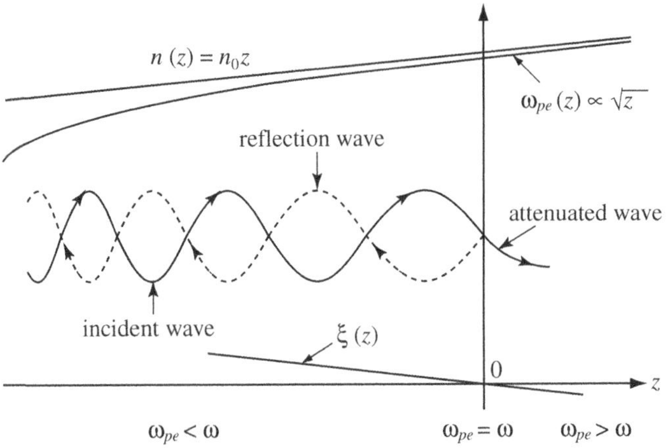

FIGURE 10.11 Propagation of electromagnetic waves into a plasma with a spatial profile $n_0 z$.

Exercise 10.4.2

When the intensity of electromagnetic waves is low enough not to perturb the bulk plasma, the plasma density is measured by using the principle that a wave propagating through a plasma has a phase shift relative to the wave propagating in a vacuum. Practically, a microwave is used in order to obtain a spatial resolution of ~cm [3]. Derive a simple relation between the phase-shift $\Delta\phi$ and and the plasma density n_e.

The phase shift between two path $\Delta\phi$(rad) is given as a function of the electron plasma frequency, $\omega_{pe}(z) = (e^2 n_e(z)/m\epsilon_0)^{1/2}$:

$$\Delta\phi = \int_l (k_0 - k_{plasma})dl = k_0 \int_l \left[1 - \left(1 - \frac{\omega_{pe}(z)^2}{\omega^2}\right)^{1/2}\right] dz. \tag{10.22}$$

Then,

$$\Delta\phi = 2.82 \times 10^{-17} \lambda_0 \int_l n_e(z) dz. \tag{10.23}$$

In particular, when the plasma is uniform in the radial direction of the ICP reactor, we obtain the plasma density $n_e(cm^{-3})$:

$$n_e = 1.18 f \frac{\Delta\phi}{L}, \tag{10.24}$$

where f(Hz) is the microwave frequency and L(cm) is the effective width of the plasma investigated.

Problem 10.4.2

The earth is surrounded with the ionospheric layer of plasma density, ~ 10^6 cm^{-3}. A shortwave broadcasting (frequency: ~MHz-30 MHz) is operated on the ground station. A satellite-based broadcasting is serviced at frequency greater than GHz. Discuss the difference of the available broadcasting-frequency.

References

[1] Ventzek, P.L.G., Hoekstra, R.J., and Kushner, M.J. 1994. *J. Vac. Sci Technol. B* 12:461.

[2] Miyoshi, Y., Petrovic, Z.Lj., and Makabe, T. 2002, 2005. *IEEE Trans. on Plasma Sci.* 30:130. Miyoshi, Y., Miyauchi, M., Oguni, A., and Makabe, T. *IEEE Trans. on Plasma Sci.* 33:362.

[3] Auciello, O. and Flamm, D.L. 1989. *Plasma Diagnostics*, Vol. 1. San Diego: Academic Press.

CHAPTER 11

Magnetically Enhanced Plasma

Magnetron plasmas and electron cyclotron resonance (ECR) plasma are used in dry plasma processing at low pressure in which an external permanent magnetic field makes a significant contribution to the plasma maintenance by reducing electron losses to walls and electrodes. Helicon wave plasma and magnetic neutral loop discharge (NLD) are also magnetized plasmas. Collisionless electron heating through an electron cyclotron or a helicon wave is the main mechanism to sustain these plasmas at very low pressure. On the other hand, an induced magnetic field operates on an inductively coupled plasma (ICP) and surface wave plasma (SWP).

11.1 DIRECT-CURRENT MAGNETRON PLASMA

Direct-current (DC) magnetron plasma has been widely used to deposit metallic film on a large area substrate in electronic and photonic device fabrication. Functional glass is prepared by material film coating. Sputter deposition of Cu atoms ejected from a copper target in a DC magnetron is practically used for interconnect in a trench or hole on SiO_2 film. See Table 11.1.

Typical DC magnetron plasma (planar magnetron) is maintained between two parallel plate electrodes named the target and the substrate. Permanent magnets are arranged radially from the center on the back of the target (cathode) in order to supply a doughnutlike magnetic field in front of the target (see Figure 11.1). DC magnetron discharge is usually sustained at about 200 V in nonreactive Ar at several mTorr under a permanent magnetic field paralllel to the target, several hundred Gauss. At the low-pressure condition, an external magnetic field is essential for the maintenance of a plasma in a DC source, and the effective lifetime of electrons in the magnetron is prolonged (see Exercise 11.1.1). In particular, the net ionization rate is greatly enhanced

TABLE 11.1 Magnetron Plasma for Sputtering

Type	DC Magnetron	rf Magnetron
Power source	DC	rf (13.56 MHz)
Magnet	Permanent	Permanent
Feed gas	Ar	Ar
Admixture	O_2, N_2	O_2, N_2
Target material	Metal	Dielectric, metal
Film depo.	Metallic	Dielectric
	Oxide metal	
	Nitride metal	
Wire depo.	Cu interconnect	

at a point where the magnetic field component parallel to the electric field is zero, as compared with the plasma under no magnetic field.

The Lamor radius of ions is on the order of centimeters in contrast to the sheath thickness with several mm in front of the target. Thus, ions in magnetron plasma are considered to be uninfluenced by the external magnetic field. Inert Ar is used as the feed gas in a magnetron sputtering plasma for several reasons. Inert Ar is chemically nonreactive on the target surface, the massiveness of the Ar^+ ion enhances sputtering on the target, and Ar is highly abundant in the earth. Oxide or nitride metal film is processed in a DC magnetron with a metal target in an admixture of a small amount of O_2 or N_2 with Ar. The process is known as a reactive sputter deposition.

Exercise 11.1.1

Both the electric $\boldsymbol{E}(=E\boldsymbol{k})$ and the magnetic field $\boldsymbol{B}(=B\boldsymbol{j})$ are applied in a gas at a number density of N. Derive the neutral density N estimated by apparent electron changes to

$$N\left(1+\left(\frac{eB}{m}\right)^2\frac{1}{R_m^2}\right)^{1/2}. \tag{11.1}$$

The momentum conservation of a single electron with mass m and velocity $\boldsymbol{v}$ is

$$\frac{d}{dt}(m\boldsymbol{v}) + m\boldsymbol{v}R_m = e(\boldsymbol{E}+\boldsymbol{v}\times\boldsymbol{B}), \tag{11.2}$$

where $\boldsymbol{E}=E\boldsymbol{k}$, $\boldsymbol{B}=B\boldsymbol{j}$, $\boldsymbol{v}=(v_x\boldsymbol{i}+v_y\boldsymbol{j}+v_z\boldsymbol{k})$, and R_m is the total collision rate of the electron. We simply consider the case of R_m independent of $\boldsymbol{v}$. Then, $\boldsymbol{v}$ is

$$\boldsymbol{v}=\left(-\frac{eB}{m}\frac{eE}{m}\frac{1}{R_m^2+\left(\frac{eB}{m}\right)^2},\quad 0,\quad -\frac{eE}{m}\frac{R_m}{R_m^2+\left(\frac{eB}{m}\right)^2}\right). \tag{11.3}$$

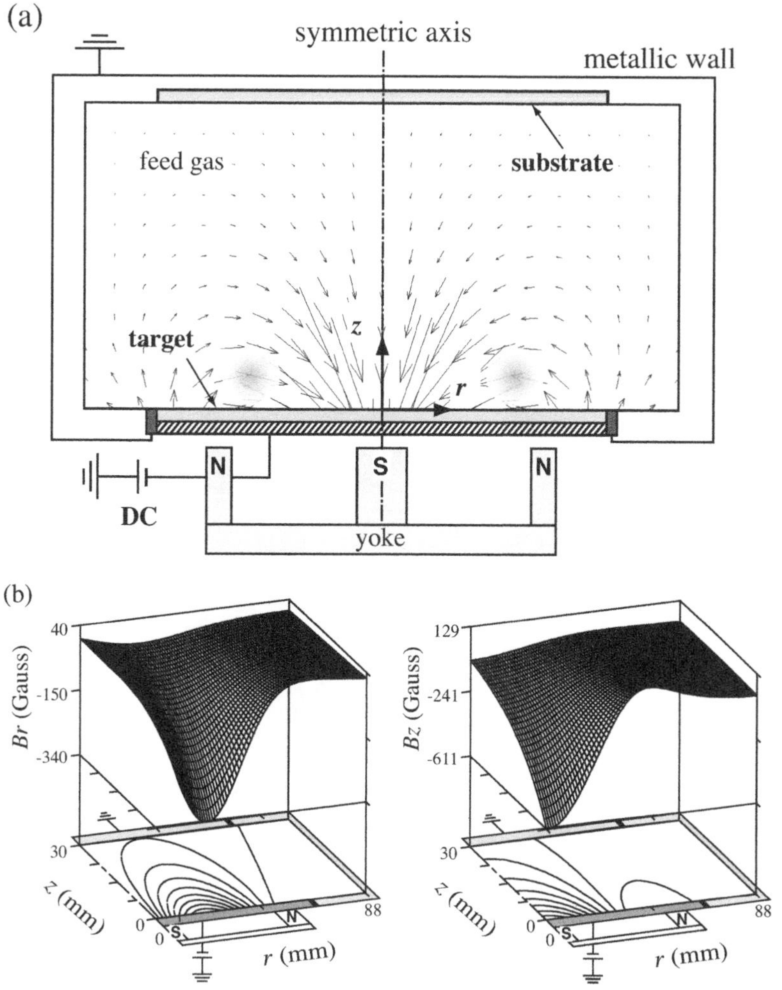

FIGURE 11.1 A typical DC magnetron plasma reactor describing a magnetic field line (a); radial and axial components of $\mathbf{B}(z, r)$ (b).

The electron mean free path λ_e under $\boldsymbol{B}$ is as

$$\lambda_e \sim \frac{v}{R_m} = \frac{eE}{mR_m} \frac{1}{R_m \left\{1 + \left(\frac{eB}{m}\right)^2 \frac{1}{R_m^2}\right\}^{1/2}}. \tag{11.4}$$

Equation 11.4 means that the gas number density changes from N to the above expression 11.1. A decisive effect of a magnetic field on plasma production is expressed under the condition of the electron cyclotron frequency (Larmor frequency) $\omega_{ce}(= eB/m) \gg R_m$.

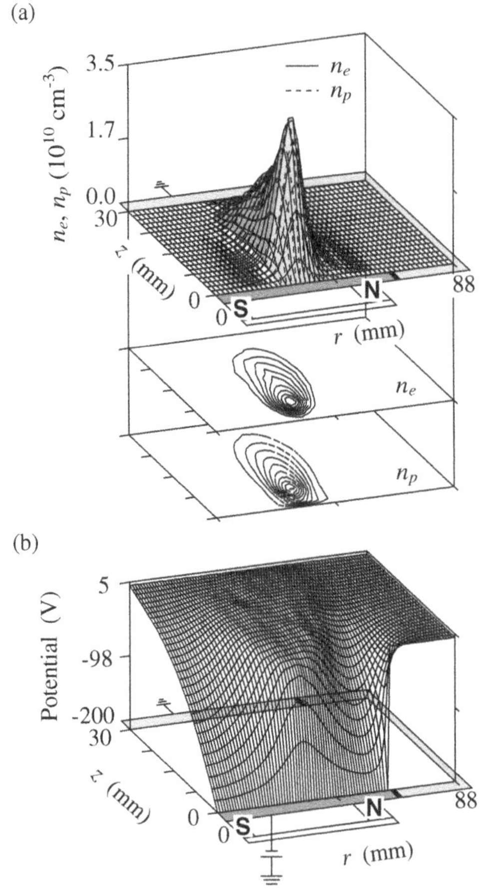

FIGURE 11.2 Plasma density (a) and potential (b) distributions in a DC magnetron in Ar at 5 mTorr driven at 200 V in the reactor in Figure 11.1.

Figure 11.2a shows a typical structure of a DC magnetron plasma in terms of the number density distribution of electrons and ions obtained by a hybrid model (see Section 7.4). The density distribution has a strong doughnutlike peak at position $\boldsymbol{r}$ (34 mm, 6 mm) with $B_z = 0$ in front of the target. This is caused by the presence of the peak net ionization rate at the position. It means that the radial nonuniformity of the plasma density (i.e., of the ion flux to the target) is the property intrinsic in a magnetron plasma, though it is capable of sustaining a plasma at lower pressure, several mTorr, as compared with a DC glow discharge or radio-frequency (rf) capacitively coupled plasma (CCP). Figure 11.2b shows the potential distribution in the reactor configuration in Figure 11.1a. A thin sheath region with a strong potential difference appears in front of the target, and the potential in the bulk plasma (plasma potential) is very low when the plasma is surrounded by the metallic reactor wall grounded to the earth. In these potential distributions, electrons are mainly produced by the collisional ionization at the radially localized region with $B_z \sim 0$ in the sheath edge, and these electrons diffuse to the bulk plasma in the very low electric and magnetic fields. On the other hand, ions produced locally as the pair of the ionization are strongly accelerated to the target with a beamlike energy, and these ions sputter the target material. It is easy to estimate a local erosion profile by ion sputtering from the local ion flux incident on a target. Magnetron plasma is usually operated in a region of current source. Removal of the external magnetic field, after the magnetron plasma is formed, introduces the discharge plasma with extinction.

Problem 11.1.1
Discuss the reasons why maximum ionization efficiency is realized at the point where the magnetic field crosses at a right angle with the electric field in front of the target (cathode) in a magnetron discharge.

Problem 11.1.2
An azimuthal drift current in the plasma ring in front of the target is on the order of a few amperes. Estimate the magnetic field generated by the loop current of electrons and compare the value with the external permanent magnetic field.

11.2 UNBALANCED MAGNETRON PLASMA

Almost all magnetic field lines from the north pole of the magnet arranged behind the target terminate at the south pole through the gas phase in the reactor. The magnetron plasma with these magnetic field arrangements is called a balanced magnetron (BM). An unbalanced magnetron (UBM) has a proper magnetic field configuration in which a finite degree of the field lines from the outer magnetic pole diverge to the substrate, though the rest of the lines finish on the inner pole behind the target (see Figure 11.3b). Sufficient plasma density and a positive ion current on a metallic substrate even at a large distance from the target can be achieved in the UBM as compared with the BM.

11.3 RADIO-FREQUENCY MAGNETRON PLASMA

Magnetron plasma with a dielectric target is sustained only by an external rf source, which is distinguished from a DC magnetron plasma with a metal target. An rf power source at 13.56 MHz having a voltage waveform,

$$V_{rf}(t) = V_0 \sin \omega t, \tag{11.5}$$

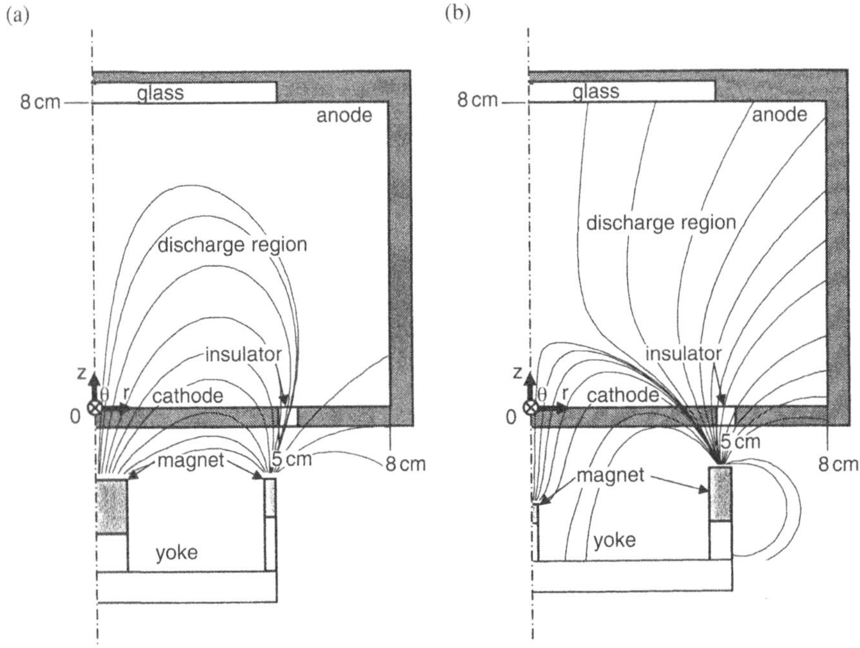

FIGURE 11.3 Comparison between a balanced and unbalanced magnetron (a) and (b).

is usually used through a blocking capacitor C_b. Therefore, the target surface is negatively biased and the surface potential takes the form

$$V_{sus}(t) = V_{rf}(t) - V_{dc}(t) \quad \text{where} \quad V_{dc}(t) = \frac{1}{C_b}\int_{-\infty}^{t} I_T(t)dt. \tag{11.6}$$

An rf magnetron sustained at 13.56 MHz has a weak temporal change of the electron density close to the target except for the region trapped deeply by the magnetic field. The other difference from the DC magnetron is the presence of the phase to release the spatiotemporal electron trapping by electromagnetic fields, based on $E(\boldsymbol{r}, t) = 0$ twice during one rf period. This allows a radially spreading distribution of the net ionization rate in front of the target (see Figure 11.4) as compared with that in a DC magnetron. The phenomena increase the efficiency both of the target utilization and of a much more radially uniform deposition on the substrate. Note that the value of C_b is carefully arranged in the case of the dielectric target, especially for a ferrodielectric target with a high dielectric constant.

Figure 11.4 shows the spatial profile of the electron density at each of the external voltage phases of $\omega t = 0$, $\pi/2$, π, and $3\pi/2$ in Ar at 5 mTorr. A local area close to the target (P_B) with $B_z \sim 0$ is weakly modulated in time,

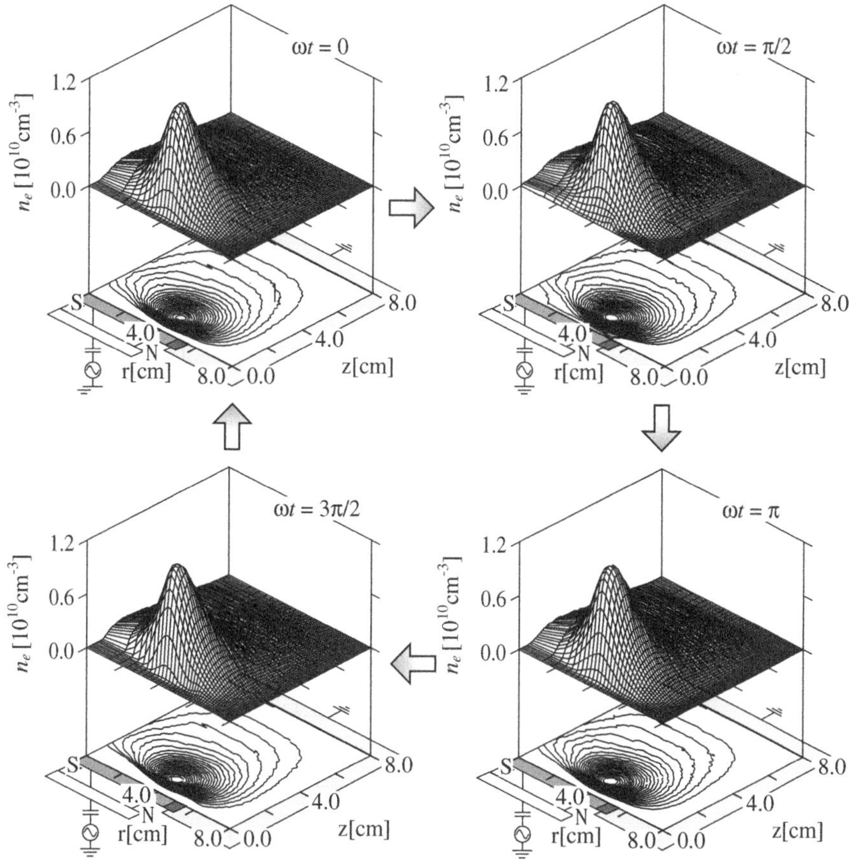

FIGURE 11.4 Typical electron density distribution in an rf magnetron sustained in Ar at 5 mTorr at 13.56 MHz and $V_0 = 400$ V.

because the electrons under a large component of B_r are spatially trapped by the $E(\boldsymbol{r}, t) \times B(\boldsymbol{r})$ fields in front of the target. As a result, the sustaining mechanism of the rf plasma is classified into two regions depending on the radial position. One is that of the conventional rf CCP, that is, the ionization multiplication in front of the target by the reflected electrons. The other is the ionization multiplication by the secondary electrons caused by the positive ion impact on the target. That is, rf magnetron plasma is maintained by mixed mechanisms between an rf CCP and a DC glow discharge. At low pressure such that rf CCP cannot be sustained, the first mechanism disappears. Figure 11.5 shows the ion flux incident on the Cu target (a) and the erosion profile (b) as a function of radial position [1].

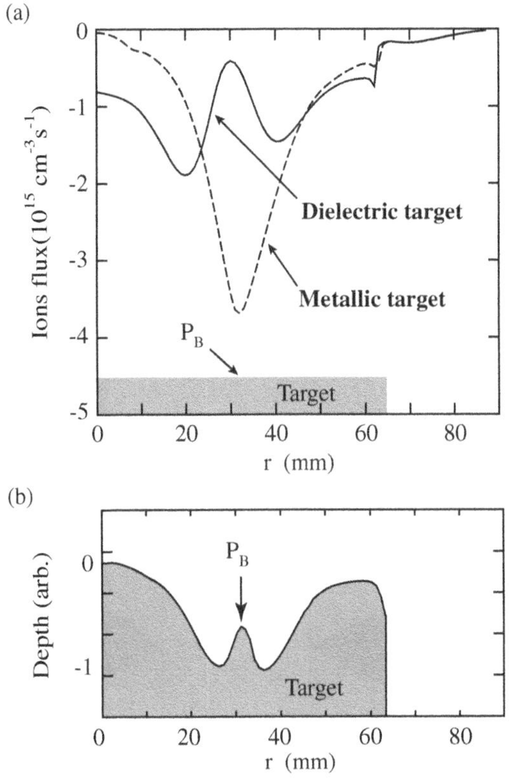

FIGURE 11.5 Ion flux incident on Cu target (a) and erosion profile (b) in an rf magnetron plasma.

11.4 MAGNETIC CONFINEMENTS OF PLASMAS

When the operating pressure of a plasma drops to the order of mTorr, electrons and ions produced in a plasma rapidly diffuse and are lost to the reactor wall by the lack of binary collisions between the electron (or ion) and neutral feed gas molecule. The multipolar magnetic field arrangement on the reactor side wall is effective for the confinement of electrons inside the reactor, when the mean free path of fast electrons is equal to or larger than the chamber size. A magnetically confined reactor is arranged by an array of permanent magnets with N-poles and S-poles alternately positioned outside the reactor as shown in Figure 11.6a. The reactor with surface cusp magnetic fields is widely used in the application of a low-pressure plasma to material processings.

The magnetic mirror effect near the cusp of two magnetic poles suppresses the electron diffusion in the direction normal to the magnetic field line in addition to the electron reflection in the wall sheath on the reactor. A set of straight magnets is arranged in the axial direction on the outer surface of the

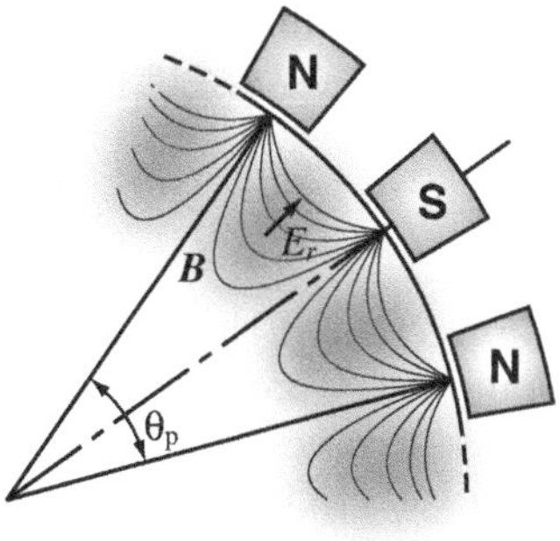

FIGURE 11.6 Magnetically confined plasma reactor. Arrangement of multipolar magnets and net ionization rate.

cylindrical reactor. The resulting line cusp magnetic field near the chamber wall confines electrons and ions. Such a field configuration has the properties of a magnetic mirror and has a very small effect on the confinement of slow electrons. That is, with increasing the number, the electron confinement becomes stronger, and the escape of electrons from the loss cone becomes significant [2]. The electron mean energy rises near the wall due to the presence of the high-energy electrons trapped by the multipolar magnetic field. The net ionization rate is expected to have a local peak. The local electron energy increases when the magnitude of the magnetic field increases due to the increase of the trapped high-energy electron. There is an optimum for the number of the magnetic poles. The magnetic field has a very small effect on the confinement of slow electrons. At pressure range where the electron mean free path is less than the reactor dimension, the magnetic field has little effect on the electron confinement by the trapping.

11.5 MAGNETICALLY RESONANT PLASMAS

We consider an elctromagnetic wave, propagating parallel to the applied magnetic field $\boldsymbol{B}$ in a cold and uniform plasma. The wave has a frequency in that ion motion is negligible, $\omega \gg \omega_{pi}$. It is known that there are two modes, R- and L-modes, for electromagnetic waves traveling along the magnetic field. Each of the refractive indexes, ξ_+ and ξ_-, are given by [3–5]

$$\xi_\pm^2 \left(\equiv \frac{k^2 c^2}{\omega^2} \right) = 1 - \frac{\omega_{pe}^2}{(\omega \pm \Omega_{ce})(\omega \pm \Omega_{cp})}, \tag{11.7}$$

where $\Omega_{ce} = eB/m$ (< 0) and $\Omega_{cp} = eB/M$ (> 0) are the cyclotron frequencies of electrons and ions, respectively. Equation 11.7 exhibits that the R- and L-modes have cut-off frequencies, ω_+ and ω_- (see Chapter 10),

$$\omega_\pm = \left(\omega_{pe}^2 + \frac{(\Omega_p - \Omega_e)^2}{4}\right)^{1/2} \mp \frac{\Omega_{cp} + \Omega_{ce}}{2}. \tag{11.8}$$

We see that the rotational direction of the R-wave corresponds to the direction of the cyclotron motion of electrons in the magnetic field. Note that the denominator of Equation 11.7 becomes zero when the frequency ω of the R-wave approaches $|\Omega_{ce}|$. Then the electrons are continuously accelerated and result in the absorption of the energy of the electromagnetic wave. It is known as the ECR between the electron and the wave. It is noted that the L-wave has no resonances with electrons but with ions.

ECR plasma is excited by the resonance mechanism of the R-mode described above. A typical ECR plasma source is driven at microwave power with 2.45 GHz in low pressure of mTorr. It is the principle to synchronize the electron cyclotron frequency $|\Omega_{ce}|$ with the UHF (microwave) frequency 2.45 GHz at B of 875 G by adjusting the external coil current. In the resonance point, electrons efficiently get the external microwave energy and a diffusive high-density plasma is produced at very low pressure in a reactor as shown in Figure 11.7a. Microwave power through a wave guide is fed into the reactor through a quartz window.

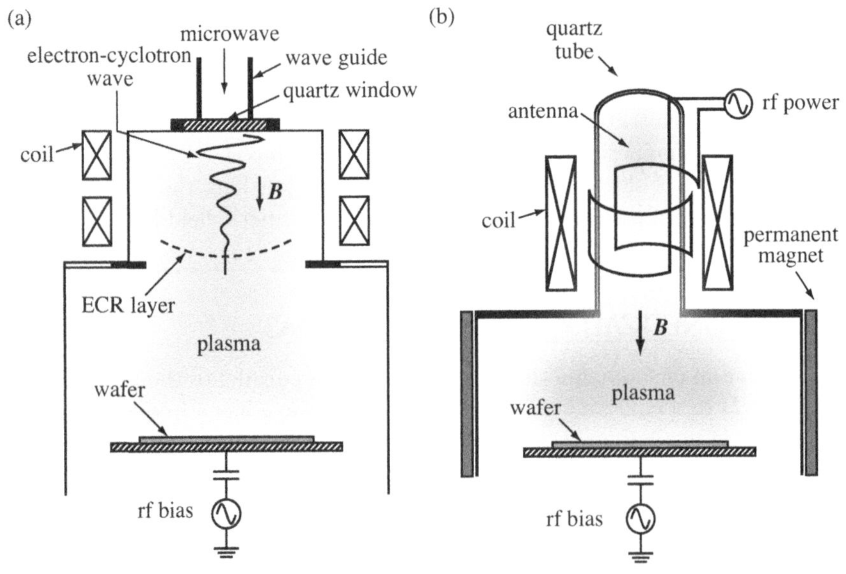

FIGURE 11.7 Representative for magnetically resonant plasma. ECR plasma source (a) and helicon source (b).

Wave heated plasmas are described in detail by Lieberman and Lichtenberg [6].

Problem 11.5.1
Calculate the magnitude of the magnetic field at the ECR condition of electrons in a collisionless plasma supplied by the microwave power at 2.45 GHz.

Problem 11.5.2
A collisionless plasma roughly satisfies the relation, $n_e/N_o \gtrsim 10^{-3}$, where n_e and N_o are the number density of the electron and the neutral feed gas molecule, respectively. Discuss the reason why the electron temperature T_e is of first importance to identify the collisionless plasma, by using the time-independent form of the collisionless Boltzmann equation, Equation 5.9 (Vlasov equation).

References

[1] Kuroiwa, S., Mine, T., Yagisawa, T., and Makabe, T. 2005. *J. Vac. Sci. Technol. B*, 23:2218.

[2] Takekida, H. and Nanbu, K. 2004. *J. Phys. D* (37:1800).

[3] Boyd, T.J.M. and Sanderson, J.J. 2003. *The Physics of Plasmas.* Cambridge: Cambridge University Press.

[4] Sturrock, P.A. 1994. *Plasma Physics.* Cambridge: Cambridge University Press.

[5] Nicholson, D.R. 1983. *Introduction to Plasma Theory.* New York: John Wiley & Sons.

[6] Lieberman, M.A., and Lichtenberg, A.J. 2005. *Principles of Plasma Discharges and Materials Processing* (2nd Edition). Hoboken: John Wiley & Sons.

CHAPTER 12

Plasma Processing and Related Topics

12.1 INTRODUCTION

Plasma processes (i.e., sputtering, deposition, etching, surface treatment, etc.) require information about the surface reactions of active species and their probability, as well as information on gas-phase collision/reaction processes and their cross sections. Physical or chemical quantities describing the plasma surface interaction are given by an effective surface reaction probability consisting of the sticking coefficient and the yield for etching or deposition.

Historically, the collision cross section of the electron or ion in the gas phase has been continuously accumulated theoretically and experimentally as a function of the impact energy, in addition to the study of the quantum characteristics of mono- and polyatomic molecules in the field of atomic and molecular physics. In the same way, we expect that the surface reaction process and the probability of neutral molecules and ions with a material surface will be rapidly elucidated and accumulated by using first-principle molecular dynamics and measurements.

12.2 PHYSICAL SPUTTERING

Metallic film deposition is usually performed by using a direct-current (DC) magnetron plasma with a metal target in pure Ar (see Figure 12.3). A relatively large number of data are available on ion sputtering in ion–surface interactions. Most of these data were collected for systems between metal targets and rare-gas ions with energy ranging from several hundred eV to several hundred keV under a clean surface [1, 2]. Very little is known about systems with low-energy ions with $\epsilon_p < 100$ eV. Reactive sputtering is a technique widely used for deposition of compound materials, for example, oxides (Al_2O_3, Ta_2O_5, etc.) and nitrides (Si_3N_4). A sputtering system of a pure metal target

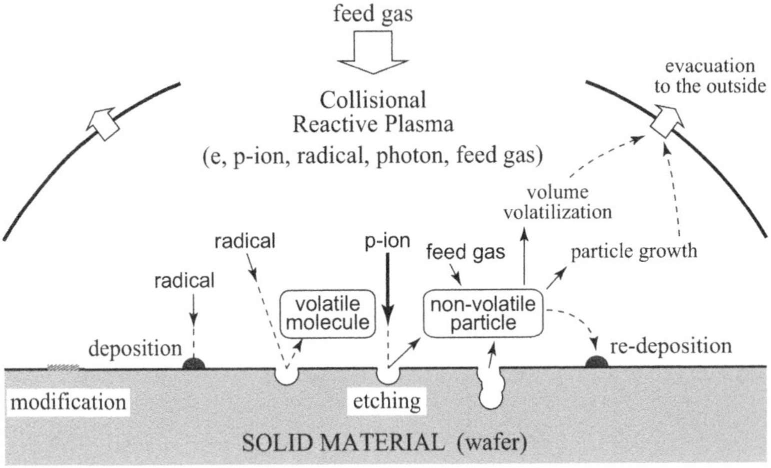

FIGURE 12.1 Schematic diagram of reactive plasma processes.

in admixture with reactive gas (O_2 or N_2) and Ar holds the advantage over a system in a compound target in Ar from the viewpoints of the stoichiometry of the film and the power density and thermal conductivity of the target. Chemical sputtering is generally considered to be a multistep process finally

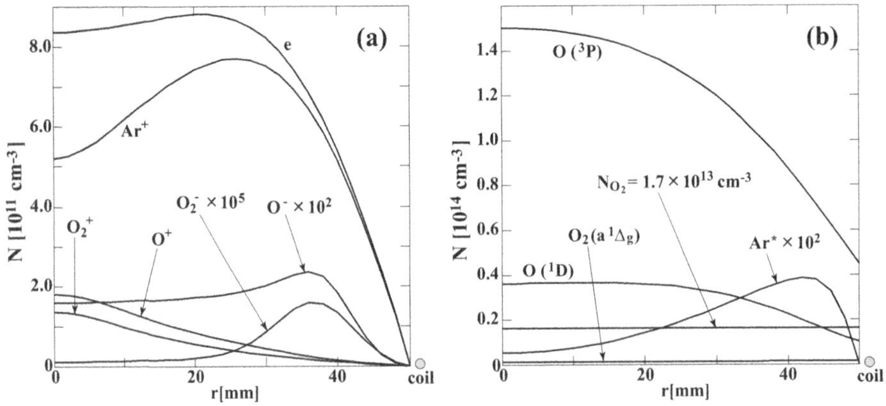

FIGURE 12.2 Density distribution of charged particles (a) and neutrals (b). ICP is driven at 13.56 MHz, 104 W at 100 mTorr in O_2(5 %)/Ar. *From Sato, T, and Makabe, T. Journal of Applied Physics , "A numerical investigation of atomic oxygen density in an inductively coupled plasma in O2/Ar mixture", Volume 41, Number 3, 2008. With permission.*

TABLE 12.1 Role of Diluent Gases in a Low Temperature Processing Plasma

Control	Diluent/Feed gas	Role
Control of gas temperature	He/Feed gas	Prevention of a glow to arc transition through high thermal conductivity
Control of residence time	He/Feed gas	Increase of diffusivity and viscosity
Control of $f(\epsilon)$ of electrons	He, Ar/Feed gas	Optimization of a collision rate through higher excitation energy of rare gases
Control of dissociation rate	He, Ar/Feed gas	Decrease with increasing atomic gases
Control of ionization rate	Ar/Ne	Enhancement by Penning effect
	O_2/Feed gas	Depletion of e by e-attachment
Control of particle growth	CO_2/Feed gas	Enhancement of volume volatilization of species through gas-phase reaction
Protection of surface profile	O_2/Feed gas	Sidewall passivation through oxidation
	N_2/Feed gas	Surface nitriding

leading to the formation of a volatile molecule that escapes into the gaseous phase. The sputtering efficiency of the system between an ion with incident energy ε and angle θ and a solid surface is described by the sputtering yield $Y(\varepsilon, \theta)$ as shown in Figure 12.4 as examples [3].

Exercise 12.2.1
A zero-dimensional model of a reactive sputtering process is successfully used for compound film deposition in a DC magnetron sputtering with a metallic target in admixture with reactive gas and Ar. Derive the governing equations of the system in which a tantalum (Ta) target is sputtered in an Ar/O_2 mixture to deposit Ta_2O_5 compound on a substrate [4].
The reaction on the surface in this system is divided into three regions: the target; substrate; and side wall with areas of A_T, A_S, and A_W (see Figure 12.5a). A fraction $(1 - X)$ of the target material sputtered by the Ar^+ ions with flux Γ_p deposits on the substrate, and X arrives at the side wall. Θ_T, Θ_S, and Θ_W give the degree of the compound (Ta_2O_5) formation on the target, substrate, and reactor wall, respectively. The values of Θ_T and Θ_S are a practical indicator of the reactive sputtering from pure metallic target ($\Theta_T = 0$) resulting in deposition of a stoichiometric compound film at the substrate ($\Theta_S = 1$). s_t, s_s, and s_w are the sticking coefficients of O_2 to the metallic part of the target, substrate, and wall, respectively. The sticking of

TABLE 12.2 Time Constants in Gas- and Surface-Phases in a Reactive Plasma for Surface process

Item	Time Constant	Surface Process	Data
Fresh feed gas	0	Clean surface	Band structure Sputter yield etc. Set of collision QS
Molecular adsorption	$\tau_{ads} < \mu$s	Covered surface	Adsorption rate
Plasma initiation	-	-	-
Ion transit by drift	$\tau_{ion} \sim \mu$s	Begining of modification	
Neutral diffusion	τ_{D} ~5(10) ms	Deposition, etching	Surface conductivity
		Change in wall	-
Ambipolar diffusion	τ_{AD} ~5(10) ms	Surface charging	
Radical formation	10(100) ms	Mono-layer etching	Q-section of impurity
Ejected particles			Yield of impurity
Gas residence time	τ_{res} ~5(500) ms		-
Process time	τ_{etch} ~1(2) min	Etching	Etch rate
	τ_{Oxi} ~1(2) s	Oxidation	-
	τ_{Nite} ~1(2) s	Nitriding	-

TABLE 12.3 Sputter Deposition

Item	Process
Advantage	
	Possible for a large-scale deposition in a simple-equipped vacuum system Possible for materials having a high melting-point, oxide film Easy to deposit multicomponent film Possible for low temperature process
Disadvantage	
	Strongly localized particle source caused by magnetron structure Low utilization efficiency of target material.

the sputtered particles on the target is assumed to be unity. Y_{Tm} and Y_{Tc} are the sputtering yields of the metal and compound material at the target. The number ratio of the reactive atom (O) between the reactive gas (O_2) and the compound molecule (Ta_2O_5) is denoted by h_r.

At the target surface, the balance equation of the compound is

$$\frac{\partial}{\partial t}\Theta_T(t)A_T = h_r\Gamma_{O_2}s_t(1-\Theta_T)A_T - \Gamma_p Y_{Tc}\Theta_T A_T. \tag{12.1}$$

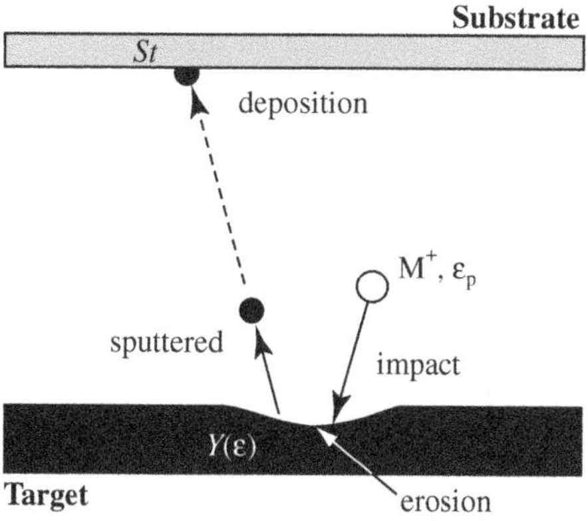

FIGURE 12.3 Physical sputtering system.

At the substrate surface, the balance equation of the compound is described as

$$\begin{aligned}\frac{\partial}{\partial t}\Theta_S(t)A_S &= b_m h_r \Gamma_{O_2} s_s(1-\Theta_S)A_S + (1-\Theta_S)b_m(1-X)\Gamma_p Y_{Tc}\Theta_T A_T \\ &- \Theta_S(1-X)\Gamma_p Y_{Tm}(1-\Theta_T)A_T, \end{aligned} \tag{12.2}$$

where b_m is the number of metal atoms in the compound molecule. In the same way, for the side wall, the surface metal balance is described as

$$\begin{aligned}\frac{\partial}{\partial t}\Theta_W(t)A_W &= b_m h_r \Gamma_{O_2} s_w(1-\Theta_W)A_W + b_m(1-\Theta_W)X\Gamma_p Y_{Tc}\Theta_T A_T \\ &- \Theta_W X\Gamma_p Y_{Tm}(1-\Theta_T)A_T. \end{aligned} \tag{12.3}$$

The total supply q of the reactive gas into the reactor is equal to the consumption on the surface and the quantity pumped out to the outside. That is,

$$\begin{aligned} q = s_t\Gamma_{O_2}(1-\Theta_T)A_T + s_w\Gamma_{O_2}(1-\Theta_W)A_W + s_s\Gamma_{O_2}(1-\Theta_S)A_S \\ + p_r V_{pump}, \end{aligned} \tag{12.4}$$

where p_r is the partial pressure of the reactive gas and related to $\Gamma_{O_2} = p_r/(2\pi MkT)^{1/2}$, and V_{pump} is the pumping speed of the gas. Here, the plasma parameters, incident ion flux Γ_p, partial gas pressure p_r, and pumping speed p_{pump} are given. Unknown parameters, X, Θ_T, Θ_S, and Θ_W can be solved in the steady state from Equations 12.1 to 12.4.

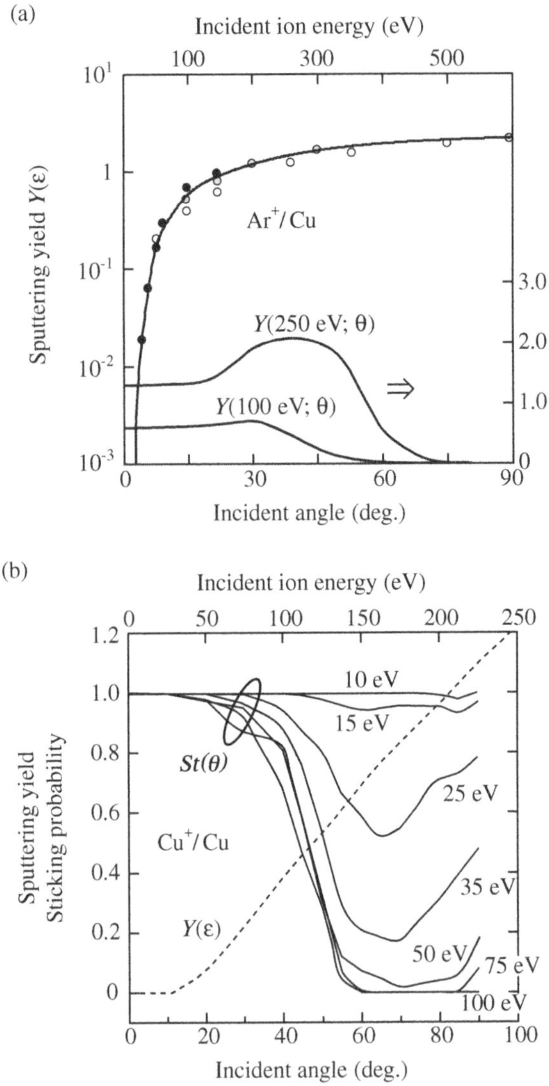

FIGURE 12.4 Sputtering yield $Y(\varepsilon)$ and $Y(\varepsilon, \theta)$ of Ar^+-Cu target (a), and $Y(\varepsilon)$ of Cu^+–Cu target and sticking coefficient $St(\theta)$ of Cu on Cu substrate (b).

The average deposition rate per unit area on a substrate R_{depo} is given as

$$R_{depo} = \Gamma_p \frac{A_T}{A_S} \{Y_{Tc}\Theta_T + Y_{Tm}(1 - \Theta_T)\}(1 - X). \tag{12.5}$$

The deposition rate R_{depo} in the reactive sputtering has a range with multi-values as a function of reactive gas flow, as shown in Figure 12.5b.

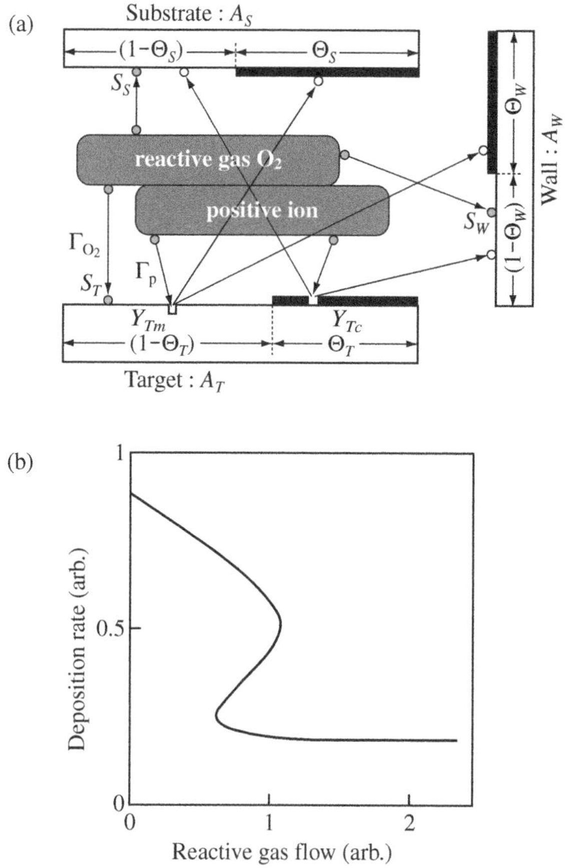

FIGURE 12.5 The 0th-order reactive sputtering model (a), and the predicted deposition rate of a compound film as a function of partial pressure (b).

12.2.1 Target Erosion

Magnetron plasma is widely applied to thin film deposition. The target in a magnetron discharge, that is, cathode, is eroded by physical sputtering by high-energy ion impact under a low-pressure and highly localized plasma condition. The target erosion is described by the sputtering yield,

$$Y(\epsilon_p; \text{target}) = \frac{\text{atoms removed}}{\text{incident ion}}, \tag{12.6}$$

for the system between an incident ion with energy ϵ_p and target material. The databases are widely available in the literature [1, 2]. The number of target atoms ejected by the sputtering of ions with velocity $\boldsymbol{v}$ incident on the target surface dS from the magnetron plasma is estimated by the velocity distribution of the incident ion, $g_p(\epsilon_p, \theta)$, as

$$\rho \; dSdl = n_p \int_\theta \int_{\epsilon_p} Y(\epsilon_p; \text{target}) \boldsymbol{v} g_p(\epsilon_p, \theta, \boldsymbol{r}) d\theta d\epsilon_p \; dt \; dS, \tag{12.7}$$

where ρ is the atomic number density of the target material, dl is the erosion depth during a small time dt, and $g_p(\epsilon_p, \theta, \boldsymbol{r})$ is normalized to unity as

$$\int_{\epsilon_p} g_p(\epsilon_p, \theta, \boldsymbol{r}) d\theta d\epsilon_p = 1. \tag{12.8}$$

The sputter rate $R_{sp}(\boldsymbol{r})$ is obtained by

$$R_{sp}(\boldsymbol{r}) \left(= \frac{dl}{dt} \right) = \frac{n_p(\boldsymbol{r})}{\rho} \int_\theta \int_{\epsilon_p} Y(\epsilon_p; \text{target}) \boldsymbol{v} g_p(\epsilon_p, \theta, \boldsymbol{r}) d\theta d\epsilon_p. \tag{12.9}$$

The erosion depth profile of the target is estimated numerically by the time development of the ion sputtering in a magnetron plasma.

Problem 12.2.1
Even in a low-pressure magnetron plasma, ions affect a target with a finite angle of incidence. Then, we must consider the angular dependence of the sputtering yield $Y(\varepsilon, \theta)$. Revise Equation 12.9 of the sputter rate by considering $Y(\varepsilon, \theta)$.

12.2.2 Sputtered Particle Transport

Next we consider the flux and energy of the atoms sputtered from the target. The ejected flux of the target atom is given at the surface $\boldsymbol{r}_t$ by

$$\boldsymbol{\Gamma}_s(\boldsymbol{r}_t) = n_p \int_\theta \int_{\epsilon_p} Y(\epsilon_p; \text{target}) \boldsymbol{v} g_p(\epsilon_p, \theta, \boldsymbol{r}) d\theta d\epsilon. \tag{12.10}$$

On the other hand, the energy spectrum of the sputtered atom is estimated by the Thompson formula [5],

$$f_s(\epsilon; \epsilon_p, \boldsymbol{r}_t) = A \frac{1 - \{(U_t + \epsilon)/\gamma\epsilon_p\}^{1/2}}{\epsilon^2(1 + U_t/\epsilon)}, \tag{12.11}$$

where A is constant. ϵ_p and ϵ are the kinetic energy of the incident ion and the sputtered atom, respectively. U_t is the binding energy of a target material. γ is the energy transfer factor in the elastic collision between the incident ion and the target atom with masses of M_p and M_s, respectively, and is given by

$$\gamma = \frac{4M_p M s}{(M_p + M_s)}. \tag{12.12}$$

Problem 12.2.2

Discuss the reason that the angular distribution of the ejected neutral is usually approximated by the $\cos\theta$-law in the physical sputtering process.

The Thompson formula is derived under the condition that the sputtered atoms come from a well-developed collision cascade in a material. The collision cascade is realized in a system between an incident heavy ion and light atom in the material. In a system between a light incident ion with low energy and a target consisting of a massive atom, however, the ion from the sheath will be easily backscattered by a massive target atom and will knock off an atom in the top layer of the target. Then, the sputtering is caused by a single knock-on mechanism. A modified formula has been proposed for the system between a light ion and massive target atom [6]:

$$f_s(\epsilon;\epsilon_p,\boldsymbol{r}_t) = A\frac{\epsilon}{(\epsilon+U_t)^{\alpha+1}}\left(\ln\frac{\gamma(1-\gamma)\epsilon_p}{\epsilon+U_t}\right)^2, \tag{12.13}$$

where $\alpha = 3/5$ for the H^+–Fe system. The angular distribution of the ejected atom is described by cosine law.

We consider the arrival flux of the sputtered atom at the substrate $\boldsymbol{r}_s$. First, the flux without collision in gas phase is given by

$$\boldsymbol{\Gamma}_s^0(\boldsymbol{r}_s;\epsilon) = \int_{S_t}\boldsymbol{\Gamma}_s(\boldsymbol{r}_t)\exp(-NQl)\,dS, \tag{12.14}$$

where l is the distance between $\boldsymbol{r}_t$ and $\boldsymbol{r}_s$. Q is the collision cross section between the sputtered atom and feed gas molecule. Here we estimate the transport of the relaxed component N_s^{rel} by collision with feed gas molecules. The neutral transport is described by the diffusion equation

$$\frac{\partial}{\partial t}N_s^{rel}(\boldsymbol{r},t) = D_s\nabla^2 N_s^{rel}(\boldsymbol{r},t) + \Lambda_s^{rel}(\boldsymbol{r},t), \tag{12.15}$$

where D_s is the diffusion coefficient of the sputtered atom in the feed gas molecule. The production rate of the relaxed component Λ_s^{rel} is given under the assumption that the sputtered atom is randomized in energy and direction after one collision with feed gas molecules. Then, from the difference of the ejected flux distribution at small distance $(l,\ l+dl)$,

$$\Lambda_s^{rel}(\boldsymbol{r}) = -\nabla\cdot\Gamma_s^0(\boldsymbol{r}). \tag{12.16}$$

In a steady-state $(\partial/\partial t = 0)$, the spatial transport of the sputtered atom under relaxation is obtained by

$$D_s\nabla^2 N_s^{rel}(\boldsymbol{r}) - \nabla\cdot\Gamma_s^0(\boldsymbol{r}) = 0. \tag{12.17}$$

The arrival flux of the sputtered atom, collisionally relaxed in the gas phase, is given by

$$\Gamma_s^{rel}(\boldsymbol{r}_s) = -D_s\nabla\cdot N_s^{rel}(\boldsymbol{r}_s). \tag{12.18}$$

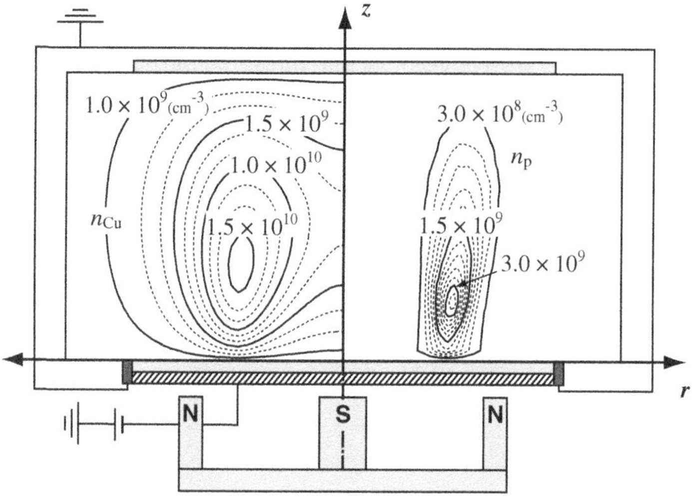

FIGURE 12.6 Relations of the distribution between the ion density $n_p(\mathbf{r})$ and sputtered atom $N_s(\mathbf{r})$ in a DC magnetron plasma.

As a result, the total flux arriving at the substrate at $\boldsymbol{r}_s$ is

$$\Gamma_s^{total}(\boldsymbol{r}_s) = \Gamma_s^0(\boldsymbol{r}_s) + \Gamma_s^{rel}(\boldsymbol{r}_s). \tag{12.19}$$

Figure 12.6 shows the spatial density distribution of Ar^+ ions and sputtered Cu neutrals in a DC magnetron sputtering system operated between axisymmetric parallel plates at 5 mTorr in Ar sustained at V_{sus} of −200 V and maximum $B_{\|}$ of 300 G on a Cu target surface.

12.3 PLASMA CHEMICAL VAPOR DEPOSITION

Chemical vapor deposition (CVD) is a process for fabricating thin films or particles using chemical reactions of gas molecules. CVD is classified according to the various forms of the external energy source into plasma CVD, thermal CVD, photo CVD, and catalytic CVD. In these different processes, chemically active species (precursors) produced by dissociation of gas molecules have a strong influence on the deposition. To distinguish it from CVD, a deposition through a physical vacuum evaporation or sputtering is termed a physical vapor deposition (PVD). The quality of the film deposited is influenced by the properties of the precursor molecules, which depend on the process and the external condition. There are several different methods for depositing Si films (see Table 12.4): low-temperature plasma processing, thermal decomposition, photo decomposition, and catalytic decomposition on a hot tungsten filament.

TABLE 12.4 Chemical Vapor Deposition of $SiH_4(+H_2)$

Deposited Film	Process	Radicals	Conditions	Ref.
a-Si:H, μc-Si:H	Plasma process	H, Si, SiH_2, SiH_3	400 K, 1–10^3 Pa	[7]
Poly-Si	Thermal process	SiH_2, H_2,	900 K, 10–10^3 Pa	—
a-Si:H, μc-Si:H	Catalytic process	H, Si	2000 K, 0.5–10 Pa	[8]
a-Si:H	Photo process	—	Room temperature	—

12.3.1 Plasma CVD

Capacitively coupled parallel plates plasma is widely used for hydrogenated amorphous silicon (a-Si:H) deposition. A good-quality a-Si:H film with a deposition area of up to a few m^2 can be manufactured at a very low-power condition, 50 $mWcm^{-2}$, in SiH_4/H_2 in parallel plate capacitively coupled plasma (CCP). Plasma damage to the film quality caused by an ion impact during deposition is prevented by a wafer arrangement on the grounded electrode and by using a high-frequency source at 13.56 MHz and more efficiently at very high frequency, 100 MHz. Recently, microcrystalline silicon (μc-Si:H) thin film has been developed into the basic materials of large-area solar cell and thin film transistor (TFT) for a large-area flat panel display. Hydrogenated amorphous carbon (a-C:H) or diamondlike carbon (DLC) is the other application of the low-temperature plasma CVD. a-C:H film is hard and wear resistant. The head and disk surfaces in magnetic disk drives are usually coated with a thin layer of DLC to minimize wear and corrosion.

Thin films deposited by plasma polymerization (surface grafting) have a large variety of applications. The plasma has a high ability to modify the surface. These materials are widely used as biocompatible films with hydrophilicity and are suitable for various biomedical and pharmaceutical materials, for example, optical lenses, implants, and drug delivery devices. Plasma polymerization is usually performed in a glass reactor excited by helical coils driven at 13.56 MHz (i.e., inductively) in carrier gas Ar at several hundred mTorr with monomer vapor. Excitation of the carrier gas takes place in radiofrequency (rf) Ar plasma, generating active species (ions and radicals) in the gas phase. These active species interact with the substrate and generate reactive sites (mainly free radicals) on the wafer surface. The monomer vapors in the plasma chamber readily react with these radicals, yielding grafted surfaces.

Exercise 12.3.1

Diamond thin film grows on a diamond substrate in a thermal plasma in CH_4/H_2. The surface under deposition is divided into two sites: the diamond lattice growth with probability D and nth layers of amorphous carbon with probability C_n. Atomic H and C, dissociated from the feed gases CH_4/H_2, in thermal plasma diffuse to the substrate surface. The thin film growth is highly selective on the site and highly competitive between deposition and etching and is modeled by the birth and death process in the stochastic process [9–11]. That is, on the substrate surface, the evolution equation is given in the form of the 0th-order simultaneous rate equations

$$\frac{dD}{dt} = k_c \Gamma_H C_1 - k_s \Gamma_C D + k_{etch} \Gamma_H C_1 \tag{12.20}$$

$$\frac{dC_1}{dt} = -k_c \Gamma_H C_1 + k_s \Gamma_C D - k_{etch} \Gamma_H C_1 - k_s \Gamma_C C_1 + k_{etch} \Gamma_H C_2 \tag{12.21}$$

$$\frac{dC_n}{dt} = k_s \Gamma_C C_{n-1} - k_{etch} \Gamma_H C_n - k_s \Gamma_C C_n + k_{etch} \Gamma_H C_{n+1} \quad (n > 1), \tag{12.22}$$

where k_s is the sticking coefficient of the C atom. k_{etch} is the etching probability of a-C:H by an H atom. k_c is the conversion probability of a-C:H to diamond lattice by an H atom, and

$$D + \sum_{n=1}^{\infty} C_n \equiv 1. \tag{12.23}$$

i. Derive the surface structure (C_1, C_n, and D) in the steady-state condition.

ii. Derive the growth rate of the diamond film R_{depo}.

By summing up Equations 12.20 to 12.22 in the steady state, we obtain the relation

$$\frac{dD}{dt} + \sum_{n=1}^{\infty} \frac{dC_n}{dt} = -k_s \Gamma_C C_n + k_{etch} \Gamma_H C_{n+1} = 0.$$

Therefore,

$$C_n = \chi C_{n-1} \quad \text{where} \quad \chi = \frac{k_s \Gamma_C}{k_{etch} \Gamma_H}.$$

We obtain C_n as

$$C_n = \frac{(1 - \chi^n)}{(1 - \chi)} C_1. \tag{12.24}$$

By substituting Equation 12.24 into Equation 12.23, we have

$$D + \lim_{n \to} \left\{ \sum_{n=1}^{\infty} \frac{(1 - \chi^n)}{(1 - \chi)} C_1 \right\} = 1 \quad \text{and} \quad D = 1 - \frac{C_1}{(1 - \chi)}. \tag{12.25}$$

Also, from Equation 12.20 we have

$$C_1 = \frac{\chi D}{\left(1 + \frac{k_c}{k_{etch}}\right)}.$$

Equations 12.24 and 12.25 give D , C_1, and C_n as

$$\begin{aligned} D &= \frac{(1-\chi)(1+\frac{k_c}{k_{etch}})}{1+(1-\chi)\frac{k_c}{k_{etch}}}, \\ C_1 &= \frac{(1-\chi)\chi}{1+(1-\chi)\frac{k_D}{k_{etch}}}, \\ C_n &= \frac{(1-\chi^n)\chi}{1+(1-\chi)\frac{k_C}{k_{etch}}}. \end{aligned} \tag{12.26}$$

The deposition rate of diamond R_{depo} is

$$R_{depo} = k_c \Gamma_H C_1 - k_{etch}^D \Gamma_H D,$$

where the etching rate coefficient k_{etch}^D is less than the other rate constant, and

$$R_{depo} \sim k_c \Gamma_H C_1 = \frac{k_s(1-\chi)\frac{k_c}{k_{etch}}\Gamma_c}{1+(1-\chi)\frac{k_c}{k_{etch}}}. \tag{12.27}$$

At $k_c \gg k_{etch}$,

$$R_{depo}(\chi) = \frac{k_s(1-\chi)\Gamma_C}{(1-\chi)+\frac{k_{etch}}{k_c}} \sim k_c \Gamma_C. \tag{12.28}$$

At $k_c \ll k_{etch}$,

$$R_{depo}(\chi) = k_s(1-\chi)\frac{k_c}{k_{etch}}\Gamma_C. \tag{12.29}$$

12.3.2 Large-Area Deposition with High Rate

High-quality thin film of hydrogenated amorphous silicon (a-Si:H) is fabricated under a low-energy ion impact in a very high frequency (VHF) plasma. A VHF source provides a higher density plasma with low-energy ions based on a high degree of spatial trap of electrons as compared with that at 13.56 MHz. However, inhomogeneity of plasma arises from the nonuniform surface potential on the powered electrode caused by a standing wave of the VHF voltage. Figure 9.9 shows the potential nonuniformity of the rod electrode as a function of position in a plasma of 2×10^9 cm^{-3} sustained at 100 MHz at 13 Pa in Ar. The numerical value, simulated by the transmission-line model (TLM) (see Section 7.5.3) reproduces the experimental observation. Numerical design of the surface potential uniformity is achieved by time-averaging the surface potential through the control of the modulation rate of the phase of the VHF source at the feed position in a one- or two-dimensional electrode system.

Problem 12.3.1

Excess radicals produced dissociatively in a reactive plasma will proceed to a particle growth around a neucleus in gas phase in advance of the surface deposition. Discuss the particle growth process by considering the governing equation. The detail of the mechanism will be found in [15].

TABLE 12.5 Typical Example of Plasma Chemical Vapor Deposition. (g): gas-phase, and (s): solid-phase

Volume Reaction	Feed Gases	Products in Plasma	Thin Film
Dissociation	SiH_4 (g) + H_2(g)	Si(s) + H_2(g)	a-Si:H
Reduction	$SiCl_4$(g) + H_2(g)	Si(s) + HCl(g)	poly-Si
Oxidation	SiH_4(g) + O_2(g)	SiO_2(s) + H_2O(g)	SiO_2
Nitride	SiH_4(g) + NH_3(g)	Si_3N_4(s) + H_2(g)	Si_3N_4
Oxynitride	SiH_4(g) + N_2O(g)	SiON(s) + NH_3(g) + H_2(g)	SiON
Carbonization	$TiCl_4$(g) + CH_4(g)	TiC(s) + HCl(g)	TiC
Hydrolysis	$AlCl_3$(g) + CO_2(g) + H_2(g)	Al_2O_3(s) + HCl(g) + CO(g)	Al_2O_3
Org.-metal comp.	$(CH_3)_3Ga$(g) + AsH_3(g)	GaAs(s) + CH_4(g)	GaAs

12.4 PLASMA ETCHING

Dry etching for micro- and nanoelectronic device fabrication should be highly selective of one material over others and highly competitive with the deposition process. A competitive surface between deposition and etching is described by two physical quantities: the etching yield $Y_{etch}(\epsilon)$ and the sticking coefficient $S_t(\theta)$ of incident active particles. Plasma etching has two extremely different phases, that is, isotropic chemical etching by neutral radicals and anisotropic etching assisted by energetic ion impacts (i.e., reactive ion etching). It is much more complicated in a practical etching that has a surface process between a purely physical sputtering and a spontaneous chemical etching as well as deposition as a function of radical-to-ion ratio incident on the wafer. The practical etching in ultra-large scale integrated (ULSI) manufacturing is mainly devoted to two processes: Si-gate etching and SiO_2-contact hole (trench) etching (see Table 12.2 and Figure 12.7). In general, the atomic scale mechanisms that enhance the surface process by ion impact are

i. Sticking coefficient of radicals;

ii. Diffusion of etchant (i.e., surface coverage);

iii. Impact damage; and

iv. Volatile- and nonvolatile-molecule productions, and the like.

Together, these effects mean that ion-assisted etching in a plasma process produces an increase in the local surface coverage of the etchant and enhances the effective etching yield compared with chemical etching under the same number of radicals incident on the surface.

The etching yield $Y_{etch}(\varepsilon)$ is defined as the ratio between the number of removed substrate molecules and the number of the incident species on the substrate. More specifically, the etching yield of ions is a function of the impact energy ε_p and the incident angle θ, $Y(\varepsilon_p, \theta)$. The etching rate $R_{etch}(\boldsymbol{r})$ of ions with velocity distribution $g_p(\boldsymbol{v}, \boldsymbol{r})$ incident on the substrate $\boldsymbol{r}$ is related to the etching yield $Y(\varepsilon_p, \theta)$ when we define $R_{etch}(\boldsymbol{r})$ as

$$R_{etch}(\boldsymbol{r}) = \frac{n_p(\boldsymbol{r})}{\rho_n} \int Y(\varepsilon_p, \theta)\boldsymbol{v} g_p(\boldsymbol{v}, \boldsymbol{r}) d\boldsymbol{v} \Big/ \int g_p(\boldsymbol{v}, \boldsymbol{r}) d\boldsymbol{v}, \qquad (12.30)$$

where $n_p(\boldsymbol{r})$ is the number density of ions close to the wafer. ρ_n is the atomic number density of the substrate material.

Problem 12.4.1

Reactive ion etching (RIE) is based on a first adsorption of radical species on a surface and successive ion impact with high energy. Discuss the necessary condition of radicals and ions incident on the surface in the RIE.

TABLE 12.6 Feed Gases in Plasma Etching for Semiconductor and MEMS/NEMS Fabrications (○: usable, △: existent, -: nonexistent)

Etching Process	Chemistry	Typical Gases	Set of QS	Volatile products	Typical Source
Poly-Si gate etching	Chlorine-	$HBr/Cl_2/O_2$	○	$SiCl_x$(x=1,4)	ICP
SiO_2 contact hole etching	Fluorine-	$C_4F_8/O_2/Ar$	△	SiF_x(x=1,4), CO_2	2f-CCP
		$C_4F_6/O_2/Ar$	-		-
		CF_4/Ar	○		-
Organic low-k etching	Hydrogen-	H_2/N_2	○	CH_x, C_2N_2, HCN	2f-CCP
		NH_3	△		-
Non-organic low-k etching	Fluorine-	$C_4F_8/Ar/N_2$	△	SiF_x(x=1,4), CO_2	2f-CCP
Metal(TaN) gate etching	Bromine-	$BCl_3/O_2/Ar$	△	$TaCl_x$(x=1,5)	Biased-ICP
High-k (HfO_2) gate etching	Bromine-	BCl_3/Ar	△	$HfCl_4$	Biased-ICP
Resist pattern trimming	Oxygen-	$SO_2/O_2/He$	○	CO, CO_2	ICP
Poly(Single)-diamond	Oxygen-	O_2/Ar	○	CO, CO_2	ICP
Deep-Si etching	Fluorine-	SF_6/O_2	○	SiF_x(x=1,4)	2f-CCP
Deep-Quartz etching	Fluorine-	CF_4/O_2	○	SiF_4, CO_2	2f-CCP
Resist ashing	Oxygen-	O_2	○	CO, CO_2	ICP
Chamber cleaning					
for gate etch	Fluorine-	SF_6/O_2	○		ICP
for contact hole	Fluorine-	$C_4F_8/O_2/Ar$	△		CCP

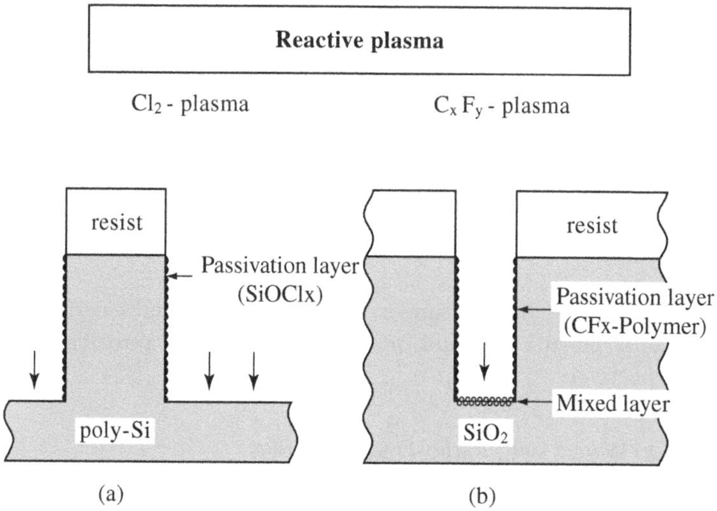

FIGURE 12.7 Typical etching profile in Si-ULSI: Si-gate etching (a) and trench or hole etching of SiO_2 (b).

TABLE 12.7 Basic Rule of Main Two Plasma Etchings. ϵ_a and ϵ_b are the chemical energy released from a radical incident on a surface and the bond energy of a material, respectively.

Item	Chemical Plasma Etching	Reactive Ion Etching
Criteria	$\epsilon_a > \epsilon_b$	$\epsilon_a < \epsilon_b$
Etchant	Radicals	P-ions with high energy (ϵ_{ion}) + radicals
Transport	Diffusion	Drift + diffusion
Surface reaction	Chemical activation	Chemical reaction assisted by ion kinetic energy
Anisotropy	under sidewall passivation	at ϵ_{ion} of 50 eV - 500 eV
Precursor	Non-volatile (SiF_k(k=1-3))	Non-volatile (Si, SiF_k(k=1-3))
Etch product	Volatile molecule ($AlCl_3$, SiF_4, CF_4)	Volatile molecule (SiF_4)
Examples	Al by Cl-chemistry Si by F-chemistry C by F-chemistry	SiO_2 by F-chemistry & Ar^+

12.4.1 Wafer Bias

The wafer to be etched is set on the substrate holder, which is usually electrically isolated from the reactor base potential. The simple relation between the wafer potential and the impact ion energy is briefly described in advance.

12.4.1.1 *On Electrically Isolated Wafers (without Radio-Frequency Bias)*

When the wafer is set on an electrically insulated holder from the system, the wafer surface is kept at floating potential V_{fl}, which provides the ambipolar diffusion with $e\Gamma_e = e\Gamma_p$ on the surface in a time-averaged fashion. In a collisionless sheath, Bohm's sheath criterion predicts the relation (see Section 3.6),

$$V_{plasma} - V_{fl} = \frac{kT_e}{2e} \ln\left(\frac{M_p}{2.3m}\right), \tag{12.31}$$

where V_{plasma} and T_e are the time-averaged potential and electron temperature in the bulk plasma. M_p and m are the mass of the positive ion and the electron, respectively.

12.4.1.2 *On Wafers with Radio-Frequency Bias*

In the case of a high-energy ion-assisted etching, we must apply an rf bias voltage $V_{bias}(t)$ on the holder. For sophisticated modern plasma etching, an ion-assisted etching with very high energy, 500 eV to 1 keV, is practical. Then, it is essential to perform the functional separation between plasma production and ion acceleration, because some degree of ionization is unavoidable in the active sheath in front of the biased wafer. As already described in Chapter 8, a negative DC self-bias voltage V_{dc} appears on the wafer under the ambipolar diffusion in the steady state. Accordingly, in a time-averaged fashion, the wafer surface is irradiated by positive ions with energy $\langle\varepsilon_p\rangle$, which is defined as

$$\langle\varepsilon_p\rangle = e(V_{plasma} - \langle V_{dc}\rangle). \tag{12.32}$$

As a result, the impact energy of ions on the wafer is roughly estimated by the relation with/without bias voltage.

$$\begin{aligned} \langle\varepsilon_p\rangle &= \frac{kT_e}{2} \ln\left(\frac{M_p}{2.3m}\right) : \quad \text{without bias (floating);} \\ &= e(V_{plasma} - \langle V_{dc}\rangle) \quad : \quad \text{with rf bias;} \\ &= eV_{plasma} \quad : \quad \text{on the ground.} \end{aligned}$$

Another important particle on the surface exposed to the plasma etching is the chemically active radicals. The neutral radical transport is usually treated by the random motion in the gas phase. That is, the radical flux incident on a wafer with/without rf bias voltage is estimated by using the Maxwellian velocity distribution in the quasithermal condition

$$\Gamma_{radical} = \frac{N_r}{4}\left(\frac{8kT_g}{\pi M_r}\right)^{1/2}, \tag{12.33}$$

where T_g is the gas temperature. N_r and M_r are the number density and mass of the neutral radical, respectively.

12.4.2 Selection of Feed Gas

Dry etching in a low-temperature plasma utilizes a physical and chemical reactivity on material surfaces based on ions and dissociated neutrals (radicals) in a plasma. Feed gases for plasma etching are carefully prepared under the following conditions (see Table 12.2):

i. The reactive product caused by the surface reaction must be in gas phase (volatile) in order to remove the surface material, and thus the vapor pressure will be high for the reactive product (see Table 12.3);

ii. The binding energy of reactive products must be lower than that of the material to be etched; and

iii. It is preferable that the feed gas, the dissociated molecules, and the reactive products are all nontoxic and all have low global warming potentials.

Admixture gases are usually used for plasma etching for the following purposes:

i. Density control of the dissociated chemically active species (i.e., radicals);

ii. Control of the absorption site on the material surface to be etched;

iii. Control of the surface reactivity;

TABLE 12.8 Reactive Product in Plasma Etching and the Boiling Point and Vapor Pressure

Materials	Reactive Product	Boiling Point (C)	Vapor Pressure
Al	AlF_3	1291	—
	$AlCl_3$	183*	0.6
	$AlBr_3$	255*	—
Si	SiF_4	–90	≤ 760
	$SiCl_4$	57	—
	$SiBr_4$	154	—
Fe	FeF_2	≤ 1000	—
	$FeCl_3$	319	—
	$FeBr_2$	684	—
Ni	$NiCl_2$	1000	—
	$Ni(CO)_4$ -25	—	—
W	WF_6	17*	—
	WCl_6	346*	—
Co	CoF_2	1200	—
	$CoCl_2$	1050	—

*Sublimation temperature.

TABLE 12.9 Global Warming Potential (GWP) and Lifetime of Gases for Plasma Etching

Feed Gas	GWP (Units of 500 yr)	Lifetime (yr)
CO_2	1.0	1.8×10^2
CF_4	8.9×10^3	5×10^4
CHF_3	1×10^4	2.6×10^2
C_3F_8	1.2×10^4	2.5×10^3
SF_6	3.2×10^4	3.2×10^3
c-C_4F_8	9100	3200
l-C_4F_8	100	1.0
C_3F_6	5×10	1.0
l-C_4F_6	5×10	1.0

iv. Improvement of heat transfer in gases and of surface cooling; and

v. Control of the plasma density (impedance) by way of the plasma structure.

The influence of each of the gases on surface etching is very complicated. Most of the feed gases used for plasma etching are greenhouse gases. That is, greenhouse gases absorb infrared radiation and trap energy in the atmosphere. The atmospheric lifetime and global warming potential (GWP) characterize the effect of the greenhouse gas. The GWP of a greenhouse gas is the ratio of global warming from one unit mass of a greenhouse gas to that of one unit mass of CO_2 over a period of time. Hence this is a measure of the potential for global warming per unit mass relative to CO_2 (see Table 12.9).

12.4.3 Si or Poly-Si Etching

Si and poly-Si material are etched by halogen and halogen compounds. A fluorine atom can be used to etch the material without the assistance of ion impact, and isotropic etching is naturally realized. F atoms incident on the clean Si surface saturate the dangling bond and insert into the Si-Si bond, resulting in an ejection of volatile particles, SiF_4, as the major etching product at room temperature:

$$4F + Si \rightarrow SiF_4. \tag{12.34}$$

The chemical etching probability (yield) of F atoms and Si crystals is sensitive to the surface condition, contamination, roughness, and so on, and the value distributes between 0.025 and 0.064. With increasing substrate temperature ($T_S > 500$ K), the fraction of SiF_4 gradually decreases and that of SiF_2 increases rapidly. On the other hand, the etching probability (yield) of crystalline Si is 0.005 for a Cl atom at room temperature, and the dominant etching product is $SiCl_4$. With increasing surface temperature, the dominant etching product becomes $SiCl_2$:

$$\begin{aligned} 4Cl &+ Si \rightarrow SiCl_4 \quad \text{at } T_S < 500 \text{ K}, \\ 2Cl &+ Si \rightarrow SiCl_2 \quad \text{at } T_S > 750 \text{ K}. \end{aligned} \tag{12.35}$$

Ion energy-assisted etching of Si-crystal or poly-Si enhances anisotropic etching in an rf plasma in chlorine or bromine compounds. Therefore, the anisotropic etching of Si or poly-Si is performed on the wafer biased at less than 100 V in a high-density chlorine plasma in inductively coupled plasma (ICP) or rf-magnetron, and so on. It should be noted that the etching probability of crystalline Si by molecular F_2 and Cl_2 will be very small.

Problem 12.4.2
The volume density of neutral etching products is simply estimated without redeposition in a plasma as

$$N_{etch} = \frac{1}{V} \frac{R_{etch} \rho S_{eff}}{\Sigma R_j n_e + k_{pump}}, \tag{12.36}$$

where S_{eff} is the effective area of a substrate exposed to etching, V is the plasma volume, R_j is the rate of the destruction processes of the species in gas phase, and k_{pump} is the pumping speed. Derive the expression 12.36.

Plasma etching is used for fine-pattern transfer in ULSI processing. The target of the present technology is the fabrication of sub-45 nm rule or less. The validity of the model depends directly on knowledge of both the gas-phase collision processes and the physics/chemistry of the surface exposed to the plasma. Figure 12.8 shows the etching yield $Y(\varepsilon)$ of Si as a function of incident ion energy [12]. Note here that in a typical plasma etching, the Cl flux is larger than the ion flux by a factor of 100. Attention should be paid to the difference of the yield between the nonreactive Ar^+ ion and reactive Cl_2^+ ion. As described in Equation 12.12, the magnitude of the physical etching is the direct result of the mass ratio, M_p/M_s. Judging from the fact that the mass ratio of Cl_2^+ in Cl_2 plasma is similar to that of Ar^+ in Ar plasma a predominant element comes from the chemical reactivity, that is, chemical etching, of the neutral radical species, Cl. In fact, this is confirmed from the values of the yield of Cl_2^+/Cl_2 as compared with that in Ar^+/Ar in Figure 12.8. Active fluxes of ions and neutral radicals incident on the surface and passive ejected flux of neutral etch products can be, in principle, extracted from a self-consistent modeling of the total system of the dry etching. However, it may be difficult to perform the modeling due to the lack of an available database.

12.4.4 Al Etching

Al etching is generally performed by chlorine and the compounds. However, fluorine compound is not valid for this purpose, because the reactive etching

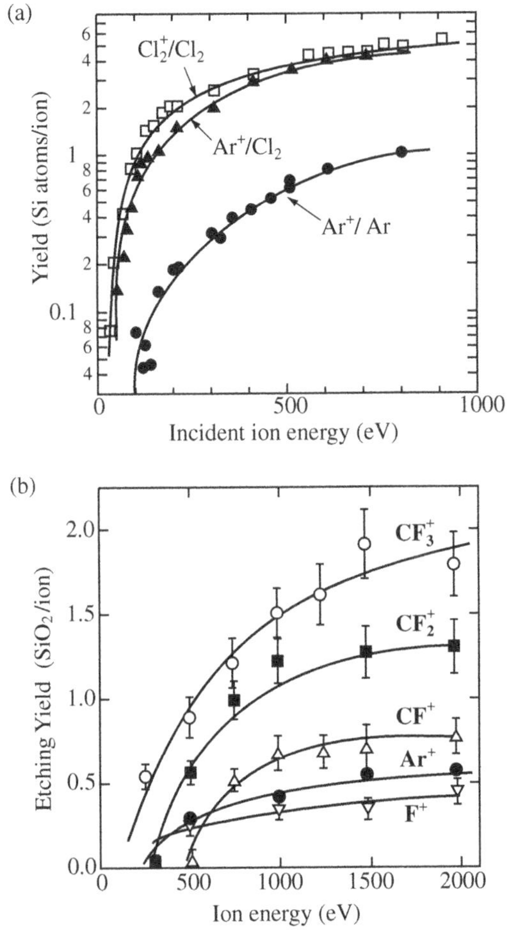

FIGURE 12.8 Etching yield of Si (a) and SiO_2 (b) as a function of ions from plasmas.

product has a high boiling point and is usually nonvolatile (see Table 12.8). The surface etching is described by the direct reaction between Cl_2 and Al:

$$3Cl_2 + 2Al \rightarrow 2AlCl_3. \tag{12.37}$$

As a result, the etching speed depends on the gas number density of Cl_2. On the other hand, in BCl_3 the plasma density controls the etching speed because the surface reaction proceeds by way of the dissociated Cl_2 in BCl_3 plasma:

$$\begin{aligned} e(plasma) + BCl_3 &\rightarrow (Cl, Cl_2, BCl, BCl_2) + e \\ 3Cl_2 + 2Al &\rightarrow 2AlCl_3. \end{aligned} \tag{12.38}$$

That is, the etching speed in rf plasma in Cl_2 is independent of the dissipated power, and the speed is proportional to the power in BCl_3 plasma. This means that ion bombardment from the plasma has no effect on the etching rate, and isotropic etching is performed. The native surface oxide is removed by the ion impact in BCl_3 plasma.

12.4.5 SiO_2 Etching

SiO_2 etching for electrical contact holes or trenches in ULSI circuits is usually performed by a CCP. This process requires a high-energy ion-assisted etching with several hundred eV or 1 keV in order to maintain a high etching rate and high selectivity of SiO_2 to the Si surface. For this purpose, a two frequency CCP with a different frequency source, functionally separated for sustaining a high-density plasma and for biasing the wafer, at each of two parallel electrodes has been adopted by a combination of VHF and low-frequency sources.

SiO_2 thin film is generally etched by an rf plasma in a fluorocarbon C_jF_k gas system. In SiO_2 the characteristics of a fluorocarbon polymer deposition as well as the high-energy ion-assisted etching determine the feature profile of the hole or trench. The control of the profile with a high aspect ratio is performed by the sidewall passivation film of the polymer. The selective etching of SiO_2 over Si and photoresist is due to the selective formation of protective fluorocarbon polymer film over the Si surface. In a steady plasma etching, a mixed amorphous interfacial layer, $Si_lC_mF_n$, is formed on SiO_2 under impact of energetic CF_3^+ ions. In actual practice, the apparent etching yield of SiO_2 in the continuous plasma irradiation is given for the value of a mixed interfacial layer rather than for the pure SiO_2 surface. Carbon in the form of CF_i radicals and CF_j^+ ions from fluorocarbon plasma reacts with the oxygen on the SiO_2 surface under energetic ion impact to form volatile products, whereas fluorocarbon is deposited on the sidewall to form a protective layer of C_iF_j polymers that inhibit lateral etching. This leads to "anisotropic etching." Low-k materials with a relative permittivity ϵ_r smaller than 3.8 of SiO_2 are used as the dielectric in a multilayer interconnect system in ULSI. CCP maintained in admixture of H_2 and N_2 is used to etch organic low-k materials. The etching yield is shown in Figure 12.9.

Problem 12.4.3
Discuss the effects of additive gases (Ar, O_2, CO, H_2) on SiO_2 etching in fluorocarbon plasma (see Table 12.11).

12.4.6 Feature Profile Evolution

The space and time evolution of a patterned wafer surface exposed to particles of positive ions and neutral radicals from a plasma as shown in terms of the velocity distribution of ions in Figure 12.11 is estimated by the Level Set method based on the Hamilton–Jacobi-type equation under a moving boundary [13]

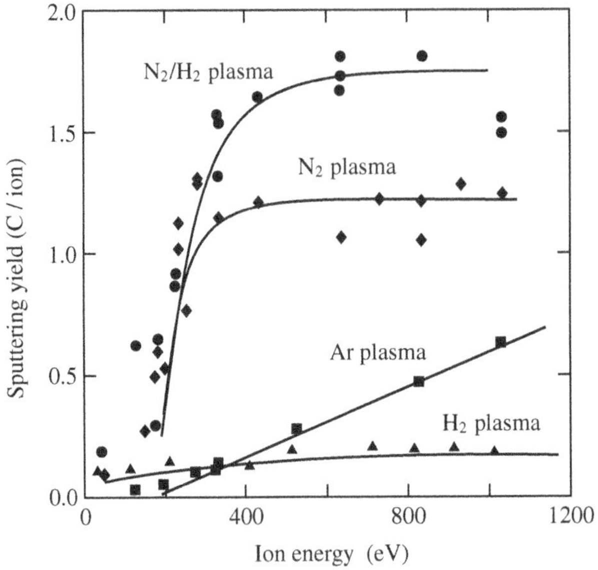

FIGURE 12.9 Etching yield of organic low-k as a function of ions from plasmas.

(see Chapter 8):

$$\frac{\partial}{\partial t}\Phi(x,z,t) = R_{etch}(x,z,t)|\nabla\Phi|. \tag{12.39}$$

Here, R_{etch} is the etching rate (speed function) of a material surface. The surface as a function of Cartesian coordinates (x, z) and time t in Figure 12.12 is defined as

$$\Phi(x,z,t) = 0. \tag{12.40}$$

That is, the gas and solid phase are given at $\Phi > 0$ and $\Phi < 0$ at t, respectively. The dielectric surface $\Phi(\boldsymbol{r}, t) = 0$ is usually locally charged by ions and electrons incident on the surface during etching, and the metallic surface is kept at equipotential.

A brief description of the Level Set method is given below. First, we investigate the relation between the time constant τ_0 for one monolayer etching and the time step to trace an ion trajectory Δt_{tr} in order to estimate a surface evolution Δt under the following physical requirement:

$$\Delta t_{tr} \ll \tau_0 < \Delta t.$$

Also Δt is adjusted to satisfy the condition that a renewal of the surface during Δt does not exceed the spatial mesh size, Δx and Δz; that is, $R_{etch}\Delta t \ll \Delta x$(and Δz).

Step 1: At a position $\boldsymbol{r}_{intl}$ unaffected by a topographically and locally charged wafer surface, the velocity distribution $g_{p(e)}(\boldsymbol{v}, \boldsymbol{r}_{intl})$ and the density $n_{p(e)}(\boldsymbol{r}_{intl})$ of positive ions and electrons are given from the plasma structure in the reactor simulated, in advance, as the initial conditions of ions and electrons

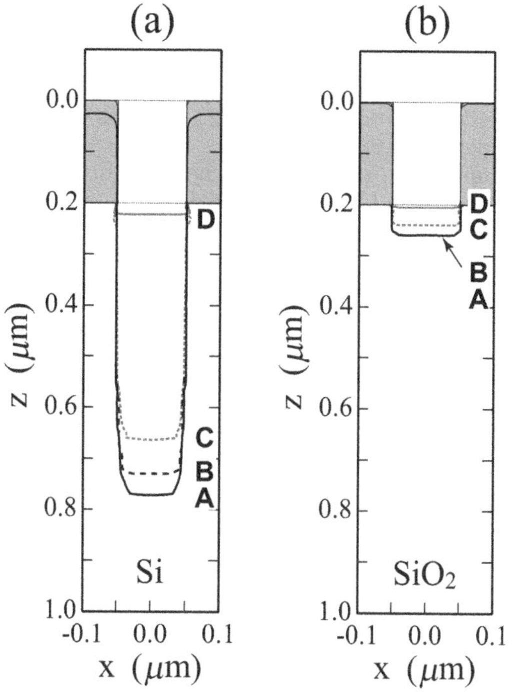

FIGURE 12.10 Comparison of the feature profile evolution on a patterned Si (a) and SiO_2 (b) in 2f-CCP at 50 mTorr in CF_4(5%)/Ar, driven at 100 MHz, 300 V and biased at 1 MHz, 100 V. Active species are Ar^++CF_3^++CF_3+F (A), Ar^++CF_3^++CF_3 (B), Ar^+ (C), and F (D).

incident on the wafer surface (see Figure 12.12). Then, the ion (and electron) trajectory incident in a trench from $\boldsymbol{r}_{intl}$ is traced by Monte Carlo simulation with time step Δt_{tr} by using Poisson's equation and Newton's equation under the consideration of collisions with the gas molecule and on the surface. At each position $\boldsymbol{r}_p(t_{tr})$ of the ion trajectory, $\Phi(\boldsymbol{r}_p)$ is interpolated from $\Phi_{i,j}$ at four grid nodes surrounding $\boldsymbol{r}_p$ (see Figure 12.12) as

$$\begin{aligned}\Phi_p(\boldsymbol{r}_p) &= (1-r)(1-s)\Phi_{i,j} + (1-r)s\Phi_{i,j+1} \\ &+ r(1-s)\Phi_{i+1,j} + rs\Phi_{i+1,j+1},\end{aligned} \tag{12.41}$$

where $r = (x_p - x_i)/(x_{i+1} - x_i)$ and $s = (z_p - z_i)/(z_{i+1} - z_i)$. When $\Phi_p(\boldsymbol{r}_p) > 0$ is satisfied, the trace of the ion flight is continued at $t_{tr} + \Delta t_{tr}$. Otherwise, if $\Phi_p(\boldsymbol{r}_p) < 0$ is satisfied, go to Step 2.

Step 2: We estimate the point $\boldsymbol{r}_0$ that intersects with the surface, $\Phi(\boldsymbol{r}_0) = 0$, from $\Phi(\boldsymbol{r}_p)$ and the value one step before $\Phi(\boldsymbol{r}_{p-1})$ at $t_{tr} - \Delta t_{tr}$ as

$$\boldsymbol{r}_0(x_0, y_0) = \left(\frac{\Phi_p x_{p-1} - \Phi_{p-1} x_p}{\Phi_p - \Phi_{p-1}}, \frac{\Phi_p y_{p-1} - \Phi_{p-1} y_p}{\Phi_p - \Phi_{p-1}} \right). \tag{12.42}$$

TABLE 12.10 Methodology in Modeling of Surface Profile Evolution

Method (authors)	Advantage & Disadvantage
String method: Jewett et al(1977)	
	Surface: string of nodes $P_{node}(\boldsymbol{r},t)$ connected by straight line segments (no mesh)
	Limitation: step-size $\leq$ segment length
(a)**Perpendicular bisector algorithm**	
	Node-shift: along bisector of angle between two adjacent segments
	Suitable: isotropic etching and deposition with uniform rate
	Simple: direct tracing of $P_{node}(\boldsymbol{r},t)$ with little memory
	Difficulty: at sharp corner
	Warning: loop formation at segment comparable to step-size
(b)**Segment advancement algorithm:**	
	Segment-shift: parallel to itself by a distance under local rate
	New node: given by the intersection of the adjacent segments
	Conservation: segment slope at initial shape
Method of characteristics: Shaqfeh & Jurgensen(1989)	
	Improvement over perpendicular bisector method
	Node trajectory: given by $R_{etch}(\theta, \boldsymbol{r}_{surf})$ & ∇R_{etch}
	Expressible for the bend of node trajectory
Level set method: Osher & Sethian(1988)	
	Surface: traced by shortest distance btwn surface & lattice using Hamilton-Jacob equ.
	It provides accurate geometrical property under etching/deposition
	Capability: multilayer structure with different physical property
	Interpolation: essential for surface profile estimation
	Accuracy: dependent on number of meshes (lattice).
Cell-based method: Hwang & Giapis(1997), Osano & Ono(2005)	
	Set of cell: positioned on the boundary btwn gas and materials
	Cell-size: selected from atomic scale up to tens of nanometers
	Cell: data structure during computation
	Requirement: more CPU resource than other methods
Molecular dynamics: Schoolcraft & Garrison(1990), Barone & Graves(1995)	
	Advantage: Increase of physical accuracy under simple algorithm.
	Disadvantage: Lack of interaction potential, and huge CPU resource

TABLE 12.11 Function of Additive Gas in SiO_2 Etching in Fluorocarbon Plasma

Gas	Property
O_2	Decelerates polymerization through the production of volatile CO, CO_2, COF_2
H_2	Scavenges [F] to form volatile HF, increasing the [C]/[F] ratio, and shifting the chemistry from an etching to a polymerization
CO	Tunes gas to increase the selectivity in SiO_2/Si Increase of relative CF
Ar	Serves as a buffer gas for control of the dissociation of C_jF_k

When the ion with energy ε_p is reflected at the material surface having a reflection coefficient $\alpha_{ref}(\varepsilon_p)$, that is, when the condition

$$\xi \leq \alpha_{ref}(\varepsilon_p)$$

is satisfied, the trace of the ion trajectory in gas phase is continued at Step 1. Here, ξ is a uniform random number distributed over [0, 1].

In the case of $\xi > \alpha_{ref}(\varepsilon_p)$, the information of the flux velocity $\mathbf{\Gamma}(\boldsymbol{v}_0, \boldsymbol{r}_0)$ at the surface is transcribed into the adjacent two-grid nodes (see Figure 12.12):

$$\mathbf{\Gamma}(\boldsymbol{v})_{i+1,j} = \left[1 + \left(\frac{s^2 + (1-r)^2}{r^2 + (1-s)^2}\right)^{1/2}\right]^{-1} \mathbf{\Gamma}(\boldsymbol{v}_0, \boldsymbol{r}_0),$$

$$\mathbf{\Gamma}(\boldsymbol{v})_{i,j+1} = \left[1 + \left(\frac{r^2 + (1-s)^2}{s^2 + (1-r)^2}\right)^{1/2}\right]^{-1} \mathbf{\Gamma}(\boldsymbol{v}_0, \boldsymbol{r}_0). \tag{12.43}$$

Then, the incident angle of the transcribed flux vector θ with respect to the normal vector $\mathbf{n} = \nabla\Phi/|\nabla\Phi|$ at both nodes is obtained by $\Phi(\boldsymbol{r})$ and the

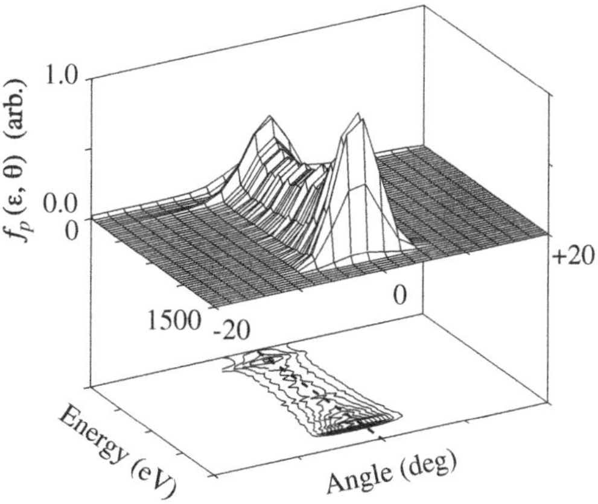

FIGURE 12.11 Example of the energy and angular distributions of ions incident on a wafer as a function of radial position.

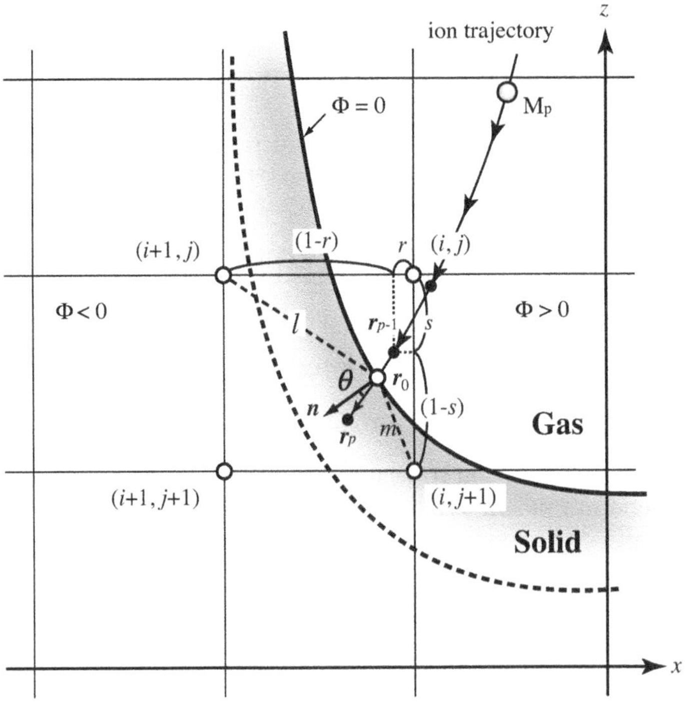

FIGURE 12.12 Schematic diagram of the Level Set method and the surface evolution. From Makabe, T, and Yagisawa, T. Plasma Sources Science and Technology, "Nonequilibrium radio frequency plasma interacting with a surface", Volume 18, Number 1, 2009. With permission.

derivative of the nodes:

$$\theta = \cos^{-1}\left(\frac{\frac{\partial \Phi}{\partial x}\cdot v_x + \frac{\partial \Phi}{\partial y}\cdot v_y}{\parallel \nabla\phi \parallel \cdot \parallel \boldsymbol{v} \parallel}\right). \tag{12.44}$$

Then we have the component of the etching yield at the node as a function of the ion energy ε_p and incident angle θ as $Y(\varepsilon_p, \theta)$. The above procedure is carried out for the number of ions at initial position $\boldsymbol{r}_{intl}$ by considering $g(\boldsymbol{v}, \boldsymbol{r}_{initl})$ and $n_p(\boldsymbol{r}_{intl})$. Then, the flux velocity $\boldsymbol{\Gamma}(\boldsymbol{v}, \boldsymbol{r})$ at each node adjacent to the surface $\Phi = 0$ is accumulated during Δt.

Step 3: After tracing the ion trajectories incident on a trench structure from a plasma during $\Delta t = m\Delta t_{tr}$ ($m \sim 100$), the accumulated flux velocity $\Gamma(\varepsilon_p, \boldsymbol{r})$, energy ε_p, and angle θ at each of the grid nodes adjacent to the surface $\Phi = 0$ enables us to estimate the new surface $\Phi = 0$. That is, at each of the nodes, the etching rate (speed function) is given as

$$\begin{aligned} R_{etch}(\boldsymbol{r}) &= \frac{n_{p_{intl}}}{\rho} \int \boldsymbol{v} g_p(\boldsymbol{v}, \boldsymbol{r}) Y(\varepsilon_p, \theta) d\varepsilon_p d\theta / \int \boldsymbol{v} g_p(\boldsymbol{v}, \boldsymbol{r}) d\varepsilon_p \\ &= \frac{1}{\rho} \int \boldsymbol{\Gamma}(\varepsilon_p, \boldsymbol{r}) Y(\varepsilon_p, \theta) d\varepsilon_p d\theta \end{aligned} \tag{12.45}$$

where ρ is the number density of the material.

As a result, the new value of $\Phi(\boldsymbol{r}, t)$ is obtained at each node from Equation 12.39

$$\Phi(\boldsymbol{r}, t + \Delta t) = \int_t^{t+\Delta t} R_{etch}(\boldsymbol{r}, t) |\nabla \Phi(\boldsymbol{r}, t)| dt. \tag{12.46}$$

Step 4: By using the information of the renewed surface ($\Phi(\boldsymbol{r}, t) = 0$), each grid surrounding the surface $\Phi(\boldsymbol{r}, t)$ is redefined. At the same time, the local charge distribution on the new surface is also renewed by using the previous ones. Successive simulation of the etching is performed by repeating the loop from Step 1 to Step 4.

The same procedure is adopted for chemical etching by neutral radicals incident on the surface. In general, the neutral radicals have a Maxwellian velocity distribution with a temperature of T_g. Plasma etching is a more-or-less competitive process between deposition and etching. As described in Section 12.3, neutral radicals are the predominant species for deposition, and etching is achieved by positive ion impact in addition to the chemical etching of neutral radicals. In particular, the surface reaction of neutral species is a function of surface temperature as predicted by Arrhenius' equation. It is possible to carry out the simulation of the competitive process by considering a surface evolution by neutral radicals and ions using the Level Set method under the database of $Y(\varepsilon_p, \theta)$ of ions and $S_t(\theta)$ of radicals. Figure 12.13 exhibits self-consistent results of a feature profile evolution based on plasma reactor-scale and feature-scale simulation.

Problem 12.4.4

Discuss the string model of the feature profile evolution in etching in comparison with the Level Set method.

12.4.7 Plasma Bosch Process

Plasma etching of a large hole/trench with high aspect ratio (depth-to-width) in Si is a basic process to fabricate integrated components for micro-electromechanical systems (MEMS) [14]. Holes or trenches in MEMS range from 1 μm to 100 μm in width and from 10 μm to several 100 μm in aspect ratio. The etching needs a high etch rate with anisotropy and selectivity to the mask material. Fluorocarbon plasma, maintained in CF_4, C_4F_8, SF_6, and so on, in admixture with O_2 is used for the processes under the basic reactions that F isotropically etches Si wafer with rapid chemical etching, and a

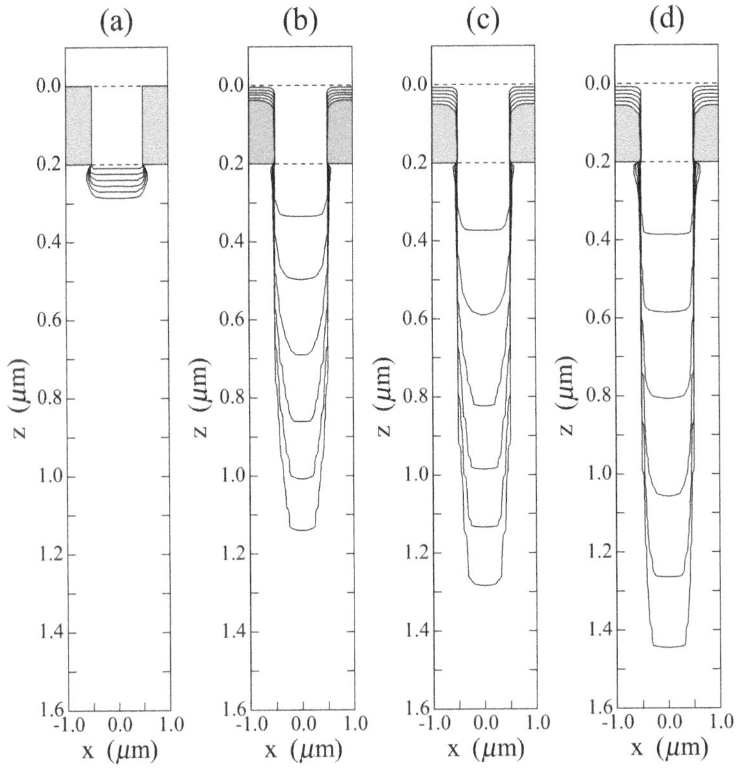

FIGURE 12.13 Effect of incident particles on a feature profile evolution on a patterned SiO_2, predicted by VicAddress. Active species are F (a), Ar^+ (b), Ar^++CF_3^++CF_3 (c), and Ar^++CF_3^++CF_3+F (d). *From Makabe, T, and Yagisawa, T. Plasma Sources Science and Technology, "Low-pressure nonequilibrium plasma for a top-down nanoprocess". Volume 20, Number 2, 2011. With permission.*

fluorinated silicon oxide passivating layer is formed on the sidewall to keep the anisotropic etching.

A series of alternating processes of etching and depositing is effective in forming MEMS construction. The time-multiplexed deep etching is known as the Bosch process. One of the effective and rapid plasma processes consists of three steps: isotropic etching of Si/polymer in SF_6, polymer passivation in C_4F_8, and polymer depassivation in O_2. The resulting feature profile keeps a high anisotropy, though the etching is chemically isotropic. The Plasma Bosch process is widely carried out by using a high-density ICP reactor at pressure between 1.33 Pa (10 mTorr) and 13.3 Pa (100 mTorr). The dimension of the MEMS structure is usually comparable with the sheath thickness or larger.

TABLE 12.12 Spatial Limitation for Anisotropic Plasma Etching

Loading Effect	Phenomena and Origin
Microloading (RIE lag)	<Phenomena> With decreasing a pattern width, the etch rate decreases. <Origin> Lack of radicals and ions incident on the exposed pattern bottom, and difficulty of escape of volatiled particles from the bottom
Micro Masking	<Phenomena> A grass-like whisker grows locally inside the surface of etching. <Origin> Residues, non-volatiles, ejected from an etching and wall surfaces.
Macroloading	<Phenomena> With increasing etchable area on wafer, the etch rate decreases. <Origin> Depletion of radicals and ions incident on each of patterns
Plasma Molding	<Phenomena> With increasing a pattern width, an isotropic etching develops. <Origin> At a hole/trench width greater than the sheath thickness, and incursion of the sheath into the hole/trench

It means that the bulk plasma and sheath structure exposed to the substrate will change in time (plasma molding).

Exercise 12.4.1

In a plasma for deposition or etching, a large size polyatomic molecule may grow up in circumstances of high degree of dissociation of the feed gas. The generated species is a named particle, dust, or powder [15]. Discuss the tool for detection of these large polyatomic molecules in a plasma.

The plasma potential is always a positive value during the cw operation. Therefore, positive ions and neutrals are detected through an orifice on the electrode or reactor wall. However, it is not easy to detect negative ions from

TABLE 12.13 Detection of Polyatomic Molecule or Cluster in Plasmas

Method	Molecular Size	Mass Resolution	Ref.
Quadrupole mass spectrometer	$M/e < 10^3$	—	[16, 17]
Time-of-flight mass spectrometer	$M/e < 10^5$	5×10^3	[18]
Laser Mie scattering	$d_0 > \mu m$	—	[19]

the electrode. Technically, a pulsed operation of the plasma is introduced to measure negative ions in the afterglow phase. Mie scattering is an in situ, active procedure (see Table 12.13).

12.4.8 Charging Damage

We have basically two types of materials to be etched in ULSI. One is metal or poly-Si with adequate electric conductivity. The other is dielectric SiO_2 or low-k materials. These are known as Si-gate etching and the contact hole (via or trench) etching in SiO_2, which is affected by an anomalous etching caused by a local charging of the surface exposed to a plasma as shown in Figure 12.14. The surface resistivity changes with the flux composition from the plasma, that is, ions, electrons, and neutral radical species.

12.4.8.1 Surface Continuity and Conductivity

A number of experiments have emphasized that the etching of a high aspect ratio structure of SiO_2 is a highly competitive process between etching of SiO_2 and deposition of a thin CF_x polymer, and that a thin CF_x polymer is deposited on the active surface exposed to photons, neutral radicals, electrons, and ions. These facts imply that electrons may be conducted on the surface of the thin polymer layer (passivation film) under a field distribution due to a local accumulation of electrons and ions and photo-irradiation from the plasma. Then, the polymer film would prevent the wall from charging by a recombination process of the accumulated positive ions through a fast electron transport.

The surface conduction in conjunction with the incident local fluxes of positive ions and electrons is described by a simple surface continuity equation (see Figure 12.15):

$$\left.\frac{\partial n_e(\boldsymbol{r},t)}{\partial t}\right|_{surf} + \mathrm{div}\boldsymbol{\Gamma}_e(\boldsymbol{r},t) = -\mathrm{div}[\boldsymbol{j}_{e_{surf}}(\boldsymbol{r},t)/e] - R_r n_e n_p, \quad (12.47)$$

$$\left.\frac{\partial n_p}{\partial t}\right|_{surf} + \mathrm{div}\boldsymbol{\Gamma}_p(\boldsymbol{r},t) = -R_r n_e n_p, \quad (12.48)$$

where $\boldsymbol{\Gamma}_e$ and $\boldsymbol{\Gamma}_p$ are the instantaneous local fluxes of electrons and positive ions incident on the surface, respectively. $\boldsymbol{j}_{e_{surf}}(\boldsymbol{r},t) = \sigma(\boldsymbol{r},t)\boldsymbol{E}_{surf}(\boldsymbol{r},t)$

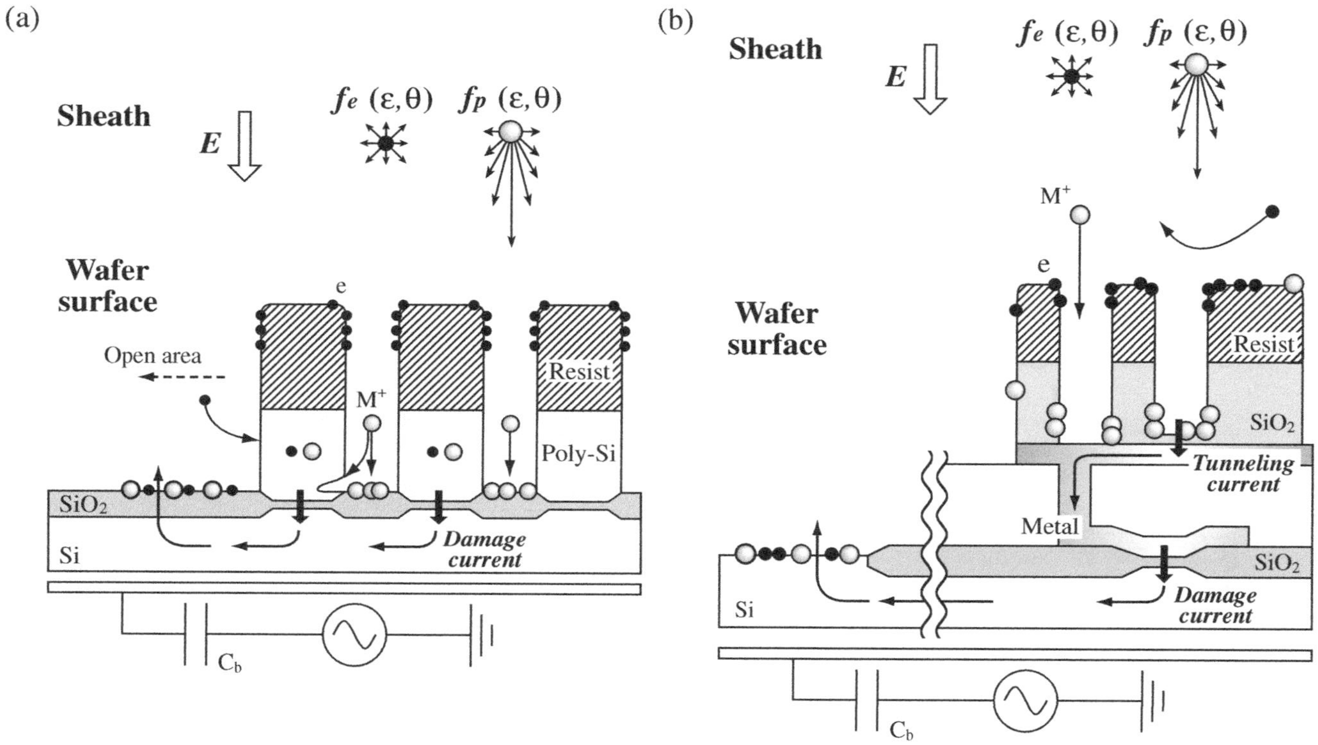

FIGURE 12.14 Schematic diagram of the local charging damage during plasma etching: gate-Si etching (a) and SiO_2 trench etching (b).

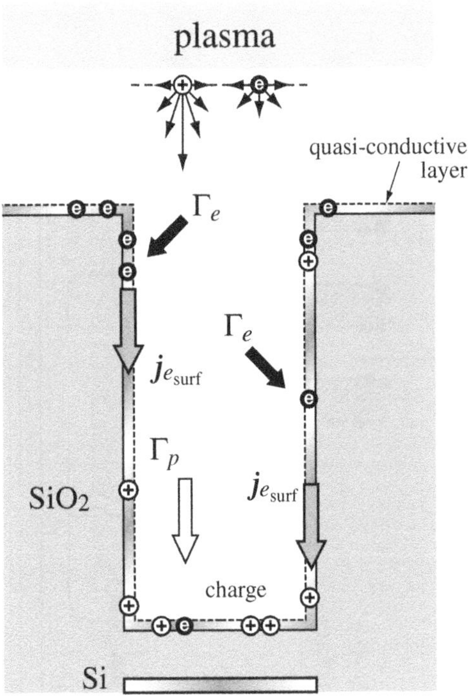

FIGURE 12.15 Local surface charges and conduction in the trench (hole) of SiO_2 during etching.

is a surface current density of the electron conduction, σ is the electrical conductivity of the CF_x thin polymer, and R_r is the surface recombination coefficient. The time evolution of the local surface potential $V_{surf}(\boldsymbol{r}, t)$ is solved at t in the system including the continuity Equations 12.47 and 12.48 and Poisson's equation under consideration of the incident fluxes of electrons and ions from the plasma through the sheath as described in the previous section.

The resulting potential distribution across the gas, surface, and solid phase is calculated by solving

$$\begin{aligned} \nabla^2(\boldsymbol{r}, t) &= -\frac{\rho_s(\boldsymbol{r}, t)}{\varepsilon_0 \varepsilon_r} \; : \text{surface of } SiO_2 \\ &= 0 \; : \text{inside of } SiO_2 \\ &= -e\frac{n_p(\boldsymbol{r}, t) - n_e(\boldsymbol{r}, t)}{\varepsilon_0} \; : \text{in gases,} \end{aligned} \qquad (12.49)$$

where ρ_s is the surface charge density on SiO_2, and ε_0 and $\varepsilon_0 \varepsilon_r$ are the permittivities of a vacuum and SiO_2, respectively. That is, the overall potential from the gas phase to the bulk SiO_2 is simultaneously solved by changing

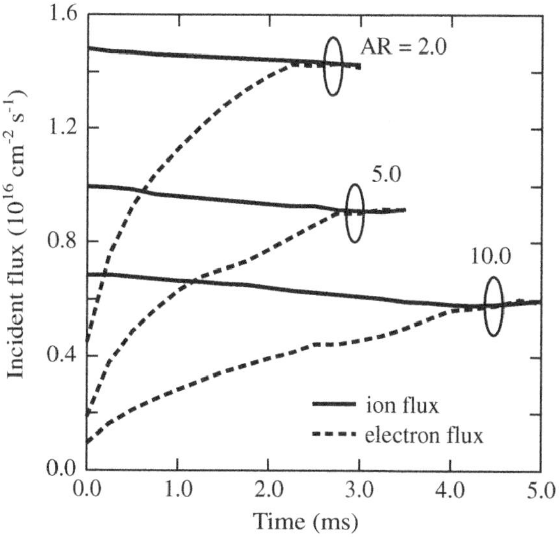

FIGURE 12.16 Time constant for local charging as a function of aspect ratio in typical trench etching of SiO_2.

the mesh size in gas and solid. Then, the distribution of the surface charge at the boundary between SiO_2 and lower poly-Si (or metal) in Figure 12.15 is iteratively calculated under the principle that the equipotential of the surface of lower poly-Si (or metal) must be maintained. The origin of the local charging of electrons and positive ions arises from the significantly different velocity distribution. In a steady state, the velocity distribution of ions incident on a wafer with a beamlike component is quite different from an isotropic distribution of electrons in the positive ion sheath, and both of the charged fluxes incident on a flat surface exposed to the plasma have the same magnitude in a time-averaged fashion. The great difference of the velocity distribution leads to a local accumulation of electrons at the upper part of the trench (hole) and ions at the lower part and bottom, that is, charging on the inside wall of a hole or trench with a high aspect ratio.

Figure 12.14 demonstrates the schematic diagram of the charging damage in the gate-Si etching by Cl_2 plasma (a) and SiO_2 trench etching by CF_4/Ar plasma (b). Both of the chargings seriously damage the profile and lower-level device elements (i.e., the thin gate etc.).

Exercise 12.4.2

Discuss the relationship among the time constants of charging, radical deposition, and ion etching in a typical SiO_2 etching (see Figure 12.13.)

Typical fluxes of radicals and ions incident on a wafer are 10^{18} cm^{-2} s^{-1} and 10^{16} cm^{-2} s^{-1}, respectively. Also, the time constant for local charging is usually ms and the time for effective monolayer etching needs 100 ms.

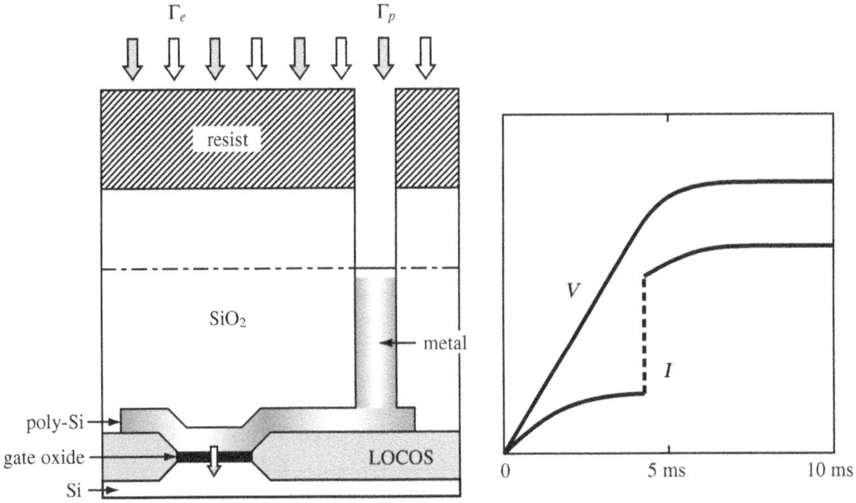

FIGURE 12.17 Current damage at the gate oxide during plasma etching.

12.4.8.2 Charging Damage to Lower Thin Elements in ULSI

The current damage through thin gate oxide is typical of a high-density plasma processing (see Figure 12.14). The current densities of the direct tunnel and Fowler–Nordheim tunnel, J_{DT} and J_{FNT}, are expressed, respectively [20], as

$$J_{DT} = \frac{e^2}{2\pi h d_{ox}^2}\left(\phi_B - \frac{V_{ox}}{2}\right)\exp\left(-4\pi d_{ox}\sqrt{2em_{DT}^*\left(\phi_B - \frac{V_{ox}}{2}\right)}\frac{1}{h}\right), \quad (12.50)$$

$$J_{FNT} = \frac{e^3 V_{ox}^2}{8\pi h d_{ox}^2 \phi_B}\exp\left(-\frac{4\pi\sqrt{2m_{FNT}^*\phi_B^{3/2}}}{3\pi e h}\right), \quad (12.51)$$

where d_{ox} is the thickness of the oxide film, and V_{ox} is the applied voltage between both sides of the thin oxide. h is the Planck constant, ϕ_B is the barrier height, and m^* is the effective mass of the electron. Plasma current damage is typical in a metal gate etching as shown in Figure 12.14a, and besides, in a multilayer interconnect system we take care of the current damage during trench or via etching of SiO_2 (see Figure 12.17).

12.4.9 Thermal Damage

High-energy ions are indispensable for the etching of SiO_2. There exists a high-energy deposition at a very short time within narrow spots near the material suface. The ion irradiation induces a rapid local heating and gradual cooling with high-temperature gradients, leading to thermal damage such as thermal stress. The thermal damage appears on the lower-level device elements, for example, gate oxide in addition to the etching surface. In the case of the elastic binary collision at the surface, the fractional energy of the incident ion with ε_p,

$$\Delta\varepsilon = \frac{4M_pM_s}{(M_p + M_s)^2} \cos\theta^2 \varepsilon_p, \qquad (12.52)$$

is transferred into an atom in the surface layer of the wafer. Here, θ is the incident angle of the ion to the wafer. Except for the case of $M_p = M_s$ and $\theta = 0$, some of the energy of the incident ion is dissipated in the form of excitation of the substrate atoms. After a short relaxation time of the highly nonequilibrium local state, a high-temperature local spot with $T(\boldsymbol{r}_0, t)$ diffuses thermally to the circumference. The thermal diffusion, except for the highly nonequilibrium state during a very short time, is described by

$$\rho c(T)\frac{\partial T(\boldsymbol{r}, t)}{\partial t} = \nabla_r \left(k(T)\frac{\partial T}{\partial \boldsymbol{r}}\right) + S(\boldsymbol{r}_0, t_0) \qquad (12.53)$$

where ρ, $c(T)$, and $k(T)$ are the mass density, the heat capacity, and the heat conductivity of the substrate material, respectively. S is the volume power density of the heat source, that is, the fractional ion energy dissipated to the surface.

In a typical feature-scale etching by using the 2f-CCP reactor, the positive ion flux to the wafer has the magnitude on the order of 10^{16} cm^{-2} s^{-1}. It corresponds to the one ion impact every 1 μs. Under these conditions, the time constant of thermal diffusion is on the order of 0.1 μs to 1 μs. That is, in the micro- or nanometer scale etching, the etching by the aid of ion impact is a discontinuous process in time subject to the stochastic process. Also note that when the scale of the material to be etched approaches nanometer scale, we should pay attention to the change of the physical quantity, $c(T)$ and $k(T)$ and so on.

12.4.10 Specific Fabrication of MOS Transistor

12.4.10.1 Gate Etching

The fabrication of the poly-Si gate electrode of the MOS transistor in Figure 1.3 requires a precise control of the critical dimension (CD) and very high selectivity to the gate oxide, SiO_2 in the plasma etching. After the bottom anti-reflection coating (BARC) film is etched in CF_4/O_2 plasma as shown in Figure 12.18(a), the resist pattern is trimmed by O_2 containing plasma to adjust the gate length (see Figure 12.18(b)). Then, a thin SiO_2 hard mask on the top of the poly-Si is etched in CHF_3/O_2 in Figure 12.18(c). After the removal of the resist mask in O_2 in Figure 12.18(d), the poly-Si is finally etched by using $HBr/Cl_2/O_2$ plasma as shown in Figure 12.18(e). The dissociation of by-products ($SiBr_x$) in the plasma must be controlled as well as the reduction of the sticking probability on the poly-Si side wall to suppress the variation

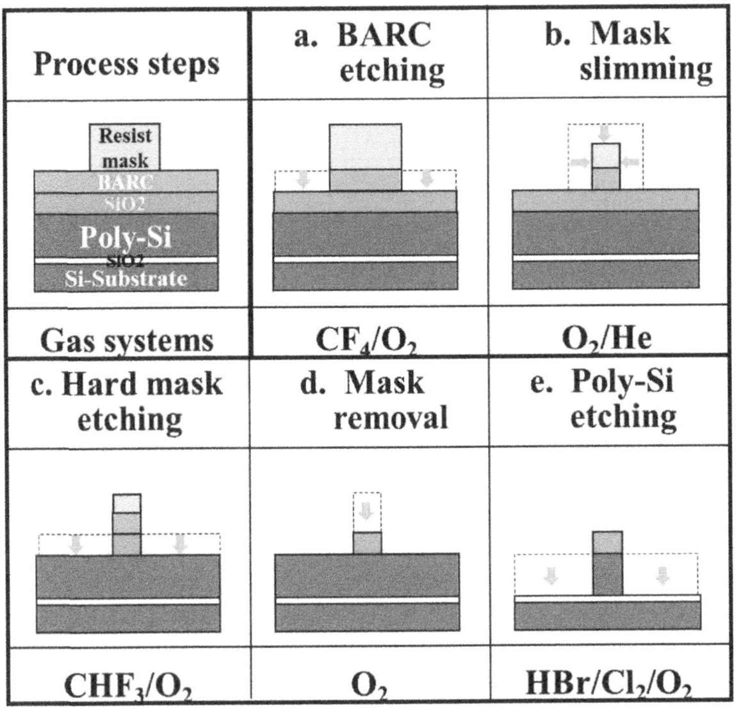

FIGURE 12.18 Schematic of typical plasma etching of poly-Si gate [21]. *From Makabe, T, and Tatsumi, T. Plasma Sources Science and Technology "Workshop on atomic and molecular collision data for plasma modelling: database needs for semiconductor plasma processing", Volume 20, Number 2, 2011. With permission.*

of CD. Furthermore, the ion energy incident on the Si surface through the ion sheath should be carefully controlled to realize high selectivity, because the thickness of the underlying gate oxide is very thin. The dissociated hydrogen atoms from HBr plasma will penetrate into the Si substrate through gate SiO_2, and this penetration will cause damage. The damaged layer will be removed after oxidization by wet treatment using HF solutions, called Si recess. The series of plasma processes for poly-Si gate etching are carried out in an ICP. The ICP is functionally superior to any other plasma sources in the performance of high plasma density and low-energy ions.

12.4.10.2 Contact Hole Etching

A very high aspect ratio contact hole (HARC) has to be prepared by using plasma etching in order to connect the transistor electrodes to the interconnect

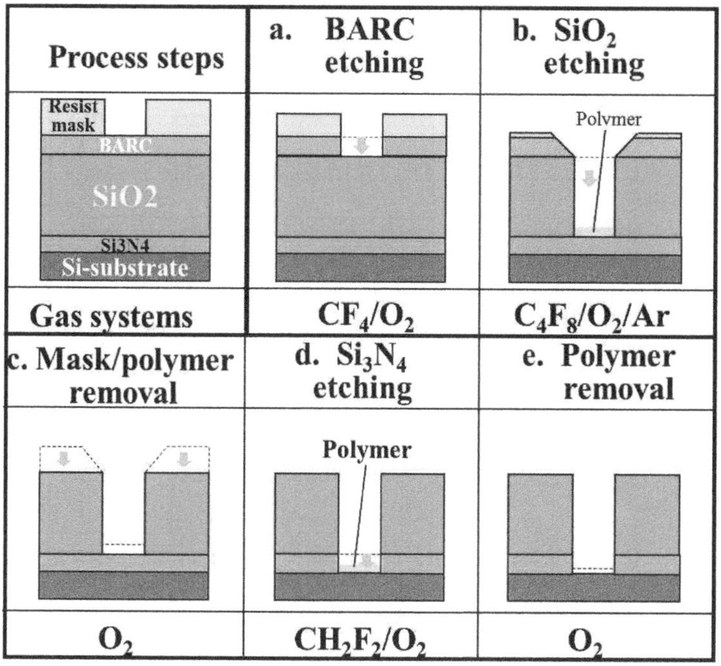

FIGURE 12.19 Schematic of typical plasma etching of contact hole [21]. *From Makabe, T, and Tatsumi, T. Plasma Sources Science and Technology "Workshop on atomic and molecular collision data for plasma modelling: database needs for semiconductor plasma processing", Volume 20, Number 2, 2011. With permission.*

layer (see Figure 1.3). It is key to establish high selectivity to both the resist mask and the underlying Si_3N_4 film. After the BARC etching in CF_4/O_2 shown in Figure 12.19(a), SiO_2 HARC is opened by using $C_4F_8/Ar/O_2$ plasma in Figure 12.19(b). A very high energy ion flux with a beam-like profile of 1 keV - 1.5 keV is required in order to realize the high etch rate of SiO_2. The etch rates of both resist mask and underlying Si_3N_4 increase with the excess F atoms, generated from the dissociation of C_4F_8. The excess amount of F in the reactor has to be carefully suppressed during the short residence time (τ_r ¡ 10 ms) of the feed gas. After removing the resist mask in O_2 in Figure 12.19(c), Si_3N_4 film is etched by using CH_2F_2 plasma in Figure 12.19(d). The surface cleaning of the Si substrate is performed by the treatment in O_2 plasma as shown in Figure 12.19(e). A 2f-CCP is usually employed in SiO_2 etching under a low degree of dissociation of F and a very high energy flux of ions.

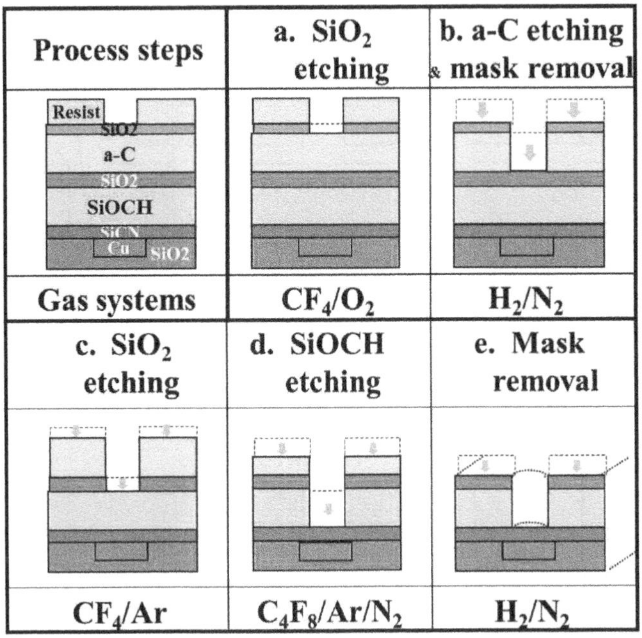

FIGURE 12.20 Schematic of typical plasma etching of low-k to fabricate a via [21]. *From Makabe, T, and Tatsumi, T. Plasma Sources Science and Technology "Workshop on atomic and molecular collision data for plasma modelling: database needs for semiconductor plasma processing", Volume 20, Number 2, 2011. With permission.*

12.4.10.3 *Low-K etching*

Low-k films are used in interconnect layers to reduce the capacitance between Cu wirings in order to realize high-speed and low-power-consumption devices. Figures 12.20 and 12.21 show a typical process flow to fabricate dual damascene (DD) interconnects for a via and a trench. First, a multi-stacked mask for a via hole is formed on the SiO_2/SiOCH in Figures 12.20(a)-(c). In particular, H_2/N_2 plasma is employed for the a-C etching and mask removal as shown in Figure 12.20(b). During the etching of the SiOCH hole pattern, high selectivity to SiC is required to reduce the loss of the bottom stopper material (see Figure 12.20(d)). $C_4F_8/Ar/N_2$ plasma is used to etch SiOCH. The process window is very narrow for the selective etching of SiOCH/SiC, because the oxygen content of SiOCH is lower than that in SiO_2. It is known that a slight fluctuation of the plasma itself or the film composition will induce etch stopping, residue, or changes in the bottom or top diameter of via holes. Next, the via hole in Figure 12.20(e) is filled with organic polymer(a-C), and a resist mask for the trench pattern is formed on the SiO_2 layer (see Figure

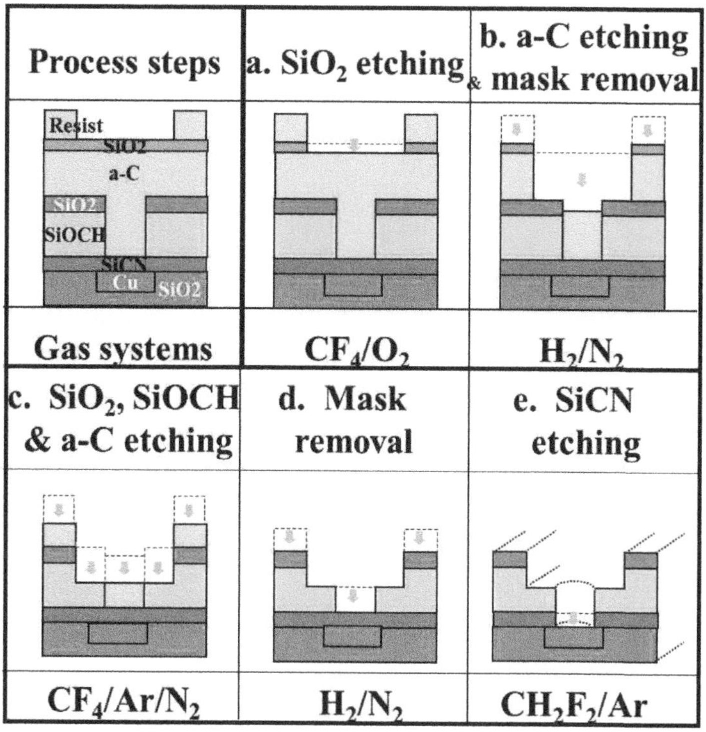

FIGURE 12.21 Schematic of typical plasma etching of low-k to fabricate a trench [21]. *From Makabe, T, and Tatsumi, T. Plasma Sources Science and Technology "Workshop on atomic and molecular collision data for plasma modelling: database needs for semiconductor plasma processing", Volume 20, Number 2, 2011. With permission.*

12.21). Then, followed by the etching of SiO_2 in CF_4/O_2 in Figure 12.21(a), the a-C and the mask are removed by N_2/H_2 plasma in Figure 12.21(b). The SiO_2/SiOCH trench is formed in $CF_4/Ar/N_2$ plasma as shown in Figure 12.21(c). After the resist mask removal by N_2/H_2 plasma, the SiCN stopper layer is finally opened on the underlying Cu wiring in Figure 12.21(e).

It is notable that SiOCH film has Si-CH_x bonding in the film network to reduce the film density. The Si-CH_3 bond, in particular, is weaker than the Si-O bond, and is easily oxidized during O-containing plasma treatment. The damage increases the k-value (i.e., the permittivity) and/or reduces the angle of the side profile of the via after removal. The damage could be decreased by changing the ashing plasma from O_2 to H_2/N_2 plasma.

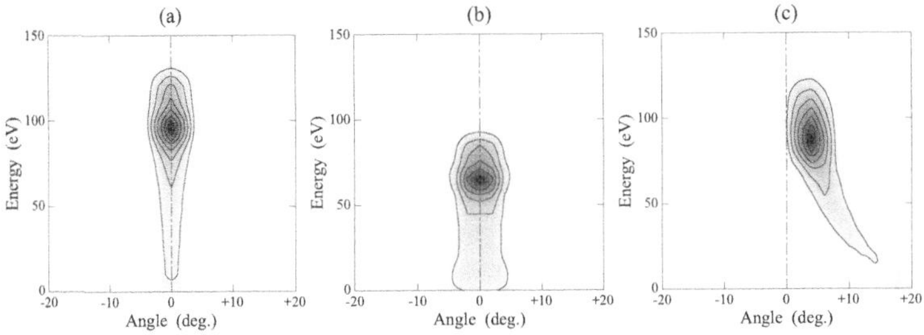

FIGURE 12.22 Time-averaged flux velocity distribution of SF_5^+ incident on a structural wafer (diameter:500 μm, depth:500 μm). Without hole (a), center of a hole (b), and hole edge (c). 2f-CCP, driven at VHF (100 MHz, 300 V) and biased at LF (1 MHz, 100 V), is maintained at 300 mTorr in SF_6(83%)/O_2 [22]. *From Makabe, T, and Yagisawa, T. Plasma Sources Science and Technology, "Low-pressure nonequilibrium plasma for a top-down nanoprocess". Volume 20, Number 2, 2011. With permission.*

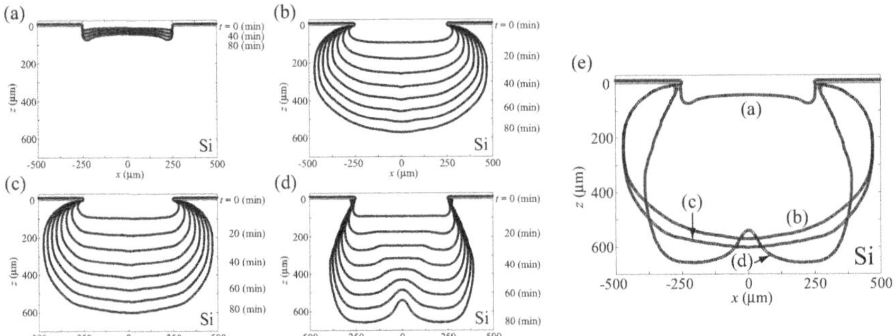

FIGURE 12.23 Feature profile evolution of 500-μm-wide trench under a plasma molding: Active species are SF_5^+ (a), F (b), SF_5^+ and F (c), and RIE with passivation layer by O (d). (e) shows the feature profiles among (a)-(d). External conditions are in Figure 12.22 [23]. *From Hamaoka, F, et al. Plasma Science, IEEE Transact, "Modeling of Si Etching Under Effects of Plasma Molding in Two-Frequency Capacitively Coupled Plasma in SF6/O2 for MEMS Fabrication". Volume 35, Issue 5, 2007. With permission.*

12.4.11 MEMS Fabrication

Reactive ion etching (RIE) is applied to a process of several 10 μm or 100 μm pattern etching, i.e., a micro-electro-mechanical system (MEMS). In a large scale of etching, the etched profile influences the sheath configuration (thickness and potential distribution) in front of the wafer. It is named "plasma molding" (see Table 12.12). As is shown in Figure 12.22, the incident distribution of the ion flux is strongly influenced by the presence of the etched hole profile. It will be possible to estimate the plasma molding's local characteristics by modeling. Figure 12.23 shows the feature profile of Deep-Si etching under plasma molding in 2f-CCP in SF_6(83%)/O_2 at 300 mTorr, where O_2 is mixed to prepare a passivation layer on the inside wall. The removal of the layer at the bottom corner is strengthened by the distorted ion by plasma molding. Thus, the etching is enhanced at the bottom corner. As the pattern width decreases and the oxygen mixture ratio increases, the etching profile becomes anisotropic by the formation of the passivation layer at the sidewall. In addition, the bottom profile becomes flat because of the uniform ion flux under the reduced influence of the plasma molding.

References

[1] Behrisch, R., Ed. 1981. *Sputtering by Particle Bombardment I. Topics in Appl. Phys.* 47. Berlin: Springer Verlag.

[2] Behrisch, R., Ed. 1983. *Sputtering by Particle Bombardment II. Topics in Appl. Phys.* 52. Berlin: Springer Verlag.

[3] Kress, J.D., Hansen, D.E., Voter, A.F., Liu, C.L., Liu, X.-Y., and Coronell, D.G. 1999. *J. Vac. Sci. Technol. A* 17:2819.

[4] Engelmark, F., Westlinder, J., Nyberg, T., and Berg, S. 2003. *J. Vac. Sci. Technol. A* 21:1981.

[5] Thompson, M.W. 1968. *Philos. Mag.* 18:377.

[6] Kenmotsu, T., Yamamura, Y., and Ono, T. 2004. *Jpn. Soc. Plasma. Sci. Nucl. Fusion Res.*

[7] Matsuda, A. 2004. Thin-film silicon (Inv. Review paper), *Jpn. J. Appl. Phys.* 43:7909.

[8] Matsumura, H., Umemoto, H., and Masuda, A. 2004. *J. Non-Cryst. Solids* 19:338–340.

[9] van Kampen, N.G. 1981. *Stochastic Processes in Physics and Chemistry.* Amsterdam: North-Holland.

[10] Gardiner, C.W. 1983. *Handbook of Stochastic Methods for Physics, Chemistry and the Natural Sciences.* Berlin: Springer Verlag.

[11] Ford, I.J. 1995. *J. Appl. Phys.* 78:510.

[12] Balooch, M., Moalem, M., Wang, W-E., and Hamza, A.V. 1996. *J. Vac. Sci. Technol. A* 14:229.

[13] Sethian, J.A. and Strain, J. 1992, 1993. *J. Comp. Phys.* 98: 231; Sethian, J.A. and Chopp, D.L. 1993, *J. Comp. Phys.* 106:77.

[14] Esashi, M., and Ono, T. 2005. *J. Phys. D, Topical Review* 38:R223.

[15] Bouchoule, A., Ed. 1999. *Dusty Plasmas:Physics,Chemistry and Technological Impacts in Plasma Processing* John Wiley & Sons, Chichester 1999.

[16] Auciello, O. and Flamm, D.L. 1989. *Plasma Diagnostics,* Vol. 1. San Diego: Academic Press.

[17] Bruno, G., Capezzuto, P., and Madan, A. 1995. *Plasma Diagnostics of Amorphous Silicon-Based Materials.* San Diego: Academic Press.

[18] Saito, N., Koyama, K., and Tanimoto, M. 2003. *Jpn. J. Appl. Phys.* 42:Part 1, 5306.

[19] Bohren, C.F., and Huffman, D.R. 1983. *Absorption and Scattering of Light by Small Particles.* New York: John Wiley & Sons.

[20] Hirose, M. 1996. *Mater. Sci. Eng.* 41:35.

[21] Makabe, T., and Tatsumi, T. 2011. *Plasma Sources Sci. & Technol.* 20:024014.

[22] Makabe, T., and Yagisawa, T. 2011. *Plasma Sources Sci. Technol.* 20:024011.

[23] Hamaoka, T., Yagisawa, T., and Makabe, T. 2007. *IEEE Trans, on PS.* 35:1350.

CHAPTER 13

Atmospheric-Pressure, Low-Temperature Plasma

13.1 HIGH PRESSURE, LOW-TEMPERATURE PLASMA

13.1.1 Fundamental Process

In an atmospheric-pressure plasma, we have to consider the three-body collision and the effect of high-density excited species in addition to the collision process in a low-pressure plasma. Typical reactions are shown in Table 13.1 as an example in Ar. The radiation from polyatomic molecules excited to the rotational and vibrational levels is absorbed by the lattice or electron gases in a metallic wall. The excited electron gas transmits the energy to the metallic lattice through an inelastic electron-phonon scattering. The microscopic phonon excitation results in the heating of the metallic wall.

Problem 13.1.1
A plasma irradiation produces an increase in wall temperature of the reactor. It is known that the metallic wall is heated up by a heavy particle collision and by a radiated photon from short-lived excited molecules. Practically, the kinetic energy of the heavy particle and the emission from a vibrational and rotational transition will contribute to the heating.

Discuss the reason why (1) the potential energy of ions and metastables incident on a metallic wall makes few contributions to the wall heating, and (2) the visible and UV radiation from the transition between the electronic levels of a molecule is ineffective to the heating (see Table 4.4).

TABLE 13.1 Collisional Reactions in a High-pressure, High-density Plasma in Ar

Process	Reaction Scheme	Rate constant
three-body dimer ion formation	$Ar^+ + Ar + Ar \rightarrow Ar_2^+ + Ar$	$10^{-31} cm^6 s^{-1}$
dissociative recombination	$e + Ar_2^+ \rightarrow Ar^j + Ar$	$5\times 10^{-7} cm^3 s^{-1}$
(cf. radiative recombination	$e + Ar^+ \rightarrow Ar^j + h\nu$	$\sim 10^{-12} cm^3 s^{-1}$)
stepwise ionization	$e + Ar^* \rightarrow e + Ar^+$	Boltzmann eq.
metastable pooling	$Ar^* + Ar^* \rightarrow e + Ar^+ + Ar$	$6.4\times 10^{-10} cm^3 s^{-1}$
	$\rightarrow e + Ar_2^+$	$4\times 10^{-10} cm^3 s^{-1}$
excitation to radiative level	$e + Ar^* \rightarrow e + Ar^j$	$2\times 10^{-7} cm^3 s^{-1}$

13.1.2 Historical Development

A low-temperature plasma at atmospheric pressure generally will be produced at the initial stage of a gas discharge. With the temporal progress of the discharge, the external electrical power is consumed in order to heat up the feed gas, and the low temperature plasma changes to thermal arc plasma. In this way, atmospheric-pressure plasma usually belongs to a thermal plasma, in which the gas temperature is sufficiently high as compared to the room temperature. Historically, a number of studies have been performed for the transition from a glow discharge to arc. It is known that a low-temperature plasma at atmospheric pressure can be maintained in He or gases highly diluted by He with high thermal conductivity and large specific heat capacity, when an appropriate power source and electrode system are employed. Also the increase of the effective surface area of the electrode made of metallic mesh will lead to a stable low-temperature plasma even in a polyatomic gas. The gas discharge sustained between electrodes covered by dielectric material is traditionally named the barrier discharge or silent discharge. [1,2] The charged particles on a dielectric surface, produced by the discharge initiation, are not able to flow through the metallic electrode to the external power supply. Accordingly, at a finite time delay after the ignition, an inverse electric field is formed between the two dielectric surfaces before the convergence of the glow mode of the discharge. This means that the glow to arc transition by a growth of a local current density and the gas heating will be practically prevented by the use of the dielectric electrode. As a result, it becomes possible to maintain both the stable glow-like discharge and the low-temperature plasma even under atmospheric pressure. Typical electrode configurations are shown in Figure 13.1.

The barrier discharge on a dielectric surface is usually driven at a frequency from 50 Hz to 100 KHz before the appearance of ion-trapping. The low-frequency barrier discharge at several kHz consists of a large number of thin pulsive discharges. At high frequency under ion-trapping, the discharge will spread over the electrode. Various types of barrier discharges are applied to ozonizers, gas laser excitation systems, and plasma display panels (PDP), etc.

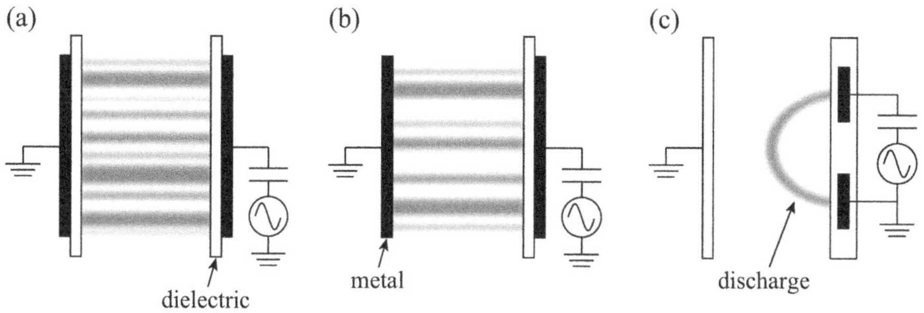

FIGURE 13.1 Schematic diagram of typical barrier discharges (a), (b), and (c).

13.2 MICRO PLASMA

13.2.1 Radiofrequency Atmospheric Micro-Plasma Source

An opened or closed system of microcell plasmas driven at radiofrequency (rf) under atmospheric pressure has been investigated experimentally and theoretically for applications to surgical plasma knives, skin and dental treatments, semiconductor processes, and environmentally decontaminating processes of polluted materials, etc., based on a small but high-density source such as those available for the charged-particle beam, neutral radical source, or illuminant [3–7].

As the size of a reactor decreases, the ratio of the surface (S) to the volume (V), S/V, increases. As the pressure in the reactor increases, the spatial diffusion of the plasma is restricted to a small area. As a result, the microcell plasma will be sustained in a high density mode by both direct and stepwise ionizations in the presence of highly accumulated long-lived excited molecules (i.e., metastables) (see Table 13.1). These facts mean that the surface wall will act significantly as an absorber of charged particles and of heat from the plasma. The heating of the neutral gas and the wall surface is caused mainly by collisions with high-density, energetic ions in atomic gases. This means that in a micro-plasma confined in a small volume at high pressure, the local heating of gas molecules must be considered, as it will be influential to the plasma structure itself and to the plasma stability. Secondly, electrons having a mid-range of energy may play an important role in the maintenance and the structure of the microcell plasma through the stepwise ionization of the metastable.

13.2.2 Gas Heating in a Plasma

The spatiotemporal heat transfer in the feed gas and the solid vessel of the microcell plasma is numerically estimated along with a consideration of the electrons, ions, long-lived metastable atoms, and potentials by using the relaxation continuum model in the two-dimensional cylindrical coordinates. The governing equation of the heat transfer in a gas and metal,

$$\frac{\partial}{\partial t}T(z,r,t) = \frac{1}{\rho c}\nabla\Big\{\kappa(T)\nabla T(z,r,t)\Big\} + S(z,r,t), \tag{13.1}$$

is joined with the relaxation continuum model. Here, $T(z,r,t)$ is the temperature at position $\boldsymbol{r}(z,r)$ and time t. $\kappa(T)$, ρ, and c are, respectively, the thermal conductivity, mass density, and thermal capacity [8, 9]. The physical quantities in a system of Ar gas in Al-microcell are summarized in Table 13.2.

The heat source $S(z,r,t)$ is given by the Joule heating in the gas phase and by the kinetic energy of ions on the metallic surface in atomic gases. The radiation heating from Ar-plasma is not taken into account, because the metallic crystal lattice is transparent to the emission wavelength from the excited atomic Ar. Under very different time constants for the transport among the charged particles, long-lived excited atoms, and gas temperature, the spatiotemporal structures of electrons, ions, and potential are first calculated during several tens of periods of the rf source. By using the cycle average of each of the internal plasma parameters, the space and time development of the metastable is calculated 100 times at $\Delta t_m = 10^{-10}$s. After the procedure is repeated during several thousand periods of the rf source, the gas temperature is calculated for 300 times at $\Delta t_T = 10^{-6}$s. The series of calculations is repeated until all the internal plasma parameters attain a periodic steady-state distribution. When the heat-transfer time is much faster than the residence time of the feed gas, the local gas number density $N_g(z,r)$ is derived from the ideal gas law $p = N_g(z,r)k_B T_g(z,r)$, where k_B is the Boltzmann constant.

TABLE 13.2 Example of the Physical Quantity in Heat Conduction Equation

Physical quantity	ρ (gcm^{-3})	c (Jg^{-1}K^{-1})	κ (Jcm^{-1}s^{-1}K^{-1})
Ar gas	1.67×10^{-3}	0.52	$1.77\times10^{-5}T^{1/2}$
Al metal	2.73	0.90	2.04
Air(N_2 gas)	1.17×10^{-3}	1.006	2.71×10^{-4}

TABLE 13.3 Elements of Gas/Metal Heating

Mechanism	Expression	Comment
Joule heating	$\boldsymbol{J}\cdot\boldsymbol{E}$	in gases (main)
ion impact	kinetic energy ϵ_p	at wall (major)
photo irradiation		
in atomic gas	$h\nu$ (UV,visible)	at wall (minor)
in polyatomic gas	$h\nu$ (infrared)	at wall (major)
deexcitation of A*	potential energy ϵ_j	at wall (minor)

13.2.3 Effect of Local Gas Heating

Figure 13.3 shows a typical 2D profile of the steady gas temperature, $T_g(z,r)$, in a microcell (Figure 13.2), driven at 13.56 MHz at $V_o = 500$ V and dissipated power of 160 Wcm^{-3} in Ar at 760 Torr [10]. In the temperature distribution $T_g(z,r)$, the plasma structure has a periodic steady state synchronized with the driving frequency, 13.56 MHz. As is expected from the deep negative value of the self-bias voltage, the gas temperature is locally heated with a peak value of 740 K in front of the powered electrode. The outer gas temperature is uniform and is 300 K at 1 mm from the outer surface of the microcell. The spatial profile of $T_g(z,r)$ and the temperature in the metal $T_M(z,r)$ in Figure 13.3 reflects the large difference in thermal conductivity between gas and solid metal. Under the local gas heating, the feed gas number density $N_g(z,r)$ has a local and broad minimum in front of the powered electrode as shown in Figure 13.4. This means that the reduced field, defined as $E(z,r,t)/N_g(z,r)$,

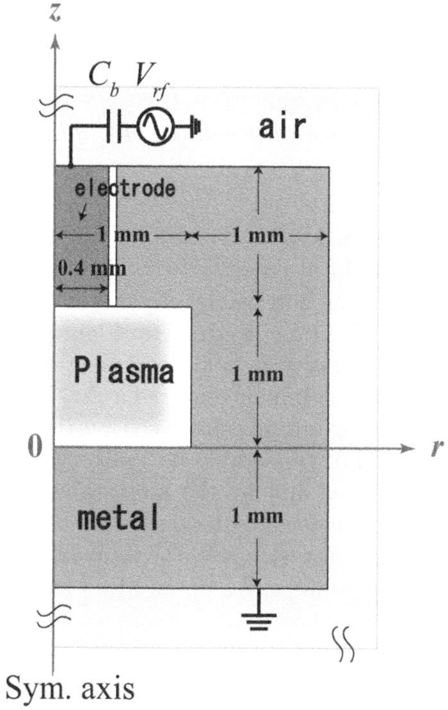

FIGURE 13.2 Schematic diagram of the plasma (Ar), microcell (Al), and surroundings (air). *From Yamaski, M, et al. "Effects of local gas heating on a plasma structure driven at rf in a microcell in Ar at atmospheric pressure". Spring 2014. JJAD JSAP Publishing. With permission.*

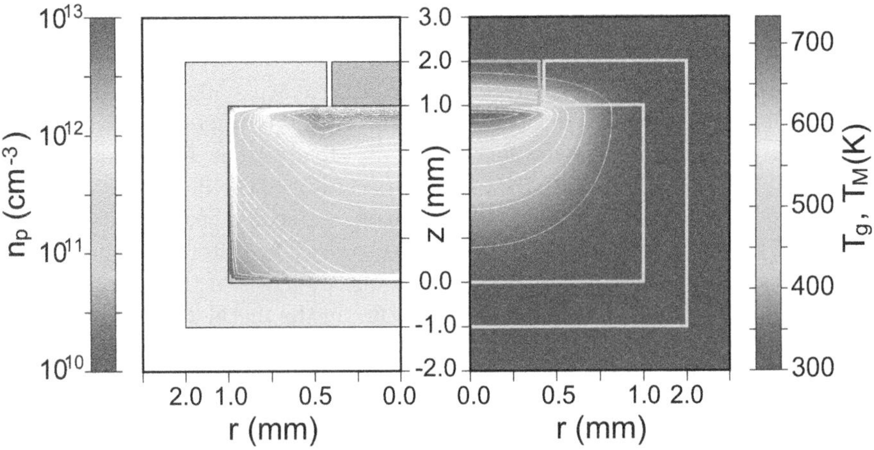

FIGURE 13.3 2D-distribution of $T_g(z,r)$ and $T_M(z,r)$ (right), and total positive ions $n_p(z,r)$ (left) in a microcell-plasma at 13.56 MHz, V_o= 500 V at atmospheric pressure in Ar. *From Yamaski, M, et al. "Effects of local gas heating on a plasma structure driven at rf in a microcell in Ar at atmospheric pressure". Spring 2014. JJAD JSAP Publishing. With permission.*

will be locally strengthened close to the rf-electrode as compared to that at constant $T_g(z,r)$, and the local energy of the electron will further increase, thus enhancing plasma production by the electron impact ionization.

The ion density at the time-average in Figure 13.3, having a peak $7\times10^{12}\text{cm}^{-3}$ close to the rf electrode and decreasing exponentially in the axial direction, consists of the dominant molecular Ar_2^+ and a small amount of atomic Ar^+. That is, Ar^+ is initially produced by the electron impact both of the ground state Ar and of metastable Ar^* with the threshold energy, 15.7 eV and 4.21 eV, respectively, and by the metastable pooling with the rate constant, $6\times10^{-10}\text{cm}^3\text{s}^{-1}$, under a high density Ar^* (see Table 13.1).

Finally, Figure 13.4, in contrast, shows the time-averaged number densities of positive ions $n_p(z)$, metastables $N_m(z)$, and feed gases $N_g(z)$ as a function of axial position z at $r=0$.

Exercise 13.2.1

In a high density plasma, long-lived excited species (i.e., metastables) and/or dissociated neutral radicals frequently have a number density N^* comparable to that of the ground-state of the feed gas molecule N_o. The metastables (radicals) may have a high reactivity on a solid surface. We should pay careful attention to the influence of the metastable on the plasma structure and the sustaining mechanism, judging from the great difference of the ionization energy between the metastable and the ground-state molecule, and from the

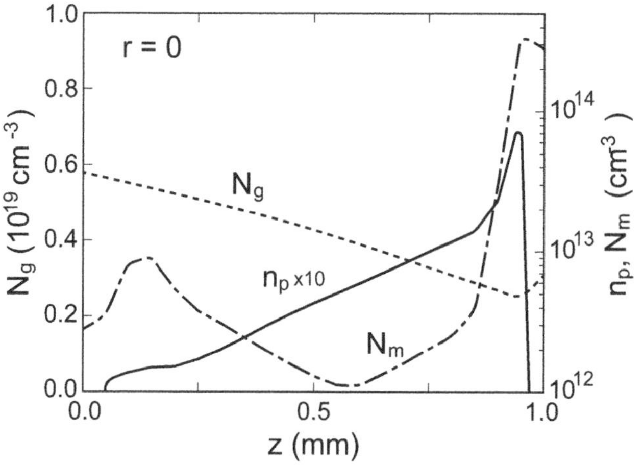

FIGURE 13.4 Time-averaged number density of total ions, metastables, and feed gas Ar at $r=0$ as a function of axial position. The external plasma conditions are described in Figure 13.3. *From Yamaski, M, et al. "Effects of local gas heating on a plasma structure driven at rf in a microcell in Ar at atmospheric pressure". Spring 2014. JJAD JSAP Publishing. With permission.*

number density ratio, N^*/N_o. Also we have to consider an electron and ion pair formation by way of the heavy particle collision between metastables, because the collision cross section at thermal energy is generally high (see Figure 4.38). Find the relative magnitude between both electron-ion pair formations, based on the stepwise ionization from the metastable and the metastable-metastable collision (i.e., metastable pooling).

The degree will be evaluated both from the ionization rate and from the initial energy of electrons:

(a) for the ionization rate by way of the electron impact,

$$\frac{\int_{\epsilon_{i_s}} N^* Q_{i_s}(v) v g(\boldsymbol{v}) d\boldsymbol{v}}{\int_{\epsilon_{i_d}} N_o Q_{i_d}(v) v g(\boldsymbol{v}) d\boldsymbol{v}} \sim \frac{N^*}{N_o} \times \frac{Q_{i_s}(\sim \epsilon_{i_s})}{Q_{i_d}(\sim \epsilon_{i_d})} \times \left(\frac{\epsilon_{i_s}}{\epsilon_{i_d}}\right)^{1/2} \times \frac{g(\epsilon_{i_s})}{g(\epsilon_{i_d})}, \tag{13.2}$$

(b) heavy particle ionization,

$$\frac{\int_{\epsilon_{th}} N^* N^* Q_{mp}(V_r) V_r G_M(\boldsymbol{V}_r) d\boldsymbol{V}}{\int_{\epsilon_{i_d}} N_o n_e Q_{i_d}(v) v g(\boldsymbol{v}) d\boldsymbol{v}} \sim \frac{N^*}{N_o} \times \frac{N^*}{n_e} \times \frac{Q_{mp}(\epsilon_{th})}{Q_{i_d}(\sim \epsilon_{i_d})} \times \left(\frac{\epsilon_{th}}{\epsilon_{i_d}}\right)^{1/2} \times \frac{G_M(\epsilon_{th})}{g(\epsilon_{i_d})}, \tag{13.3}$$

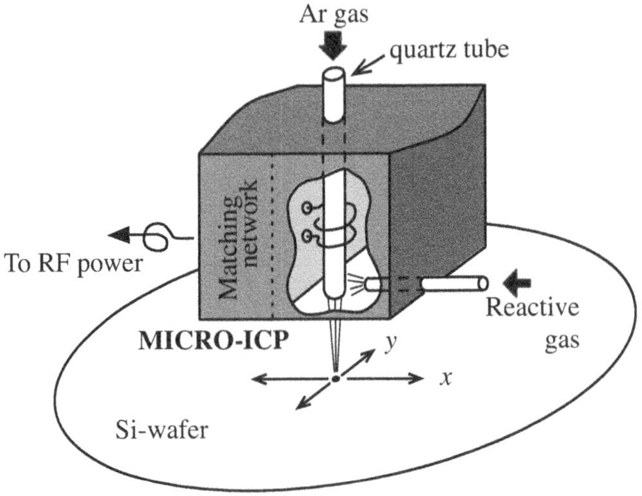

FIGURE 13.5 Schematic diagram of a scanning microplasma etcher.

by considering the fact that the ionization mostly occurs just after the threshold energy ϵ_i. Here, Q_{i_d} and Q_{i_s} are the cross sections for the direct and stepwise ionization having the threshold energy, ϵ_{i_d} and ϵ_{i_s}, respectively. Q_{mp} is the cross section for the metastable pooling at thermal energy ϵ_{th}. $g(\boldsymbol{v})$ and $G_M(\boldsymbol{V})$ are the velocity distributions normalized to unity, respectively. The influence will be evaluated from each of the physical quantities in both equations.

Exercise 13.2.2
Manufacturing plasma etcher is a large system providing for the uniformity of a high density plasma flow to a large wafer in order to achieve high efficiency and throughput. On the contrary, a prototype of a scanning microplasma etcher operated at atmospheric pressure is proposed (see Figure 13.5). Discuss the advantage of the microplasma system for surface processing.

The first advantage will be the possibility of the maskless pattern etching by using a microplasma jet. A low-temperature microplasma jet driven at 13.56 MHz in Ar/SF_6 is applied for a maskless pattern etching of Si by the aid of a drive unit of computer assisted numerical two-dimensional data. A high etch rate of Si, 500 μmin^{-1} has been reported [11].

References

[1] von Engel, A. 1983. *Electric Plasmas: Their Nature and Uses*. London: Taylor & Francis Ltd.

[2] Hippler, R., Pfau, S., Schmidt, M., and Schoenbach, K. H. 2001. *Low Temperature Plasma Physics.* Berlin: Wiley-VCH.

[3] Park, GY., Park, SJ., Choi, MY., Koo, IG., Byun, JH., Hong, JW., Sim, JY., Gollins, G. J., and Lee, JK. 2012. *Plasma Sources Sci. Technol.* 21: 043001.

[4] Kushner M. J. 2005. *J. Phys. D.* 38: 1633.

[5] Sakiyama, Y., Graves, D. B., and Stoffels, E. 2008. *J. Phys. D.* 41: 095204.

[6] Sakiyama, Y. and Graves, D. B. 2009. *Plasma Sources Sci. Technol.* 18: 025022.

[7] Ito, Y., Sakai, O., and Tachibana, K. 2010. *Plasma Sources Sci. Technol.* 19: 025006.

[8] Cherrington, B. E. 1979. *Gaseous Electronics and Gas Lasers.* Oxford: Pergamon Press.

[9] Eden, J.D. and Cherrington, B. E. 1973 *J. Appl. Phys.* 44: 4920.

[10] Yamasaki, M., Yagisawa, T., and Makabe, T. 2014. *Jpn. J. Appl. Phys.* 53:036001.

[11] Ideno, T. and Ichiki, T. 2006. *Thin Solid Films* 506/507: 235.

Index

V

W

Printed and bound by CPI Group (UK) Ltd, Croydon, CR0 4YY
18/10/2024
01776256-0014